WRITING THE CITY

WRITING THE CITY

Eden, Babylon and the New Jerusalem

Edited by Peter Preston
and Paul Simpson-Housley

London and New York

First published 1994
by Routledge
2 Park Square, Milton Park, Abingdon, Oxfordshire OX14 4RN
Simultaneously published in the USA and Canada
by Routledge
711 Third Avenue, New York, NY 10017
First issued in paperback 2014
Routledge is an imprint of the Taylor and Francis Group, an informa company
Transferred to Digital Printing 2005

© 1994 Peter Preston and Paul Simpson-Housley

Typeset in Garamond by Solidus (Bristol) Limited

British Library Cataloguing in Publication Data
A catalogue record for this book is available from the British Library

Library of Congress Cataloging in Publication Data
Writing the city: Eden, Babylon, and the New Jerusalem / edited by
Peter Preston and Paul Simpson-Housley.
p.   cm.
Includes bibliographical references and index.
1. Cities and towns in literature. 2. Literature, Modern—20th
century—History and criticism. 3. Literature, Modern—19th
century—History and criticism. I. Preston, Peter
II. Simpson-Housley, Paul.
PN56.C55W75   1995
809'.93321732—dc20   93-44177

ISBN 13: 978-0-415-10667-2 (hbk)
ISBN 13: 978-0-415-75637-2 (pbk)

CONTENTS

CONTENTS

This book is dedicated
by Peter Preston
to his parents

and by Paul Simpson-Housley
to Arely Ayala, Carol Davison and Bridget Sine

LIST OF MAPS
AND FIGURES

NOTES ON CONTRIBUTORS

Gary Brienzo teaches at Nebraska Wesleyan University, Lincoln, USA. He has published a number of articles on Bernard Mac Laverty and Willa Cather and is co-editor of a forthcoming annotated bibliography of Cather scholarship from 1984 to 1992; his Ph.D. dissertation on Willa Cather is being prepared for publication.

Jonathan Crush is Associate Professor of Geography at Queen's University, Ontario, Canada. He has a special interest in South Africa and has published articles on customary rights in late colonial Swaziland, migrancy and the South African Labour Commission, and black miners.

Pierre Deslauriers is Lecturer in the Department of Geography, Concordia University, Montreal, Quebec, Canada, and has previously taught at l'Université de Montréal. His main interest lies in the processes of growth of cities, particularly in rural–urban fringes, and he has published several articles, alone and in collaboration, on issues concerning land use around cities.

Lorne Foster received his Ph.D. in sociology in 1984. He is the author of many articles and publications on various aspects of urban life and culture.

Jacqueline Gibbons is Assistant Professor of Sociology at York University, Ontario, Canada. She has recently published papers on the semiosis of Art Nouveau and contributed "The north and native symbols: landscape as universe" to P. Simpson-Housley and G. Norcliffe (eds.), *A Few Acres of Snow: Literary and Artistic Images of Canada* (1992).

Emily Gilbert is a graduate student in the Department of Geography at York University, Ontario, Canada. She holds, among other scholarships, the Graduate Fellowship for Academic Distinction.

Pauli Tapani Karjalainen received his Ph.D. in 1986 and is now Docent in Humanistic Geography at the University of Joensuu, Finland, where his main fields of research are humanistic and cultural/social geography.

Yossi Katz is Professor and Chairperson of the Department of Geography at Bar Ilan University, Jerusalem, Israel. His main interest lies in the historical geography of Palestine and the settlement processes of the Jewish people throughout the world. His most recently completed book, *Zionist Private Enterprise in Palestine 1990–1914*, will be published soon.

Anna Makolkin received her Ph.D. in Comparative Literature from the University of Toronto in 1987 and is now Research Fellow at the Centre for Russian and Eastern European Studies and Adjunct Professor at Victoria College, University of Toronto, Canada. She has published extensively on the poeticity of biography and the semiotics of literature and culture.

Anssi Paasi received his Ph.D. from the University of Joensuu, in 1986 and is now Professor of Geography at the University of Oulu, Finland. His research fields are the history of geographical thought, social theory and regional geography and regional consciousness and identity.

Deborah Carter Park is a doctoral student in Geography at York University, Ontario, Canada. She has had papers published in *Brontë Society Transactions* and the *Journal of Historical Geography*.

Alec Paul is a Professor of Geography at the University of Regina, Saskatchewan, Canada. He has published in the fields of geography and literature and climatology, and his regional specializations are North America (especially Canada) and Europe.

Peter Preston is Lecturer in Literature in the Department of Adult Education and Convenor of the D. H. Lawrence Centre at the University of Nottingham, UK. He has published articles and reviews on D. H. Lawrence, Arnold Bennett and Dickens, is co-editor (with Norman Page) of *The Literature of Place* (1993) and has recently completed *A D. H. Lawrence Chronology*.

John P. Radford is Associate Professor of Geography at York University, Ontario, Canada, and editor of the *Journal of Historical Geography*. His special interests are urban and historical geography and he has recently published articles on the control of the feeble-minded and identity and tradition in the post-Civil War South.

Susan J. Rosowski is Adele Hall Professor of English, University of Nebraska–Lincoln, USA. Her numerous essays on Cather have appeared in a wide variety of journals and she is the author of *The Voyage Perilous: Willa Cather's Romanticism*, editor of the series Cather Studies and co-general editor of the Cather Scholarly Edition.

Jamie S. Scott is Associate Professor of Humanities and Director of Religious Studies at York University, Ontario, Canada. His special interests are in Bonhoeffer and Thomas More and his books include *Cities of God: Faith, Politics and Pluralism in Judaism and Islam* and *Sacred Places and Profane Spaces: The Geographics of Judaism, Christianity and Islam* (co-edited with P. Simpson-Housley).

Paul Simpson-Housley is Associate Professor of Geography and Director of Graduate Geography at York University, Ontario, Canada. His interests include Chile, natural disasters and literary landscapes and among his many publications are *Geography and Literature* (co-edited with W. Mallory) and *Antarctica: Exploration, Perception and Metaphor*.

Rana P. B. Singh is Associate Professor in Geography at Banaras Hindu University, Varanasi, India, and Director of the Varanasi Studies Foundation. He has written many research papers and monographs and co-edited eighteen anthologies, including *Where Cultural Symbols Meet: Literary Images of Varanasi* and *Varanasi: Cosmic Order, Sacred City and Hindu Traditions*.

ACKNOWLEDGEMENTS

The editors are glad to acknowledge the help of the following people in the preparation and publication of this book: Carolyn King and Carol Randall for cartographic assistance; Tristan Palmer, Commissioning Editor, and Moira Taylor, Senior Desk Editor, at Routledge; Robert Peden, the Copy Editor; and the contributors for their unfailing encouragement and commitment to the project.

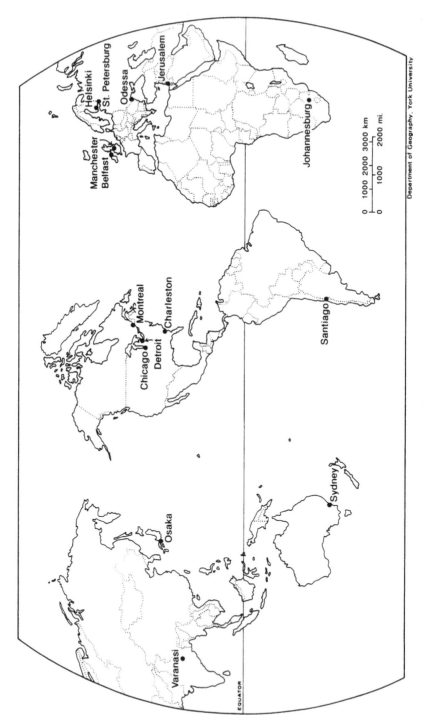

Map 1 The world

Department of Geography, York University

0   1000 2000 3000 km
0   1000   2000 mi.

Manchester
Belfast
Helsinki
St. Petersburg
Odessa
Jerusalem
Johannesburg

Montreal
Charleston
Chicago
Detroit
Santiago

Osaka
Varanasi
Sydney

EQUATOR

1

INTRODUCTION
Writing the city

Peter Preston and Paul Simpson-Housley

Saul Bellow calls the city "the expression of the human experience it embodies, and this includes all personal history."[1] This is to see the city from a viewpoint different from that of classic geographical descriptions. Urban geographers have developed models to predict phenomena such as growth or decline, demographic patterns, traffic flows and economic potential. Some have taken a mechanistic view of cities and used mathematical formulae to solve urban problems. Others, notably Mumford, have been more conscious of cities as places of human habitation, while some, such as Griffith Taylor, view cities in terms of historical development.[2] Different ideological viewpoints have informed these descriptive–predictive analyses of the city; logical empiricism has dominated the debate, but Marxist-inspired work, such as that by David Harvey on the bid–rent curve, has recently made its impact.[3]

Such empirically conceived models, however, fail to accommodate that aspect of city life emphasized by Saul Bellow: the human experience, both individual and collective, contained by the city. Philip Larkin, meditating on the appeal of a country church for a secular, agnostic age, asks if he is drawn to the church

> because it held unspilt
> So long and equably what since is found
> Only in separation – marriage, and birth,
> And death, and thoughts of these

and concludes that the building itself contains something that acts as an irresistible magnet even for the casual visitor:

> A serious house, on serious earth it is,
> In whose blent air all our compulsions meet,
> Are recognised, and robed as destinies.[4]

The city's air, too, may be blent, composed of the hopes, aspirations, disappointments and pain of those who live in it; like Larkin's church, it is a kind of vessel, filled with human experience. The city is an aggregation or accumulation, not just in demographic, economic or planning terms, but also

1

in terms of feeling and emotion. Cities thus become more than their built environment, more than a set of class or economic relationships; they are also an experience to be lived, suffered, undergone.

The principal impulse behind this collection of essays, therefore, is to concentrate on the city as seen through the eyes of novelists and poets and their characters, in order to offer a particular kind of witness to the challenges, opportunities, stresses and frustrations of city life. This is not to deny the value of the contribution made by established geographical approaches – many of the contributors are themselves professional geographers. Rather, the collection seeks to complement some of their conclusions by concentrating on insights, attitudes and values derived from a personal perspective. Before we turn to the nature of those insights, it is necessary to look briefly at the history of the city in literature.

The city has always been an important literary symbol, and the ways in which a culture writes about its cities is one means by which we may understand its fears and aspirations. The New Jerusalem of the Book of Revelation or St. Augustine's City of God are types of the ideal city whose perfections only serve to highlight the shortcomings of all earthly cities. In Sir Thomas More's *Utopia* (1516) or Thomas Campanella's *The City of the Sun* (1623) we find representations of the ideal Renaissance city-state, which themselves look back to Plato's ideal republic. Indeed, the frequency with which the city appears as the physical embodiment of the Utopian community reminds us of its perceived potential to achieve a kind of contained perfection, always in a desirable and sustaining equipoise and forming a refuge from the chaos that lies outside its walls. Such powerful fictional images have their roots in the real politics of earlier centuries, when there may indeed have been a pressing need to place clear boundaries around the state; boundaries which also served as defenses and were therefore a means of maintaining political power as well as protecting the inhabitants against aggressors and controlling the movement of people and goods. It is rare to find a utopia – even at its most idealistic – that does not specify or imply this element of containment and control.

Negative literary images of the city also often have their roots in its political nature. In a wide variety of texts from the classical period to the present day the urban and the rural have been placed in diametric opposition, to the advantage of the latter. In this view the city is seen as the site of guile, corruption, intrigue and false values, as against the positive, natural, straightforward values of the countryside.[5] To escape from the city and live far from courts and princes is to make a choice in favor of a kind of authenticity that can only be found in the pastoral life. Yet the pastoral contains its own ambivalence, as is famously exemplified by Shakespeare's *As You Like It* (c.1599). In that play, Shakespeare adds to the established terms of the pastoral argument – here represented by the court and the forest – a third term which stands back from and offers a critique of both. The view of the radically alienated figure of Jacques, for whom "All the world's a stage,/ And all the men and women merely players," can hardly be taken as normative; but the verse he offers for Amiens to sing serves as a corrective to any temptation to see the forest as an ideal retreat:

> If it do come to pass
> That any man turn ass,
> Leaving his wealth and ease,
> A stubborn will to please,
> Ducdame, ducdame, ducdame.
> Here shall he see
> Gross fools as he,
> An if he will come to me.[6]

For Jacques, the pastoral gesture is in its way as false as the life of the court and city, no more likely to lead to an authentic way of life than the world that has been rejected. The forest is populated by actors.

Implicit in Jacques' attitude is a profound ambivalence towards the life of city and court. The city comedies of Shakespeare's contemporaries offer savage attacks on the vanity, greed, malice and deceit of city-dwellers, particularly those who represent a decaying courtier class and those who belong to a burgeoning trader-*rentier* class. Yet, in the work of Ben Jonson, for instance, it is undeniable that the experience of city life – its variety of character and language, the never-failing resourcefulness of its inhabitants in finding new ways of gulling and exploiting their neighbours – provides a kind of energizing force for the drama. Bartholomew Fair, like Vanity Fair, has its allure. The words of another Johnson – that a man who is tired of London is tired of life – are relevant here. Samuel Johnson is a sharp satirist of the human vanity he found in eighteenth-century London; yet it is impossible to imagine his career as a writer having developed outside the metropolis, for his imagination is fired and nourished by the very material it excoriates.

Such ambivalence is carried through into the following periods of literature. As industrialization developed in Europe, what might be called the pastoral debate – city vs. country, culture vs. nature, mechanical vs. organic – took on a greater urgency as the city came to be regarded as the most obvious symbol of a new and pressing reality. But the ways in which that symbol may be employed and its potency in imaginative terms vary enormously. Wordsworth, for instance, in Book VII of *The Prelude*, finds in London some of the delights of variety and energy, but ultimately finds no difficulty in rejecting its *mélange* of sights, sounds and inhabitants. His images for London are sometimes of confusion; he writes of its "Babel din" (line 157), or "the thickening hubbub" (227). Equally significant, however, is the sense of falsity of so much that he sees in London: "those mimic sights that ape/ The absolute presence of reality ... imitations, fondly made in plain/ Confession of man's weakness and his loves" (247–8, 254–5). Wordsworth concedes the allure of such sights and the other entertainments, from theater to lawcourts, that London offered him, culminating in the description of the crowded variety of Bartholomew Fair (648–94). But his summing-up of the Fair becomes a summing-up of what the city means to him:

> Oh, blank confusion! and a type not false
> Of what the mighty City is itself
> To all except a straggler here and there,
> To the whole swarm of its inhabitants;
> An undistinguishable world to men,
> The slaves unrespited of low pursuits,
> Living amid the same perpetual flow
> Of trivial objects, melted and reduced
> To one identity, by differences
> That have no law, no meaning, and no end.
>
> (695–704)

This may be "[b]y nature an unmanageable sight" but can be managed by one who has "among least things/ An under-sense of greatest," to whose mind "[t]he mountain's outline and its steady form/ Gives a pure grandeur," even when living "in that vast receptacle" (708, 710–11, 722–3, 734). Strengthened by "The Spirit of Nature," he can find, even among "the press/ Of self-destroying, transitory things,/ Composure, and ennobling Harmony" (735, 738–40).[7]

Ultimately, it is not difficult for Wordsworth to reject the city and return to a different way of life. For a writer like Dickens, however, the city may be a location both threatening and alluring, menacing and exciting. The very aggregation of experience that finally appals Wordsworth, offers to Dickens rich material, and his characters' experience of the city may be a mixed one. Nicholas Nickleby's first sight of London, for instance, offers a vision of chaos as telling as anything in *The Prelude*:

Streams of people apparently without end poured on and on, jostling each other in the crowd and hurrying forward, scarcely seeming to notice the riches that surrounded them on every side; while vehicles of all shapes and makes, mingled up together in one moving mass like running water, lent their ceaseless roar to swell the noise and tumult.

As they dashed by the quickly-changing and ever-varying objects, it was curious to observe in what a strange procession they passed before the eye. Emporiums of splendid dresses, the materials brought from every quarter of the world; tempting stores of everything to stimulate and pamper the sated appetite and give new relish to the oft-repeated feast; vessels of burnished gold and silver, wrought into every exquisite form of vase, and dish, and goblet; guns, swords, pistols, and patent engines of destruction; screws and irons for the crooked, clothes for the newly-born, drugs for the sick, coffins for the dead, churchyards for the buried – all these jumbled each with the other and flocking side by side, seemed to flit by in motley dance like the fantastic groups of the old Dutch painter, and with the same stern moral for the unheeding restless crowd.[8]

Like Wordsworth in his description of Bartholomew Fair, Dickens achieves his effect in part by accumulation, by the sheer number of people and things to be seen and acquired in the city; but although the paragraph ends with a reminder of "the ... stern moral for the unheeding restless crowd," Dickens's excitement at the city's sense of promise is evident in the rhetoric of his long, cumulative sentence.

It is in this London that Nicholas's hapless father looks "among the crowd without discovering the face of a friend,"[9] and it is in this London that Nicholas, too, will experience a sense of loss and alienation, finding that the plenitude of London is formless and meaningless. Yet Nicholas is discovered by the tirelessly benevolent Cheeryble brothers in the midst of this same London – a place where the reader would hardly expect two strangers to notice each other, let alone that one should offer the other a job. This is in a sense special pleading on Dickens's part; he is beginning to occlude the portrait in the first half of the novel of a hostile dehumanizing city and to replace it by one in which he can create small havens of retreat and redemption. On the one hand, there is Golden Square, where the miserly Ralph Nickleby has his office, the window of which looks out on "a melancholy little plot of ground," a piece of "unreclaimed land," a "desolate place," which no one thinks of "turning ... to any account," and containing nothing but rubbish and a "distorted fir tree" rotting in a pot.[10] On the other hand, there is City Square, where the Cheerybles have their business:

> The City Square has no enclosure, save the lamppost in the middle; and has no grass but the weeds which spring up round its base. It is a quiet, little-frequented, retired spot, favourable to melancholy and con-templation, and appointments of long-waiting; ... In winter-time, the snow will linger there, long after it has melted from the busy streets and highways. The summer's sun holds it in some respect, and, while he darts his cheerful rays sparingly into the square, keeps his fiery heat and glare for noisier and less-imposing precincts. It is so quiet, that you can almost hear the ticking of your own watch when you stop to cool in its refreshing atmosphere. There is a distant hum – of coaches not of insects – but no other sound disturbs the stillness of the square.[11]

In many respects the ambience of City Square is similar to that of Golden

Square, but Dickens, who by this point in the book has begun the process of retrenchment by which he will return Nicholas and Kate to a miraculously recreated world of their childhood, subtly alters the impression by a careful choice of language. The passage is partly ironical, a gently satirical comment on the "fervent encomiums bestowed upon [City Square] by Tim Linkinwater,"[12] the Cheerybles' clerk. But the overall impression is that of a kind of rural retreat, "retired" and "favourable to melancholy and contemplation," where the busy and indifferent life of the city is reduced to "a distant hum." It is, for Nicholas and the group that gathers around him, a suitable stopping-point on the way back to the real countryside, where the city's grosser aspects can be redeemed by a contained life of affection and companionship. Yet, as can be seen from Dickens's later works, the city was a location to which he had constantly to return, partly in order to record and interpret its rapidly changing physical appearance, partly because there is something in the rhythm and variety of city life that stimulates his imagination.

The question of what the nineteenth-century writer, and particularly the novelist, does with the experience of the city is of some importance, because of the special relationship between the rise of the city and the rise of the novel. The aggrandizing potential of the realist novel, which quickly showed itself to be a suitable vehicle for addressing the full range of contemporary issues, personal and political, local and national, may be compared to the development of the city as a location for an enormous range of people and activities. Thus, as Malcolm Bradbury points out, the urban experience may have had its effect on the form as well as the content of fiction: "one might argue that the unutterable contingency of the modern city has much to do with the rise of that most realistic, loose and pragmatic of literary forms, the novel."[13]

Bradbury's comment appears in an essay on the centrality of the city to modernist writing. Cities may have been seen as a kind of purgatory or hell, as in Eliot's "unreal city" where death has "undone so many,"[14] but they were also "generative environments" for intellectual debate and artistic experimentation, as well as "novel environments, carrying within themselves the complexity and tension of modern consciousness and modern writing."[15] Eliot may have heard

> Falling towers
> Jerusalem Athens Alexandria
> Vienna London [16]

but it was in some of those towers that the aesthetic of modernism was forged, and it is no accident that such key modernist texts as James Joyce's *Ulysses* (1922) and Virginia Woolf's *Mrs Dalloway* (1924) use the city as both setting and protagonist.

The need both to render and to comprehend the multiplicity of the city persists in contemporary fiction. Consider the following passage from *City of the Mind* (1991) by the British novelist Penelope Lively. Two of the characters, Matthew and Alice, emerge from a London restaurant:

And so, presently, they are out in the London night. It is a blaze, a swirl of light and colour, sound and smell. They pass from the intimacies of the Soho streets to the frontier of Charing Cross Road, streaming with people and traffic. Matthew takes Alice's arm; an advancing gang of tipsy youths divides around them, goes whooping into the tube station. Everyone is talking, shouting. Language hangs in the night air and throbs in giant lettering above shops and theatres. A column of buses stands pulsing in a traffic jam: Gospel Oak, Putney Heath, Clapton Pond, Wood Green. Matthew and Alice pause on the pavement and he thinks of the city flung out all around, invisible and inviolate. He forgets, for an instant, his own concerns, and feels the power of the place, its resonances, its charge of life, its coded narrative. He reads the buses and sees that the words are the silt of all that has been here – hills and rivers, woods and fields, trade, worship, customs and events, and the unquenchable evidence of language. The city mutters still in Anglo-Saxon; it remembers the hills that have become Neasden and Islington and Hendon, the marshy islands of Bermondsey and Battersea. The ghost of another topography lingers; the uplands and the streams, the woodland and fords are inscribed still on the London Streetfinder, on the ubiquitous geometry of the Underground map, in the destinations of buses. The Fleet River, its last physical trickle locked away underground in a cast iron pipe, leaves its name defiant and untamed upon the surface. The whole place is one babble of allusions, all chronology subsumed into the distortions and mutations of today, so that in the end what is visible and what is uttered are complementary. The jumbled brick and stone of the city's landscape is a medley of style in which centuries and decades rub shoulders in a disorder that denies the sequence of time. Language takes up the theme, an arbitrary scatter of names that juxtaposes commerce and religion, battles and conquests, kings, queens and potentates, that reaches back a thousand years or ten, providing in the end a dictionary of reference for those who will listen. Cheapside, Temple, Trafalgar, Quebec, a profligacy of Victorias and Georges and Cumberlands and Bedfords – there it all is, on a million pairs of lips every day, on and on, the imperishable clamour of those who have been here before.[17]

Here, Penelope Lively attempts to read – or listen to – the narrative that is inscribed in and spoken by the city. This narrative is both diachronic, in that it recounts a story of change through time, and synchronic in that it sees that story of change as existing and constantly surfacing in the contemporary city.

The city's voices therefore comprise "a babble of allusions" ranging from its mutterings in Anglo-Saxon, with place-names recording the remote topography of the land on which the city was built, to the "defiant and untamed" Fleet Street, recalling its once-full river, with "its last physical trickle locked away underground." Contemporary voices are also clamant, in the shouting passers-by and the words on the electric advertisements. This "coded narrative" requires a special kind of reading, accessible to Matthew, who is an architect specializing in the renovation of old buildings; he can sort out "the distortions and mutations of today, so that in the end what is visible and what is uttered are complementary." No judgment is made on the processes of change and development that have issued in the contemporary city – or at least no adverse judgment. The narrative inscribed in the present-day city is not necessarily one of loss or decline, and the passage speaks of no lost golden past. Then and now are simultaneously present and, so far as one can tell, equally valued; what matters is the multiplicity, the babble (babel) of voices, the "million pairs of lips." For Matthew the "charge of life" he feels in the city derives from this variety of messages that lie waiting to be decoded from "the distortions and mutations of today." Even the words "distortions and mutations" seem to be innocent of any pejorative connotations; it is almost as if the narrative takes pleasure in the change of place-names through the centuries because they offer another opportunity for reading the city. Even the "advancing gang of tipsy youths" offers no threat, but "divides around" Matthew and Alice. Individuals are pursuing their own concerns, seemingly taking little notice of each other, but again this is not seen as part of a narrative of distance or alienation, these single lives being, rather, part of the city's collective life. The passage celebrates resonance, variety, jumble, medley, profligacy and the coexistence of past and present spoken in the city's "imperishable clamour."

A different mood enters the narrative, two paragraphs later:

> Matthew, all of a sudden, is desolate. He is surrounded by people, and entirely alone. He is brushed by the indifferent glances of a hundred strangers, and looks upon them with equal disregard. Alice Cook is beside him, but he does not love Alice, nor does she love him. He is consumed with the sense of loss, of solitude. There comes surging forth the memory, not of Susan, but of love. Once, he lived within a safe warm capsule of requited feeling; now, he is adrift.[18]

But this moment of desolation, indifference, disregard, loss, solitude and of being adrift is not particularly attached to the city; it derives from Matthew's past experience, and is something he brings to the present moment rather than something that the moment imposes on him. As in the earlier passage, the novel assumes the existence of no normative city metanarrative, against which the city of today or the feelings of any of its inhabitants may be judged.

In this sense, the passage from Penelope Lively's novel may be seen as offering an example of the post-modernist view of the city, in the terms of the following definition:

> [Post-modernism relates to] a relatively widespread mood in literary theory, philosophy and the social sciences concerning the inability of these disciplines to deliver totalising theories and doctrines, or enduring "answers" to fundamental dilemmas and puzzles posed by objects of enquiry, and a growing feeling, on the contrary, that a chronic provisionality, plurality of perspectives and incommensurable appearances of the object of enquiry in competing discourses make the search for ultimate answers or even answers that can command widespread consensus a futile exercise.[19]

Boyne and Rattansi point to a key element in post-modernism, and one that is relevant to this collection, when they write of post-modernism's nervousness of "totalising theories and doctrines." In suggesting, earlier in this Introduction, that empirical accounts of the city cannot accommodate all that is implied in urban experience, we were making a similar rejection of totalizing meta-narratives, in favor of the "plurality of perspectives" to be found in literary works. Thus this book contains essays by practitioners of different academic disciplines, writing of cities on different continents as seen by radically different writers. Our domain is therefore very wide within the overall subject of the city in literature, and we hope that this variety of approach will in itself enhance understanding of the many things it means to live in a city.

A number of our contributors comment in general terms on the importance of taking into account literary renderings of city experience. Jonathan Crush, for instance, writes of Johannesburg as "a city of signs, of moral coherence, of essences. Its landscape encodes stories about its origins, its inhabitants and the broader society in which it is set. The traveler's task is to discover and write these truths."[20] As in the Penelope Lively passage quoted above, the city is a text, waiting to be read and written or rewritten in literary terms. Pauli Karjalainen and Anssi Paasi, indeed, emphasize that the essential elements of "cityness" are "manifested only through feelings: the experience of the environment is always a fusion of the external physical realm and the human being's internal capacities";[21] while for Yossi Katz, much of the interest of Agnon's fictions about Jerusalem lies in the tension between empirical description and imaginative experience. This can be linked with William Gilmore Simms's belief, shared with the English Romantic poets, that the landscape itself exists only as perceived and becomes a metaphor for the observer's state of mind. As Rana Singh argues, in relation to the work of Shivprasad Singh, we see the city through the filter of the writer's imagination, which produces a very particular and idiosyncratic way of seeing.

This individual mode of regard is the key element to emerge from these chapters. Yet even within this variety and plurality of viewpoints, and without imposing rigid and constricting taxonomies on the range of city experience, it is possible to discern some common and recurrent groups of themes or motifs.

A first group gathers around the ideas of alienation and oppression, the sense of how individual lives may be lost in the busy aggregation of the city. Of course, that very aggregation, with its opportunities for losing oneself in the crowd and living one's life untrammelled by the intrusiveness and narrowness that may exist in smaller communities, may offer a kind of freedom and possibility. But, for many of the writers discussed in this volume, that freedom proves to be illusory. Isabel Allende's characters, who move from the country to Santiago, or those who move into Helsinki in anticipation of a new life find that their initial experience of hope and liberation is quickly replaced by a sense of alienation and despair. Willa Cather's Rosicky finds an emptiness at the heart of New York life, while others, as in Singh's Varanasi, find themselves the victims of tricksters and other forms of exploitation. In a different way, the inhabitants of Elizabeth Gaskell's Manchester, only one or two generations removed from a different mode of living, have to accommodate themselves to new topographies and a new pace and rhythm of life. By presenting us with Manchester people of three generations, with varying degrees of attachment to the city, she shows how that adaptation may take place.

For many of the writers discussed here, the city is an active organism, which may prove to be a site of culture and inspiration, like Pushkin's Odessa, but is more likely to be seen as oppressing its inhabitants and creating or exacerbating divisions within individuals. The Montreal novelist Hugh MacLennan provides an exact image for this phenomenon when one of his characters contemplates "the immense empty street, and has the impression that it is his own interior life that he is contemplating." Even William Gilmore Simms, for whom the experience of living in Charleston was largely nourishing, feels a kind of resentment; as John Radford tells us "in early middle age, he was placing the blame for his failure on the society of the city he refused to leave."[22] The consequences of such oppression, even predatoriness, in the city organism, may be disastrous, resulting in break-down (Mrs. Pyy in Marja-Liisa Vartio's novel of Helsinki life), withdrawal (Goncharov's Oblomov), suicide (Christina Stead's Michael) or murder (Dostoyevsky's Raskolnikov).

The city is perhaps seen at its most violent and alienating in Elmore Leonard's Detroit and Bernard Mac Laverty's Belfast. As Lorne Foster writes, Detroit is a metaphor in the post-modern war, a city of Hobbesian brutes in business suits, where civility in its full sense had broken down and the social contract is radically breached. Only the rules of confidence tricksters and casual violence apply; cops and robbers, operators and their marks are caught in an eternal dance. In Bernard Mac Laverty's *Cal*, the eponymous central character is drawn into a world of violence by the city's sectarian conflicts, and in this sense he is both oppressed and victimized by the spirit of the city. That spirit of murderous violence is personified in its politically active inhabitants, driven by ideological arguments that are exposed as spurious and dehumanizing. The violence is also responsible for disintegration at every level – in individuals, in families and in communities – which in themselves become microcosmic symbols of the city's larger divisions. In Detroit, in Belfast, as well as in St. Petersburg and many of the other locations discussed here, the cities have, in both moral and topographical terms, lost their human scale. Far from presenting the Renaissance Utopian ideal of the protecting city state, they have become, as Rana Singh puts it, symbols of disorder that offend a sacred cosmic order.

Cities are not homogeneous places, one and indivisible. As most city inhabitants recognize, they have their zones, with boundaries, both visible and invisible. Urban geographers have long recognized the existence and importance of such boundaries. In particular, they have used a model employing the concentric circles of a central core, containing the commercial and entertainment districts, surrounded by a "twilight zone" of poor housing, which in turn succeeds to an outer suburban circle of more affluent houses with smaller pockets of commercial activity.[23] Crossing the boundaries will be easier for some individuals than others, and that crossing will often involve more than the exchange of one style of housing for another. Emily Gilbert makes a useful distinction: the spatial distance between city and country is subordinate to the social distance created by the historic and economic dialectic of city and country.[24] Although her comment is made in the context of a city/country contrast, the point is applicable to the "distances" between city zones, and almost all the writers discussed in this collection show some awareness of the existence of such zones and the effect they may have on individuals. A key theme of Pierre Deslauriers' chapter, for instance, is movement between zones and what that means in terms of ethnicity, culture, values and economic standing; while Yossi Katz's chapter also deals with the relationship between economic and social geography.

For the inhabitants of Belfast, for instance, there is a link between urban decay and inner breakdown, and the presence in a community of one family from the "wrong" side of the sectarian divide leads to persecution and violence. Elizabeth Gaskell makes clear the sharp differences between the dwellings of the rich and of the poor in Manchester in the 1840s, while other visitors to the city remarked on how the physical distances between these areas led to ignorance, incomprehension and conflict between the classes. Raskolnikov wonders why "in all great towns men are not simply driven by necessity, but in some peculiar way inclined to live in those parts of town where there are no gardens or fountains; where there are most dirt and smell and all sorts of nastiness." Yet, like more than one of the characters in the works discussed here, he is both repelled and fascinated by the miscellaneity and confusion of the underlife of St. Petersburg. That same fascination is found in the pleasure districts of Osaka, discussed in Jacqueline's Gibbons' chapter, in Singh's descriptions of the slums of Varanasi, and to some extent in the red-light areas of Sydney and Santiago.

A third area of common interest is gender. Not only are women the protagonists of many of the works covered in these essays – in Cather, the Helsinki novels, and in the novels of Elizabeth Gaskell and Christina Stead – but women may also be seen as fulfilling two key functions. The first is to suffer at the hands of the city's violence and indifference; the other is to offer some kind of redemptive escape from its worst excesses. In Mrs. Pyy, the alienated middle-class housewife, we see the damage caused to a woman who comes to the city, but plays no part in its active life and is driven into breakdown. Catherine, in Christina Stead's *Seven Poor Men of Sydney*, wanders among the city's poor and vagrant women when she herself feels confused. Rachel in Elizabeth Gaskell's *Mary Barton* finds it impossible to adjust to the move to the city and slips into its lowest depths. In *Crime and Punishment* Sonia becomes a prostitute in order to keep her family, while the self-sacrificing marriage of Raskolnikov's sister Dounia to Luzhin is seen as little more than a kind of prostitution. In the work of Shivprasad Singh and Isabel Allende, women are similarly driven into prostitution by a poverty that the indifferent city offers no way of relieving. In Detroit, many of the women seem to be no more than bit players in the larger drama of macho wiliness and violence that is being played out in Leonard's city. Willa Cather's Thea Kronberg in *The Song of the Lark* must find some way of surviving in a Chicago that is largely fuelled by what are seen as male drives of material possession and professional ambition.

Against this, however, as Thea herself finds when she visits a different kind of city in Panther Canyon, it is possible to find a kind of redemption. As Susan Rosowski shows, the alternative values of Panther Canyon are very much associated with its women, and Thea "awakens to an alternate city consciousness inside the domestic spaces that she has entered on the most personal and private terms."[25] In doing so, she sheds the competitive drives that she has acquired along with her Chicago "city consciousness." At the end of *Crime and Punishment* it is through his relationship with Sonia that Raskolnikov can hope to discover a new life. For Cal in Bernard Mac Laverty's novel, it is in the rural idyll that he enjoys with Marcella, widow of the man in whose murder he has participated, that he finds a sense of peace; but even here the logic of the novel – a combination of the city's sectarian divisions and those within Cal's own nature – makes that idyll fragile and short-lived.

The hope that can be found in Isabel Allende's novels is gender-related, and can be seen as deriving from her feminism and her belief that the true political literature of our time is writing that allows "both men and women

to become better people and to share the heavy burden of this planet." Much
of that burden throughout the world is carried in its cities, Crowded, bomb-
scarred, riven by deep divisions of race and religion, sites of the highest
culture and the most abject poverty, they are home for a high proportion of
the world's population. These chapters have been gathered in the belief that
the literary witness is a vital element in helping us to understand the variety
of city experience as it has been in the past and as it is today. Perhaps the final
words may be given to Raymond Williams, whose own work has been so
influential in showing us how to make best use of that witness:

> the pulse of ... recognition is unmistakable, and I know that I have felt
> it again and again: the great buildings of civilisation; the meeting places;
> the libraries and theatres, the towers and domes; and often more
> moving than these, the houses, the streets, the press and excitement of
> so many people, with so many purposes ... I have felt also the chaos
> of the metro and the traffic jam; the monotony of the ranks of houses;
> the aching press of strange crowds. But this is not an experience at all
> ... until it has come to include also the dynamic movement, in these
> centres of settled and often magnificent achievement.... I find I do not
> say "There is your city, your great bourgeois monument, your
> towering structure of this still precarious civilisation" or I do not only
> say that; I say also "This is what men have built, so often magnificently,
> and is not everything then possible?"[26]

NOTES

1 Saul Bellow, *More Die of Heartbreak* (London: Secker & Warburg, 1987),
   p. 124.
2 Lewis Mumford, *The City in History* (London: Secker & Warburg, 1961);
   Griffith Taylor, *Urban Geography* (orig. 1949; London: Methuen, 1951).
3 David Harvey, *Social Justice and the City* (London: Edward Arnold, 1973).
4 Philip Larkin, "Church Going," in *The Less Deceived* (Hessle: The Marvell
   Press, 1955), p. 28.
5 See Raymond Williams, *The Country and the City* (London: Chatto & Windus,
   1973) for an excellent survey and discussion of the topic.
6 William Shakespeare, *As You Like It*, II. v. 52–9, in *William Shakespeare: The
   Complete Works*, ed. C. J. Sisson (London: Odhams Press, n.d. [1953?]), p. 270.
7 William Wordsworth, *The Prelude* (1805–6 text), Book VII, "Residence in
   London," in *The Prelude: A Parallel Text*, ed. J. C. Maxwell (Harmondsworth:
   Penguin Books, 1972), pp. 258, 262, 264, 288, 290, 292, 294.
8 Charles Dickens, *Nicholas Nickleby* (1838–9; Oxford: Oxford University Press,
   1950), pp. 408–9.
9 ibid., p. 1.
10 ibid., p. 8.
11 ibid., pp. 468–9.
12 ibid., p. 468.
13 Malcolm Bradbury, "The cities of modernism," in Malcolm Bradbury and James
    McFarlane (eds.), *Modernism* (Harmondsworth: Penguin Books, 1976), p. 99.

14 T. S. Eliot, *The Waste Land* (1922) lines 60 and 63, in *Collected Poems, 1909–62* (London: Faber & Faber, 1963), p. 65.
15 Bradbury, "Cities," p. 96.
16 *The Waste Land*, lines 373–5, p. 77.
17 Penelope Lively, *City of the Mind* (1991; Harmondsworth: Penguin Books, 1992), pp. 65–7.
18 ibid., p. 67.
19 R. Boyne and A. Rattansi, *Postmodernism and Society* (London: Macmillan, 1990), pp. 11–12.
20 See page 257 of this volume.
21 See pages 62–3.
22 See page 179.
23 See Robert E. Park and Roderick D. Mackenzie, *The City* (Chicago: University of Chicago Press, 1925; repr. with an introduction by Morris Janowitz, Chicago: University of Chicago Press, 1967).
24 See page 312.
25 See page 162.
26 Raymond Williams, *The Country and the City*, pp. 6–7.

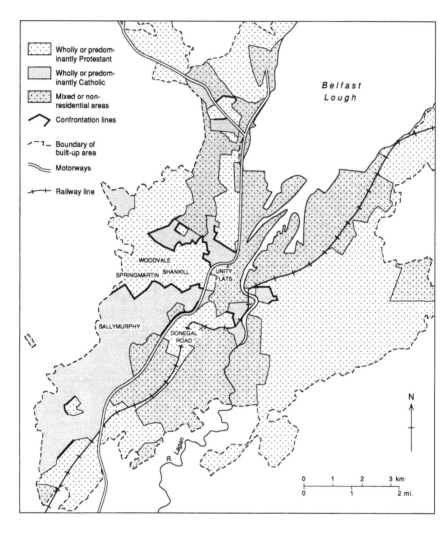

Legend:

Wholly or predominantly Protestant

Wholly or predominantly Catholic

Mixed or non-residential areas

Confrontation lines

Boundary of built-up area

Motorways

Railway line

Belfast Lough

WOODVALE
SPRINGMARTIN SHANKILL
UNITY FLATS
BALLYMURPHY
DONEGAL ROAD

R. Lagan

N

0 1 2 3 km
0 1 2 mi.

Map 2 Belfast

2

BELFAST

Bernard Mac Laverty's heart of darkness

Gary Brienzo

In the stories and novels of Bernard Mac Laverty,[1] the city – most often Belfast, but occasionally other urban centers such as London – is a powerful presence, as integral to his works as his characters, and at times even more important in being the single strongest motivator of Mac Laverty's protagonists and plots. From the direct simplicity of "Father and Son," or the even earlier "St. Paul Could Hit the Nail on the Head," to his most highly developed writing like the acclaimed *Cal*, the city, with its violence and alienation, shapes Mac Laverty's fictive vision, transforming it into what to many is "an eloquent plea for tolerance, good faith, and compassion – the virtues of an old-fashioned liberal,"[2] and to others "a primal gloom from which some sort of bright oblivion seems the only escape."[3]

Mac Laverty rejects the concept of an unremitting bleakness in his fiction, choosing as well to underscore "an ideal of good [in his stories] ... the love that exists between father and son, between boy and woman ... [the] stoicism and humour."[4] Yet Mac Laverty's refutation carries with it echoes of the darker forces at work within his tales, powers most commonly unleashed within the city and which are only countered by Mac Laverty's human connection or by the contrapuntal glimpses of a rural peace that offer his most successful characters a real, if ultimately failed, escape from urban horrors. The hope such love and country solitude hold out is a clearer definition of self, or a return to an earlier, integrated identity that is lost in the spiritual turmoil of the city as most ominously exemplified by Belfast. "He thought about how things happened to him but he brought nothing about. What he needed was self-discipline," Cal McCrystal, Mac Laverty's best-known character, derides himself as he lies in the abandoned cottage in which he has taken refuge since Protestant Loyalists burned his and his father's home (*Cal*, p. 116). And Liz O'Prey, the "Daily Woman" who is victimized by her drunken Catholic husband, her lecherous and chauvinistic Protestant employer, and by the city of Belfast itself with its public notice-boards of those killed in the troubles, thinks of herself as a "Clinchpoop ... somebody who didn't know what to do ... [with] no control over the direction of her life" ("The Daily Woman," in *A Time To Dance*, p. 117).

The believability of his characters, the sympathy Mac Laverty creates for them, the degree of self-knowledge each *seemingly* at first reveals suggest that these dark self-assessments are correct, and that what is lacking is a fortitude in Mac Laverty's victimized creations. Further insights by these same characters, however, point to a more universal heart of darkness in Mac Laverty's fiction, one that, for Cal McCrystal and Liz O'Prey, finds its fullest expression in the Ulster horror that emanates from Belfast. Cal thinks:

> To suffer for something which didn't exist, that was like Ireland. People were dying every day, men and women were being crippled and turned into vegetables in the name of Ireland. An Ireland which never was and never would be. It was the people of Ulster who were heroic, caught between the jaws of two opposing ideals trying to grind each other out of existence.
>
> (*Cal*, p. 92)

It is such insights, coupled with the clearer understanding of self Mac Laverty allows his characters when they are briefly freed from the city, that most clearly reveal the author's sympathy for his overwhelmed characters and his distrust of the urban centers in which human life often exists in indifferent isolation, if not in open peril. An understanding of the geography of Belfast, especially as the city is depicted in Mac Laverty's fiction, as well as a knowledge of Mac Laverty's own early life in Belfast, do much to illuminate the image of the city in literature as Mac Laverty envisions it. He was born into a comfortable Catholic family of artisans and tradespeople and raised among his relations in the Newington area of the city. Despite this early security, however, at the age of 12 Mac Laverty suffered the death of his father, a loss that seems seminal to his later depictions of incomplete families. As a young adult, Mac Laverty began working as a laboratory technician, a career in which he seems to have shown little early interest but which he pursued for ten years until, at the urging of his wife, he enrolled as a mature student in literature at Queen's University. His acquaintance there with lecturer Philip Hobsbaum and with such budding writers as Michael Longley, Seamus Heaney, Frank Ormsby and Paul Muldoon encouraged his own devotion to fiction, so that, by 1981, he could give his full-time efforts to writing.[5] Along the way, Mac Laverty earned Scottish Arts Council Book Awards in 1977 and 1980 and other honors, including the runner-up award in the *Guardian*'s Fiction Prize for his first novel, *Lamb*, and the literature prize in the *Sunday Independent*'s Annual Arts Award for his story collection, *A Time To Dance*.

As his career progressed, Mac Laverty became so disenchanted with his violence-torn home-place as to leave Belfast for Scotland, first as a teacher in Edinburgh and then in the Isle of Islay, and at last as a writer in Glasgow, where he now lives with his wife and children. Yet even as Mac Laverty physically left Northern Ireland, he carried Belfast with him in his art. His characters often bear the permanent stamp of Belfast, so that in one of his later stories, "Some Surrender," discussed below, an aging father and son can climb the Cave Hill that overlooks the city and look down, both with regret at the demolition that is still reshaping the inner city and with appreciation of "the blue Lough" of Belfast which rests between the Hills of Down and the Antrim Plateau (*The Great Profundo*, p. 123).

Natural settings aside, the human geography of Belfast is more powerful still in Mac Laverty's fiction, so that, with divisions of Protestant (Springmartin, Woodvale, Shankhill, Donegal Road, Sandy Row, and Ballymacarett) and Catholic (Ardoyne, Ballymurphy, Falls Road, Unity Flats, and New Lodge Road)[6] intact, the contemporary city can be defined in terms of the Inner City, the Fringe, and an expanding metropolitan region which some urban scholars depict as a series of "networks [that] link up the whole of Northern Ireland to the core of Belfast ... [an] area ... that ... extends out from the city centre for about twenty-five miles and from which many people travel in to Inner Belfast and the Fringe."[7] This connection and interrelatedness become chilling in a further description of the modern city as "expanding [its] reaches on the pioneer fringe [in which much of the action of *Cal* is set, so that the outer area is] ... not a pioneer fringe of isolation but of ready contact with the centre, thanks to the transportation and communication technology now available."[8]

This seemingly neutral assessment is echoed with deadly effect in *Cal*, when the young protagonist reflects on the futility of self defense, thinking, "His killers would be in Belfast having a cup of tea by that time [the time it would take to reach into his jacket for a gun] if they had anything above second gear" (p. 89). Mac Laverty's growing skill in expressing his themes of urban suffering and indifference can be seen from the earliest stories, where the city often exists more in asides or undertones that suggest the somber color of a work, to the writing of *Cal*, in which Ulster's violence is personified in the twisted or diminished lives of nearly every character, and, beyond *Cal*, in the short fiction that followed his acclaimed novel.

"Demolition's the thing at the moment. They're knocking down the half of Belfast," Mary, in "St. Paul Could Hit the Nail on the Head," from Mac Laverty's first collection of stories, *Secrets*, tells a visiting cousin (p. 20). In this story demolition is more than just a topic of conversation for a woman ill at ease with an uninvited guest; it is a metaphor for the strain Mary feels herself under, as she reluctantly asks her distant relation, Father Malachy, to stay with her family on his visit to Belfast, aware both of the old priest's loneliness and the lack of tolerance that her Protestant husband, Sam, has for the Roman Catholic church. Mac Laverty underscores the woman's personal discomfort in this story, which begins with the line "All that afternoon Mary's world seemed to be falling apart at the seams" (p. 18), with the larger urban decay and demolition that surround Mary's private world. He even more skilfully achieves the interplay of personal and social upheaval in

Belfast by making Mary's own family not just another witness or victim of the upheaval, but, through the work of her successful contractor husband, active agents of the urban destruction. Thinking of Sam's "business of devastation," Mary recalls the one time she had taken her boys to watch Sam work and been repelled by the bulldozers snarling in,

> crashing through kitchen walls, teetering staircases, leaving bedrooms exposed.... Mary felt she shouldn't look, seeing the choice of wall-papers ... it was too private. She rounded up the boys, "Come on, come on, this is no sight for children," and went home, remaining depressed for the rest of the day.
>
> (p. 20)

Mac Laverty provides no resolution for the muted loss and alienation of the story, unless it is in Mary's quiet and tormented sympathy which provides a tentative unity of the disparate elements at work in the tale.

In the later "The Daily Woman," Mac Laverty uses a similarly sympathetic, if less personally powerful, character, Liz O'Prey, to illustrate the malevolence of Belfast and to attempt to mollify it. "Boy, have you got problems," an American journalist Liz meets in a chance encounter says of Ireland, a remark Liz tellingly misreads by asking "Me?" (*A Time To Dance*, p. 114). Within minutes, however, after Liz, lulled by the American's encouraging conversation, has revealed much about her own life, her poverty and the drunken beatings she regularly receives from her husband Eamonn, the journalist says of Liz's own family, "You have your troubles" (p. 115). Again Mac Laverty interweaves personal and social disintegration within the urban setting, with Liz moving with no apparent recourse between Eamonn, whose life is circumscribed by a "horseshoe of wear in the pavement from one door [of the pub or the bookies] to the other" (p. 106), and Mr. Henderson, her bigoted employer who molests her and boasts of his openness in employing Catholics, while hiring Liz to serve his guests and wash his dishes "in her best dress" (p. 104).

In a larger sense, Liz is also victimized by the city itself, with its lines of demarcation never more visible than when she stops to read "the black notice-board which kept up with the death toll of the troubles" (p. 110) or is frisked, twice, in the Belfast hotel to which she has fled for temporary shelter, because, as she is told by a woman in uniform, "You have a Belfast address, you have no luggage" (p. 111). From the opening line – "She woke like a coiled spring, her head pressed on to the mattress, the knot of muscle at the side of her jaw taut, holding her teeth together" (p. 99) – to the last – "She ... tried, as she drifted off to sleep, to forget the fact that Eamonn, for the loss of her weekly wage, would kill her when he got her home, if not before" (p. 118) – Mac Laverty has created in Liz O'Prey, as in Mary before her, a rounded character as well as a personification of the fear and tension that distort her country. He also writes in "The Daily Woman" with a sense of understanding and a balanced tension that find fullest expression in *Cal*.

Other works fall between these stories and the consummate *Cal* in the intensity of the anguish and alienation they reveal, with the city and references to it often underscoring private grief. "Father and Son," the opening story in *A Time To Dance*, is thematically nearest *Cal* in depicting in a father and son's relationship the devastation of all relations when they are severed by secrecy and violence. "This is my son who let me down. I love him so much it hurts but he won't talk to me," the unnamed father relates (*A Time To Dance*, p. 9), and later, "My son is breaking my heart" (p. 10). Fearing for the safety of the son who ignores and apparently despises him, the father tells of other details he learns from his secret watching, like the presence of a gun beneath his son's pillow and the young man's involvement in the Belfast violence implied in the older man's observation, "My son rides pillion on a motor-bike. Tonight I will not sleep. I do not think I will sleep again" (p. 14).

For his part, the son conveys his own contempt of his father, thinking "Your hand shakes in the morning, Da, because you're a coward. You think the world is waiting round the corner to blow your head off" (p. 11), and

> "You live in fear. Of your own death. Peeping behind curtains, the radio always loud enough to drown any noise that might frighten you.... You undress in the dark for fear of your shadow falling on the window-blind.... By your bed a hatchet which you pretend to have forgotten to tidy away. Mice have more courage."
>
> (p. 12)

The themes of violence, father-and-son alienation, and an older man's deterioration in the face of urban terror are explained in more complexity and detail in *Cal*, but in "Father and Son" all conflict is resolved in sudden death, as the father hears a gunshot at his door and goes out to find his son "lying on the floor, his head on the bottom stair, his feet on the threshold," and realizes "The news has come to my door. The house is open to the night" (p. 14).

Two longer stories, "Between Two Shores," from *Secrets*, and "My Dear Palestrina," from *A Time To Dance*, bridge the literary gap between the simple telling of "Father and Son" and the artistry of Mac Laverty's later fiction. Though seemingly as unlike each other as they are dissimilar to Mac Laverty's Belfast stories, "Between Two Shores" and "My Dear Palestrina" reveal much about the struggle to develop and retain a sense of self that is especially jeopardized in Mac Laverty's urban settings. "Your values all belong to somebody else," his more worldly mistress tells a displaced Irish laborer in London as she gently raps his forehead and tells him to think ("Between Two Shores," in *Secrets*, p. 55). Confused by her words and not knowing what she means, the protagonist's only recourse is to recall with longing the wife, children and farm in Donegal he must leave for months at a time to find work in London: "he wanted to be at home among the sounds that he knew. Crows, hens ... the distant bleating of sheep on the hill, the rattle of a bucket handle, the slam of the back door. Above all he wanted to see the children" (p. 54). Such descriptive touches find their echo and fulfillment in the later *Cal*, in what one critic calls "a longing for the quiet life that ... characters can never take for granted."[9] In this story, too, because of the economic condition of his homeland and his own cultural impoverishment, the narrator cannot take such things for granted, and he returns home tortured by thoughts of the "tiny [syphilitic] corkscrews" he has contracted in London "boring into his wife's womb" (p. 54).

"My Dear Palestrina" is a tale of a similarly threatened quest for survival and self-knowledge, one, too, marred by ignorance and occurring on a sort of voyage, that of a young boy traveling the path between town and his music teacher Miss Schwartz's country home. Danny McErlane is almost dragged along the footpath by his mother in the story's opening scene, as he is forced to play a piano left to his family by an uncle. Danny's musical ability, however, and his growing love of independence and maturity soon turn the path, the exotic Miss Schwartz, and the politically radical blacksmith he encounters on his regular walk to Miss Schwartz's, into the centers of the boy's life. "You're coming to an age now when you've got to think. Don't accept what people tell you – even your father. Especially your father. And that includes me," the blacksmith tells the boy when he learns that Danny's father has warned him away from the smith's shop. "If you want to come back here, Danny, you come. The belt shouldn't stop you. You've got to be your own man, Danny boy" (*A Time To Dance*, pp. 60–1). The belt – and the narrow repression it represents – proves to be exactly what does keep Danny away, not only from the blacksmith but from the even more important Miss Schwartz and the larger world of self-knowledge and expression she represents. Chastised by his morally conventional parents for his loyalty to his pregnant and unwed teacher, Danny can only reflect on his loss, on never being

> allowed back again to talk and work with Miss Schwartz ... to be allowed to call in on the blacksmith and be talked to as if he were a man.... He thought of being deprived of all this ... [and] began silently in his own dark to cry.
>
> (p. 65)

The force of "My Dear Palestrina" lies in Mac Laverty's sensual depictions of the moments that make up Danny's new awareness of reality, an attention to the small detail and the ordinary delights of everyday life that has made some consider the work "his most perfect short story."[10] These scenes are also reminiscent of *Cal* and, to a lesser extent, Mac Laverty's other novel, *Lamb*, in the images Mac Laverty chooses to highlight, most often of rural scenes of peace poised against a more darkly imposing urban background which threatens to overwhelm them. While watching for a satellite to cross the night sky, for example, Danny and Miss Schwartz enjoy a night "cold, black and clear as a diamond," with a "swirl of stars [that] covered the sky so that it seemed impossible to put a finger between two of them" (p. 58). Danny is warmed by a glimpse of "the line of her arm" and the smell of Miss Schwartz's perfume, tinged with "the slightest taint of her own smell," as Miss Schwartz herself is entranced by "the pinpoint of light threading its way up the sky" and pausing above them in the absolute stillness, until she tells the boy, "The music of the spheres. Do you hear it, Danny? . . . It's a sort of silence" (pp. 58–9).

Such a reverential solitude also reigns in Miss Schwartz's garden, the place where she plays her favorite music for Danny on an old wind-up gramophone and serves him hot tea with lemon after his lessons. She later tells the boy she bought the house because of the garden, one just like the garden of her childhood home in Praszka, but when asked by Danny why she chose not to return to her native Poland, Miss Schwartz admits, "the longer you are away the more impossible it is to return," and that, although she left in "a time of fear," it is "a terrible thing" to live in exile (p. 50). Memory and rural peace ultimately cannot save Miss Schwartz from ruin at the hands of the narrow townspeople who spurn her and deny her her livelihood as teacher and church organist, and in this story of longing and loss Mac Laverty creates a fine example of the rural foil to urban horror he would create in even more graceful detail in *Cal.*

In *Lamb,* another work that preceded Mac Laverty's most finished novel, a garden image also provides one of the few beautiful memories of protagonist Michael Lamb, although, as for Miss Schwartz, memory alone proves ineffective in altering an unbearable existence. Indeed, in *Lamb,* Mac Laverty seems most nearly to write with the darkness that has earned the early novel the label of "unbearably claustrophobic,"[11] as neither in the Irish countryside, one of the few retreats from urban horror in *Cal,* nor in the depraved humanity his characters encounter in London, can salvation be found.

"Michael had never liked the city – any city," he reflects during one of his many nights of agonized sleeplessness in London, as he listens to the unceasing sounds of a distant factory (p. 84). He also feels alone and isolated among London's crowds, where he finds himself "in the strange position of wanting to see someone he knew" (p. 96), even as he and Owen Kane, the child with whom he has flown Ireland in an increasingly futile rescue attempt, continue to hide from the authorities. The pair's inescapable present is no better in their native Ireland, where Owen has fled first from a Dublin housing estate he describes as "like Ballymun – only it's rough" (p. 16), and later, with Michael Lamb's aid, from a repressive boys reform school run by the cruel Brother Benedict who boasts that his Home teaches "a little of God and a lot of fear" (p. 14). Unfortunately, both for Owen and for Michael

Lamb, who was known as Brother Sebastian when he taught at Brother Benedict's borstal, fear is the lesson learned in all the places they seek, from the London hotels they must leave often to prevent detection, to the "squat," an abandoned London building marked for demolition in which Owen is sexually accosted (p. 119), to even the rural County Donegal described by Lamb as the "wasteland of Donegal, tweed green fading into brown, fading in the pale blue of the mountains, uninterrupted by trees of any sort. A place without shelter" (p. 142).

Physically and spiritually besieged, Lamb's only escape is in memory, usually of a rural past with his father, in which the two shared fishing and rock-climbing and admiration for the very hawks that killed the father's lambs and yet which were still revered as the "rulers of the air" (p. 86). Michael's own more recent memories of maintaining the garden for the Home also offer brief comfort, as his "mind relaxed into the past" while thinking of the spears of plants "bursting through into the air" or of "the water remaining in single droplets, glittering and rattling ... like diamonds" (pp. 97–8). But memory alone fails Michael and Owen, and Michael's circumscribed garden is not the farm to which Cal McCrystal, also fleeing his persecutors, makes a real if impermanent escape, as the urban "squat" in which they take temporary shelter is not the rustic cabin in which Cal finds the only adult love he has known for anyone but his father. The single mattress, dirt-streaked windows, and empty chill of the condemned building offer no comfort, and soon Lamb and Owen are swept along by Lamb's "plan" of forced self-annihilation (p. 105) that will leave the boy dead and the man "gutted ... as if his insides and his soul had been burned out" (p. 152).

As *Cal* offers more balance than *Lamb*, and more scope than even such accomplished earlier stories as "My Dear Palestrina" or "The Daily Woman," it also surpasses these works in more powerfully depicting the urban horror and personal devastation that emanate from Belfast. From its opening scene, of Cal visiting the abattoir at which his father works, to the novel's closing line in which the youth welcomes bloody atonement for the sin which the violence of the city has forced upon him, *Cal* is shot through with loss and separation. This tension is obvious as the son reaches into his father's white butcher's coat which is "japped all over with blood and stiff with cold fat" (p. 8) to borrow a cigarette, as it is apparent in the messages painted on tin lids a religious zealot nails around the countryside, with such biblical sayings as "The Wages of Sin is Death" (p. 7). In his characters, too, Mac Laverty personifies the human indifference that can make the intellectual IRA leader Finbar Skeffington mouth such platitudes to a sceptical Cal as "This is a war, Cahal ... You have to steel yourself ... think of the issues, not the people" (pp. 25–6), or can make the brutal Crilly, once a friend of Cal's, happily commit acts of violence.

Even worse, the violence is universal in this novel, with Mac Laverty laying blame on neither side but instead on all who place ideologies before human life. "Ulstermen would die rather than live under the yoke of Roman Catholicism," a Protestant acquaintance tells Cal. "I'm serious, Cal. I would *die* rather than let that happen" (p. 123). The willingness of partisans on both sides to kill and die creates the tension that renders normal human relations impossible and that makes Cal and his father, Shamie, the last Catholics living in a Protestant-dominated housing estate, reach for a pistol and fill their bath tub with water after receiving a note that reads, "GET OUT YOU FENYAN SCUM OR WE'LL BURN YOU OUT. THIS IS YOUR 2ND WARNING, THERE WILL BE NO OTHER" (p. 29). It is the same unthinking hatred that ends in the house being burned, with Shamie being so unnerved that, by the end of the novel, he is undergoing shock treatments in a nearby hospital after having turned, in the words of a friend, from "iron to plasticine overnight" (p. 159). The same malice has given Cal, for his unwilling compliance as driver in an IRA murder, "a brand stamped in blood in the middle of his forehead which would take him the rest of his life to purge" (p. 99).

In writing with the clarity that has earned this novel praise for its considerable beauty, however, Mac Laverty both mitigates the darkness of Cal's world and in a way enhances it by his glimpses of simple human joys that are not totally obliterated even in Ulster. He accomplishes this in the characters of Marcella Morton, widow of the Protestant Reserve Policeman Cal has helped kill despite himself, and her daughter, Lucy, and in the relationship between them and Cal. Unlike most things in his world tainted by Belfast's horror, Cal finds Marcella's speech "clean and beautiful," her smile "open" (p. 82). Scenes Cal observes between Marcella and Lucy, away from the city on the farm on which they live with Marcella's parents-in-law, also provide images of a natural beauty that is only matched by the other rustic impressions Cal comes to know there. Helped by Marcella in cleaning up an abandoned cottage, after Cal's own home has been burned, Cal discovers an old pump which he and Marcella prime until water falls with "a faint cool green against the white of the enamel" of a bucket (p. 116); he watches as Marcella wipes the juice of a blackberry from Lucy's chin and then cleans her own hand on the grass (p. 114), and as, comforting the child from a briar prick, she runs to Lucy, holds her and "leaned back from the child to see her whole face [though] they remained snugly together at the waist" (p. 115); and both Cal and Marcella come to know perhaps the greatest joy given them in this novel as they go berry-picking in the country, with Lucy walking "between them, holding a hand of each," and they view countryside of "a deep winter green" falling away

> to the blue mountains of Slieve Gallon ... crossed by dark random lines of trees and hedges ... [with] here and there a red barn or a white gable ... and a window [that] shone like a diamond ... [and] cows all facing in the same direction [grazing] their way across a field.
>
> (p. 129)

Even these scenes of country peace, however, are not free of the terrors of Belfast, as the berry-picking interlude ends with the eruption of a land mine placed in a nearby field and the destruction of a Friesian cow – the rear half of which Cal finds, with its "udders, hindquarters with muscles red-raw and still jiggling" (p. 133) – and with Cal vomiting, as he has both at the abattoir and after witnessing Robert Morton's murder, as he sees the white hides of the other animals spattered and "japped with blood" (pp. 133–4), like Shamie's coat in the novel's opening paragraphs. This scene, as well as standing in starkest contrast to the quiet passages before it, also underscores the impossibility of personal fulfillment and direction in Northern Ireland and offsets Cal's sense of burgeoning content on the farm, a peace found not only with Marcella but also with such basic rhythms of farm life as the comforting lowing of cows (p. 89) and the "soft milky smell on their breath" (p. 75). Such intrusions as the mines and the relentless pryings of Crilly and Skeffington soon find Cal out, and he is delivered by their duplicity into police hands where he stands, "listening to the charge, grateful that at last someone was going to beat him to within an inch of his life" and trying to envision how he can ever seek Marcella's forgiveness or regain her love (p. 170).

It is in just the depth of this love that Cal finds completion. Like the earlier Liz O'Prey, who anticipates further beatings from her husband, Cal seems to find an end to spiritual suffering in imminent physical atonement. Unlike Liz, however, he has found a human bond that is as saving as the urban violence that surrounds him is damning. "Would you die for me?" Marcella innocently teases Cal (p. 152) before his denunciation by his former allies, and yet it is ironically only in his unspoken answer, the need to stay with her that is greater than his desire for survival, that the youth becomes most fully developed. Similarly, the feeling that Mac Laverty describes as "the love that exists between father and son"[12] is also what moves his fiction from the bitterness of the early "Father and Son" to the warm intensity of *Cal* itself or even to the simple affection described in one of his latest stories, "Some Surrender," from *The Great Profundo* collection. Cal and Shamie's relationship, though superficially like that of the equally threatened, motherless family in "Father and Son," resembles its early counterpart in no other way. While Cal demonstrates a love selfless enough to make him risk exposure simply to bring his traditional Christmas gift of after-shave and a shaving stick (p. 158) to his father, the exchanges between Mac Laverty's first father and son are marked by confused bitterness:

"Wake up, son. I'm away to my work. Where are you going today?"
"What's it to you?"
"If I know what you're doing I don't worry as much."
"Shit."

(A Time To Dance, p. 9)

The softened, more developed vision of filial relations portrayed in *Cal* continues to mark the fiction that followed the novel, in particular a story like "Some Surrender." It is the tale of love between a Protestant father and his grown son who has married a Roman Catholic woman and left Belfast for Dublin, to the dismay of his father and the absolute, unforgiving abhorrence of his mother. The reconciliation of 44-year-old Roy with his architect father, on the father's seventy-fifth birthday, is also layered with more specific geographical references to Belfast than any of Mac Laverty's other works, in part because of the older man's architectural contributions to the city. Such sections of Belfast and its surrounding area as Cave Hill, the Shore Road, the decaying Divis Flats and Belfast Lough are frequently mentioned in the story, as are the old man's multi-storey structures that, due to the corrosion of the wet Belfast climate, are fast being destroyed. "The constructive thing to get into these days is demolition" (p. 129) the father jokes, all the while making half-serious allusions to the "design fault" both of his own failing human body (p. 118) and of the city around them. Such faults aside, and despite the younger man's continued rift with his mother, the story ends in completion and love, with the son helping the father down Cave Hill, offering his hand and being "startled to feel his [father's] shoulder blades, the shape of butterfly wings, through the thin material of his jacket" (p. 129). This closing image reveals the need for redeeming love that Mac Laverty acknowledged in *Cal* and that helped make the harmony of the later "Some Surrender" possible.

Also in keeping with the graceful, detailed realism that he perfected in *Cal*, Mac Laverty's mature fiction provides more insights and instinctive appeals to compassionate reason than direct answers to the pain that is centered in Belfast. As a result, his finest writing, represented by *Cal*, becomes no mere glimpse of an unforgiving place, but, in one critic's words, a work that "opens into a world larger than itself with a confidence that makes one take that world on the novel's terms."[13] From its incarnation in the early stories, Mac Laverty's vision moves to so clear an expression in *Cal* as to justify one critic's prediction in labeling the novel the *Passage to India* of the Irish troubles:[14] "'No, not yet,'" Forster writes of his characters' desire for reconciliation and self-knowledge, "'No, not there.'"[15] These words might as aptly apply to Liz O'Prey's questions about her future, after her betrayal by her predatory employer, or to the aging architect's regret in "Some Surrender" as he seeks peace between his wife and son. Most important, however, longed-for mediation and peace seem to underscore all of Mac Laverty's best fictional creations, imbuing his artistry with the beauty and clarity that move beyond the simple telling of stories, to the sharing of something universal.

NOTES

1 The following are referred to by title only: *Cal* (New York: George Braziller, Inc., 1983); *The Great Profundo and Other Stories* (New York: Grove Press/ London: Jonathan Cape, 1987); *Lamb* (New York: Penguin Books, 1981); *Secrets and Other Stories* (1977; London: Allison & Busby, 1984); *A Time to Dance and Other Stories* (New York: George Braziller, Inc., 1982).

2 Jim Miller, "In Ireland's name," *Newsweek*, 102 (September 5, 1983), p. 68.

3 Charles Hunter, "Bernard Mac Laverty: A natural film writer," *The Irish Times* (October 18, 1986), Weekend section.

4 Paul Campbell, "In the beginning was the written word," *The Linene Hall Review*, I (Winter 1984–5), p. 5.

5 Hunter, "Bernard Mac Laverty."

6 Malcolm Falkus and John Gillingham, (eds.), *Historical Atlas of Britain* (London: Granada, 1981), p. 162.

7 J. C. Beckett and R. E. Glasscock, *Belfast, the Origin and Growth of an Industrial City* (London: BBC Publications, 1967), p. 171.

8 ibid., p. 182.

9 Michael Gorra, "Guilt and penance in Northern Ireland," *New York Times Book Review* (August 21, 1983), p. 17.

10 Hunter, "Bernard Mac Laverty."

11 Gorra, "Guilt and penance."

12 Campbell, "In the beginning."

13 Gorra, "Guilt and penance."

14 ibid.

15 E. M. Forster, *A Passage to India* (New York: Random House, 1924), p. 322.

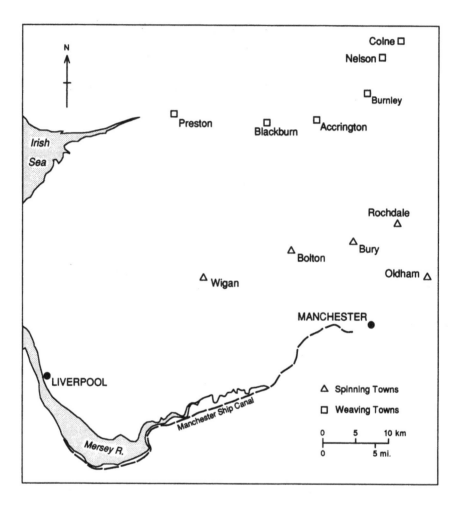

Map 3 Manchester

3

MANCHESTER AND MILTON-NORTHERN

Elizabeth Gaskell and the industrial town

Peter Preston

INTRODUCTION:
ENCOUNTERING MANCHESTER

"It is the philosopher alone who can conceive the grandeur of Manchester, and the immensity of its future. There are yet great truths to tell, if we had either the courage to announce or the temper to receive them." Benjamin Disraeli's words, published in *Coningsby* in 1844,[1] remind us that in the first half of the nineteenth century, Manchester represented a new reality, whose very existence had yet to be absorbed and whose full significance was still uncertain. For many people, Manchester symbolized both the most splendid achievements and the most abject miseries visited on British society by the Industrial Revolution. *Chambers' Edinburgh Journal* in 1858, while noting the muddiness of Manchester's streets and the density of its smoke, concluded that nothing "can prevent the image of a great city rising before us as the very symbol of civilization, foremost in the march of improvement, a grand incarnation of progress."[2] Other observers, including many of those who were as alive as *Chambers* to the excitement and energy generated by Manchester's rapid development into a major industrial and commercial center, were readier to emphasize the dangers inherent in a growth that had taken place so quickly and in so unplanned a manner. That growth had been rapid indeed: the population rose from 40,000 to 250,000 between 1794 and 1831, with a 45 per cent increase in the decade 1820 to 1830; and by the census of 1851 the figure had risen to 338,000. The growth was dependent upon individual enterprise and went ahead virtually unchecked by any govern-mental controls, either at local or at national level. Until 1832 Manchester, like Birmingham, was unrepresented in Parliament; until 1838, when it became an incorporated town, it lacked even any effective system of local government. Not that governmental interference would have been welcome, since the philosophy of Manchester – and part of its distinction was that it had its own economic and political philosophy – was to pride itself on its independence and the self-determining energy of the powerful individuals

31

who owned and directed its industrial and business concerns.

Visitors to Manchester – and there were many in the years 1830–50 – record the energy and excitement abroad in the burgeoning town, but reserve their strongest writing for the squalor and ugliness of the buildings, the constant noise of machinery, the smoky, filthy atmosphere and the failures in relationships between the social classes. What to a hard-line philosopher of the Manchester school might be the necessary consequence of progress and the creation of wealth seemed to these commentators a cause for concern and outrage and the occasion of apocalyptic images of horror and potential conflict. Hugh Miller, a Scotsman who visited the town in 1845, was struck by "the lurid gloom of the atmosphere that overhangs it ... a murky blot in one section of the sky ... [which] ... seems spread over half the firmament" with each of "the innumerable chimneys ... bearing atop its own troubled pennon of darkness." The river Irwell is "considerably less a river than a flood of liquid manure, in which all life dies ... and which resembles nothing in nature, except perhaps the stream thrown out in eruption by some mud-volcano."[3] Alexis de Tocqueville, who visited Manchester ten years earlier than Miller, emphasizes the chaotic, unplanned and unfinished appearance of the town. The unhurried, irregular variety of nature has been interrupted, but in no positive way, to create a kind of no-man's-land: "the wretched dwellings of the poor are scattered haphazard.... Round them stretches land uncultivated but without the charm of rustic nature, and still without the amenities of a town."[4] Urban development in itself is not the problem for de Tocqueville; his concern is for what happens when that development is allowed to proceed without any attempt at planning or control. For him, Manchester is a symbol of human creative powers run riot: "Everything in the exterior appearance of the city attests the individual powers of man; nothing the directing power of society. At every turn human liberty shows its capricious creative force. There is no trace of the slow continuous action of government." He goes further than Miller in his description of the Irwell, calling it "the Styx of this new Hades."[5]

Central to de Tocqueville's vision of Manchester is the contrast he found between "the palaces of industry" and the squalid dwellings of the poor, by whose labor the splendor of those "palaces" was maintained. Their houses, situated on the least desirable marshy land, seem "the last refuge a man might find between poverty and death," yet even the "damp, repulsive holes" of their cellars are "crowded pell-mell." More significantly, the palace-like factories "keep air and light out of the human habitations which they dominate; they envelope them in perpetual fog; here is the slave, there the master."[6] Here de Tocqueville touches on an aspect of Manchester that gave rise to deep concern among observers in the 1830s and 1840s. To an increasing number of visitors it seemed that the great leaders of industry, in their pride, self-reliance and single-minded pursuit of wealth, were blind or indifferent to the great suffering that lay round them. It was as if the fog that kept light and air from the dwellings of the poor also rendered them invisible to the rich. In 1839, Canon Parkinson, a local man, pointed out that "there is no town in the world where the distance between the rich and the poor is so great, or the barrier between them so difficult to be crossed."[7] Friedrich Engels, in *The Condition of the Working Class in England* (1844–5), follows de Tocqueville in linking that distance between the classes to the very geography of Manchester.

> The town itself is peculiarly built, so that a person may live in it for years, and go in and out daily without coming into contact with a working-people's quarter or even with workers, that is, so long as he confines himself to his business or to pleasure walks. This arises chiefly from the fact that, by unconscious tacit agreement, as well as with outspoken conscious determination, the working-people's quarters are sharply separated from the sections of the city reserved for the middle-class.... And the finest part of this arrangement is this, that the members of this money aristocracy can take the shortest road through the middle of all the labouring districts to their places of business, without ever seeing that they are in the midst of the grimy misery that lurks to the right and the left.[8]

Both Parkinson and Engels perceive, the latter more consciously, that Manchester is not only the symbol of a new age in terms of industrial and commercial development; it also brings into focus the existence of a new set of relationships between the industrialists and their employees. These relationships are inextricably linked to the growth in self-awareness of the urban working class, new in its size and concentration, conscious of its role in creating the wealth of the capitalists, and beginning to articulate its desire not only for a greater share in that wealth, but also for an opportunity to participate in determining the direction of society and a say in its government. Its ability to do these things was severely limited not only by the restricted electoral franchise but also by proscriptive legislation and harsh governmental responses to any manifestation of radical protest. Already, in 1819, Manchester had acquired notoriety when a peaceful demonstration at what became known as Peterloo was broken up by a brutal militia charge which left eleven dead and about four hundred injured. "The capital of an industry," as Asa Briggs puts it, "became the capital of discontent"; and he quotes *The Times*, which in reporting the massacre, said of the working class in Manchester that "their wretchedness seems to madden them against the rich, who they dangerously imagine engross the fruits of their labour without having any sympathy for their wants."[9] The situation was complicated and exacerbated by the fact that the class against which the working people were pitting themselves was also new in its consciousness of its distinct identity. The new urban aristocracy of rich industrialists and businessmen was itself seeking a political role commensurate with its contribution to the nation's economic life. If Chartism was the principal manifestation of the working-class struggle for political power, the campaign against the Corn Laws, very much centered in Manchester, shows the well-organized leaders of business flexing their muscles against metropolitan domination.

LITERARY RESPONSES TO THE INDUSTRIAL TOWN

Thomas Carlyle, who saw Manchester as "every whit as wonderful, as fearful, as unimaginable, as the oldest Salem or prophetic city," but built "upon the infinite abysses,"[10] was very alert to the problems of the new industrial towns and the dangers for society if they were ignored. As Asa Briggs points out, Carlyle's insistence that the answers to contemporary problems lay in the application of imagination as well as logic opened the way for novelists to deal with contemporary, industrial themes.[11] It was an opportunity for the novelists to act as social explorers, giving a voice, as Carlyle asked, to those who had none,

> How inexpressibly useful were true insight into it; a genuine understanding by the upper classes of society what it is that the under classes intrinsically mean; a clear interpretation of the thought which at heart torments these wild inarticulate souls, struggling there, with inarticulate uproar, like dumb creatures in pain, unable to speak what is in them![12]

Like many middle-class observers, Carlyle underestimated the power of the working class to articulate its own concerns; and when faced with the manifestations of that articulation, such as Chartist activity, he reacted with fear and rejection. Nevertheless, he saw an urgent need to make known the problems that he thought lay at the heart of the country's growing prosperity – to expose the insufficiency of the cash-nexus and to bring, particularly to a metropolitan and southern middle-class readership, some understanding of life for the masses in the expanding northern towns.

The novelists who can be seen as answering Carlyle's call were themselves largely middle-class and metropolitan, outsiders to Manchester and towns like it in both class and geographical terms. They were genuinely explorers, foreign to the experience of both the new business class and the concentrated masses of the workers; foreign, too, to the striking physical realities of the new industrial towns. It is for these reasons that Elizabeth Gaskell occupies a peculiarly interesting position among the social novelists of the 1840s and 1850s. More than any of her best-known contemporaries she writes as an insider. For her, the North was not an unknown region to be visited or "researched" in the interests of fiction. Much of her early life was spent in Cheshire, and after her marriage in 1832 she lived in Manchester itself. Manchester was, therefore, a place she knew well, part of whose development she had witnessed and whose people and problems she understood. Her husband was a Unitarian minister and she cooperated with him in a good deal of charitable work, in the course of which she visited some of the worst slums. At the same time, through her participation in Unitarian intellectual and philanthropic circles, she came to understand the employers and some of the problems they encountered. She also appreciated the potential of Manchester and the strength of its native culture, and Angus Easson has aptly described *Mary Barton* as "a novel of local pride."[13]

Elizabeth Gaskell's description of the origin of the novel in the preface to *Mary Barton* suggests how the press of local circumstances turned her from her original intention:

> I had already made a little progress in a tale, the period of which was more than a century ago, and the place on the borders of Yorkshire, when I bethought me how deep might be the romance in the lives of some of those who elbowed me daily in the busy streets of the town in which I resided.[14]

The novel was published in London in 1848 and will at first have attracted a largely metropolitan readership. Yet its point of view is unquestionably Mancunian and it is narrated from deep within the experience of the town.

Determining the narrator's standing-ground is important for our understanding of the novel, for it has a profound effect on the way Manchester is presented. When Dickens and Disraeli write about the industrial town, they do so with an air of discovery, and the rhythm and texture of their prose reflects the impact of a new vision. Their reaction may be one of shock, a recoil of horror; or there may be a more positive reaction, which leads to a rhetorical meditation on the potential at the heart of this new experience. In *The Old Curiosity Shop* (1841), for instance, Nell and her grandfather pass through the Black Country on their flight across England, spending one night in a steel mill and another in a town. Two brief passages will serve to give the flavor of Dickens's description. First, the steel mill,

> echoing to the roof with the beating of hammers and roar of furnaces, mingled with the hissing of red-hot metal plunged in water, and a hundred strange unearthly noises never heard elsewhere; in this gloomy place, moving like demons among the flame and smoke, dimly and fitfully seen, flushed and tormented by the burning fires, and wielding great weapons, a faulty blow from any of which must have crushed some workman's skull, a number of men laboured like giants.... [some] drew forth, with clashing noise upon the ground, great sheets of glowing steel, emitting an insupportable heat, and a dull deep light like that which reddens in the eyes of savage beasts.[15]

This weird, other-worldly place is guarded by the furnace-keeper, who spends his life watching and feeding the fire, and seems to identify with it: "'It has been alive as long as I have'", he tells Nell.[16]

The next day, Nell and her grandfather approach a town.

> On every side, and as far as the eye could see into the heavy distance, tall chimneys, crowding on each other, and presenting that endless repetition of the same dull, ugly form, which is the horror of oppressive dreams, poured out their plague of smoke, obscured the light, and made foul the melancholy air.[17]

They pass along a road lined with "strange engines [which] spun and writhed like tortured creatures; clanking their iron chains, shrieking in their rapid whirl from time to time as though in torment unendurable." They then encounter more chimneys "never ceasing in their black vomit, blasting all things living or inanimate, shutting out the face of day, and closing in on all these horrors with a dense dark cloud." The episode culminates with a description of the town at night, peopled by bands of desperate unemployed men, spurred by their leaders to "errands of terror and destruction," filled with the sounds of carts loaded with coffins, the shrieks of widows and the cries of orphans.[18] The overall effect is of a hell-like, tormented world, relieved by no sense of grandeur or achievement. Nature and humanity are threatened and powerless, vulnerable to injury, disease or the terrifying power of mass discontent. There is no sense of any human hand behind the industrial landscape; rather, it seems like some emanation of evil.

At the beginning of Book III of *Sybil* (1845), Disraeli describes a mining landscape. The passage is more even-toned and less metaphorical and apocalyptic than in *The Old Curiosity Shop*; it even employs a kind of irony as it describes the ways in which industry blights nature and threatens humanity.

> though the subterranean operations were prosecuted with so much avidity that it was not uncommon to observe whole rows of houses awry, from the shifting and hollow nature of the land, still, inter-mingled with heaps of mineral refuse, or of metallic dross, patches of the surface might here and there be recognised, covered, as if in mockery, with grass and corn, looking very much like those gentle-men's sons that we used to read of in our youth, stolen by the chimneysweeps, and giving some intimation of their breeding beneath their grimy livery. But a tree or a shrub, such an existence was unknown in this dingy rather than dreary region.[19]

Disraeli is more concerned than Dickens with the "social problem" aspects of the scene he describes. He goes on to discuss the scandal of women and young children working underground, leaning heavily on evidence from Government reports. But he is also conscious, in *Sybil*, of the dangers of working-class discontent. There are no marauding bands of the unemployed, but there is Devilsdust, a child of the mills, who meets with his companions in the Temple of the Muses to swear terrible oaths of brotherhood and revenge.

Disraeli reaches for his most grandiose images when he sets out to convey the true significance of Manchester at the beginning of Book IV of *Coningsby*:

> A great city, whose image dwells in the memory of man, is the type of some great idea, Rome represents conquest; Faith hovers over the towers of Jerusalem; and Athens embodies the pre-eminent quality of the antique world, Art....
>
> What Art was to the ancient world, Science is to the modern: the distinctive faculty. In the minds of men the useful has succeeded the beautiful. Instead of the city of the Violet Crown, a Lancashire town has expanded into a mighty region of factories and warehouses. Yet, rightly understood, Manchester is as great a human exploit as Athens.[20]

Disraeli then goes to make the point with which I began this chapter – that it requires a philosopher to understand the full grandeur of Manchester.

The three passages I have been discussing represent three possible ways for a novelist to respond to the new reality of the industrial towns: in Dickens we see the apocalyptic, in Disraeli a mixture of social concern and a grandiose sense of human achievement set in a long historical view. My purpose in setting these ways of writing side by side with Elizabeth Gaskell's much plainer approach is not to suggest that Dickens and Disraeli are in some way inferior or superior to her work. If we look in vain in Gaskell's novels for the kind of set-piece writing I have been discussing it is not because she is any the less imaginatively engaged with Manchester. The difference is partly to do with the quality of her imagination and her preference for the everyday, domestic and conciliatory over the sublime, grotesque, philosophical and confrontational. But a more important consideration is that she does not need to approach Manchester in the same way as Dickens and Disraeli, because, like her characters, she is *of* the city and accepts its presence. Dickens and Disraeli, coming to the industrial city as outsiders, and seeing it first from a distance, are likely to conceive of it as a set-piece; to see it, in terms of the visual arts, as a landscape; and to note its difference, as does Nell, from other regions through which they pass to reach it. In Gaskell's *North and South*,[21] where the heroine, Margaret Hale, comes to Milton-Northern from the rural, pastoral South, there is one short set-piece describing the pall of smoke over the town; and there are a number of other brief references to the shock that Margaret experiences at the appearance and conditions of Milton. But the uniqueness of that set-piece and the tone of the writing about Margaret's initial reaction, which is that of an inhabitant describing the feelings of an outsider, serve to emphasize the point. In order to construct a picture of Manchester as it appears in Elizabeth Gaskell's novels, we have to build it up from a multiplicity of small details scattered, almost incidentally, throughout her texts.

MARY BARTON

Mary Barton begins in the country, in the fields near Manchester, "the busy, bustling manufacturing town," where "the artisan deafened with the noise of tongues and engines, may come to listen awhile to the delicious sounds of rural life" (p. 39). The emphasis is on the natural rhythm of the seasons, the "old-world" farmhouse and the "old-fashioned herbs and flowers" in its garden (p. 40). The scene finally composes itself as an idyll, strong in associations with the poetry of the past:

> It was an early May evening – the April of the poets; for heavy showers had fallen all the morning, and the round, soft, white clouds which were blown by a west wind over the dark blue sky, were sometimes varied by one blacker and more threatening. The softness of the day tempted forth the young green leaves, which almost visibly fluttered into life; and the willows, which that morning had only a brown reflection in the water below, were now of that tender gray-green which blends so delicately with the spring harmony of colours.
>
> (p. 40)

Two of the most sympathetic characters in the novel are associated with country life. Alice Wilson has happy memories of her rural childhood, and her room is "oddly festooned with all manner of hedge-row, ditch, and field plants" (p. 51). Job Legh, weaver and scholar, works in a room "not unlike a wizard's dwelling" (p. 76), and is full of knowledge about the natural world.

Against these images of the ancient and rural may be set the urban and modern as found in Manchester itself. The people walk in badly lit, half-finished, identical streets, off which run mazes of houses and tiny courts. Rain has a different effect in town from in the country:

> The next evening it was a warm, pattering, incessant rain, just the rain to waken up the flowers. But in Manchester, where, alas! there are no flowers, the rain had only a disheartening and gloomy effect; the streets were wet and dirty, the drippings from the houses were wet and dirty, and the people were wet and dirty.
>
> (p. 140)

Over the whole city there lies a pall of smoke (p. 343) and at the furnace where Jem works the men seem "like demons" in the "deep and lurid red" glare of the fire (p. 276). It is hardly surprising that the people of Manchester lack the "strong local attachments" (p. 158) of the agricultural worker. Furthermore, the narrative voice of the novel seems to affirm the virtues of country life. Just after Mary burns the evidence proving that her father is Henry Carson's murderer, she goes out of the house to fetch some water:

> The hard, square outlines of the houses cut sharply against the cold bright sky, from which myriads of stars were shining down in eternal repose. There was little sympathy in the outward scene, with the internal trouble. All was so still, so motionless, so hard! Very different to this lovely night in the country in which I am now writing, where the distant horizon is soft and undulating in the moonlight, and the nearer trees sway gently to and fro in the night-wind with something of almost human motion; and the rustling air makes music among their branches, as if speaking soothingly to the weary ones, who lie awake in heaviness of heart. The sights and sounds of such a night lull pain and grief to rest.
>
> (p. 303)

This sounds very neat and tidy: natural versus human-made, human versus mechanical, country versus town. The only problem is that in order to construct such a picture I have had to ignore many aspects of the novel's real and more complex effect. The sketch of Manchester, for instance, I have taken, quite unfairly, from a few passing references: the half-finished streets are mentioned twice, the rain once, the furnace in one short paragraph. And the smoke, so central to Dickens and Disraeli's vision of the industrial town, is referred to only once, as Mary Barton leaves Manchester, when it arouses in her a nostalgia for her birthplace:

> The very journey itself seemed to her a matter of wonder. She had a back seat, and looked towards the factory-chimneys, and the cloud of smoke which hovers over Manchester, with a feeling akin to the "Heimweh". She was losing sight of the familiar objects of her childhood for the first time; and unpleasant as those objects are to most, she yearned after them with some of the same sentiment which gives pathos to the thoughts of the emigrant.
>
> (pp. 343–4)

Here Elizabeth Gaskell conflates the vision of someone seeing Manchester from a distance, perhaps for the first time, with that of the native who observes and regrets what may not be pleasant but is at least familiar.

Furthermore, what is most striking about the opening scene in the countryside and the later authorial intrusion about life in the country, is that they are – and particularly the first – so self-contained, so self-consciously literary. Any sense of an idyll of rural quietude in Green Heys Fields is quickly dispelled, for this scene is populated by townsfolk, "merry and somewhat loud-talking girls" and young men "ready to bandy jokes with

39

anyone" (pp. 40, 41); and individuals like John Barton, "a thorough specimen of a Manchester man; born of factory workers, and himself bred up in youth, and living in manhood, among the mills" (p. 41). There is even a note of partiality for townspeople as against those from the country. The young women are "not remarkable for beauty.... The only thing to strike a passer-by was an acuteness and intelligence of countenance, which has often been noticed in a manufacturing population" (p. 41). Mrs. Barton, on the other hand, "had the fresh beauty of the agricultural districts; and somewhat of the deficiency of sense in her countenance, which is likewise characteristic of the rural inhabitants in comparison with the natives of the manufacturing towns" (pp. 41–2).

The narrative is soon flooded with the onrush of the concerns of the town: Mrs. Barton's anxiety about her sister, made restless by her new life in Manchester and now missing; John Barton's anger at the gulf between the rich and the poor. The country walk is not presented as an alternative to town life; it is part of a life which may include the possibility of occasional "escape," but which cannot avoid a constant engagement with the real concerns of Manchester. Alice Wilson may have memories of a country childhood, but she also remembers the search for a better life which first brought her family to Manchester, and her life, like Job Legh's, is now inextricably of the town. Her memories are of a time and place that is irrecoverable, as in the words of the song she remembers her mother singing: "the golden hills o'heaven,/ Where ye sall never win" (p. 70).

Elizabeth Gaskell, then, accepts the reality of Manchester in a way that many of her contemporaries as novelists do not. Her knowledge of its life and people is too rich and complex to allow for easy contrasts, simple outrage, unqualified despair, fear or hope. She cannot settle for either the fear-filled image of the desperate unemployed or the grandiose notion of the Athens of industry. This is not to say that she is blind to the filth and overcrowding of parts of Manchester, or that she fails to feel outrage at these conditions. Sharp and powerful images of suffering appear in the book; but she tries always to go beyond the image in order to understand why things are as they are, why people behave as they do.

At the heart of the novel, as a kind of core-text, lies the parable of Dives and Lazarus, the rich man and the poor man, which is first referred to by John Barton in the opening scene.[22] It is this contrast, far more than that between town and country, that is central to Elizabeth Gaskell's project in *Mary Barton*. One of the novel's main strategies is to present us with a number of contrasts between the lives of the rich and poor. These are the constant subject of conversations between the more politically conscious characters, but are also dramatized, characteristically for Gaskell, in a series of scenes showing the homes of those who stand at different points in the social scale. Between the luxuriously furnished house of Carson, the mill-owner, two miles from the center of town and "almost in the country"

(p. 105), and that of the Bartons in "a little paved court" (p. 48), there lies a difference that is more than geographical, although the geography, as Engels and de Tocqueville saw, is in itself symbolic. Further contrasts exist within those homes. Early in the novel, when the Bartons are relatively prosperous, the house is warm, bright and comfortable, a place to entertain friends and family (pp. 49–50); later, after Mrs. Barton's death and when times are hard, it is "stripped of its little ornaments" (p. 159) as all but the bare essentials are pawned. The bleak picture it then presents matches the deterioration in John Barton's character and Mary's increasing unhappiness. But even comfortable homes are not immune from sadness, and against the death-bed scenes among the poor Elizabeth Gaskell sets the episode when Henry Carson's dead body is brought home (pp. 257–66).

The "social purpose" note of the novel is most evident in the scene where Barton and his friend Wilson go to visit the Davenport family, and find them living in conditions that shock even them, accustomed as they are to the conditions of the town:

> Our friends were not dainty, but even they picked their way till they got to some steps leading down into a small area, where a person standing would have his head about one foot below the level of the street, and might at the same time, without the least motion of his body, touch the window of the cellar and the damp muddy wall right opposite. You went down one step even from the foul area into the cellar in which a family of human beings lived. It was very dark inside. The window-panes were many of them broken and stuffed with rags, which was reason enough for the dusky light that pervaded the place even at mid-day. After the account I have given of the state of the street, no one can be surprised that on going into the cellar inhabited by Davenport, the smell was so foetid as almost to knock the two men down. Quickly recovering themselves, as those inured to such things do, they began to penetrate the thick darkness of the place, and to see three or four little children rolling on the damp, nay wet, brick floor, through which the stagnant, filthy moisture of the street oozed up; the fireplace was empty and black; the wife sat on her husband's chair, and cried in the dank loneliness.
>
> (p. 98)

The writing here is very detailed and engaged: the specificity of what it feels like to stand in the area and be able to touch both window and wall, and the subtle shift of viewpoint from "our friends" to "a person" to "you," suggest the narrator's knowledge of such conditions and her desire to involve the reader in the experience. Yet the tone remains level, unimpassioned, allowing the details to speak for themselves, so that by the end of the passage the "empty and black" fireplace has a kind of symbolic strength. Also by the end of the passage, Gaskell is shifting to what is always her central concern – the human consequences of living in such conditions: Mrs. Davenport experiences her home as "dank loneliness." Equally important is the effect the visit has on Barton and Wilson. Each, in a slightly different way, feels afresh the force of the contrast between the Davenports and those who are protected against the fluctuations of trade and the menace of fever. Barton notes "the contrast between the well-filled, well-lighted shops and the dim gloomy cellar" (p. 101), while Wilson, in a more subtly nuanced episode, goes to the Carson home for a hospital order and finds its atmosphere both seductive in its warmth and plenty yet chilling in its incomprehension of the Davenports' plight (pp. 105–10).

Elizabeth Gaskell is fully aware of all the problems of life in Manchester: poverty, filth and disease; the vulnerability of the workers and the protected situations of the employers; the discontent and unrest to which these conditions may give rise. The novel is set slightly in the past, in the years 1838–40, during which the Chartist petition was drafted and presented to Parliament, and which saw the beginnings of the trade depressions of the 1840s. The narrative more than once refers to these events and seeks to understand and mediate between the suffering and anger of the poor and the indifference of the rich (pp. 59–60, 125–6). The explanatory tone of voice is as crucial to the novel as the revelatory tone used elsewhere. The latter moves readers from one part of Manchester to another, introducing them to realities of English life of which they may be unaware; the former, earnest, compassionate, conciliatory, at once explains and asks for understanding of both sides of a tangled conflict.

The last part of the novel, which is taken up with a melodramatic plot of wrongful accusation, pursuit, escape and last-minute rescue, is less satisfactory – a succession of breathless chases, death-bed reconciliations and changes of heart. Like both Dickens and Disraeli, Elizabeth Gaskell found it hard to accept the value of trade-union activity (as in *Hard Times* and *Sybil*, the impetus for mass action comes largely through the intervention of an outsider, an "agitator" from London); and although her belief in the efficacy of Christian fellowship is deeply held, it may seem simplistic in the face of the real complications of this phase of industrial capitalism. It may also seem to some readers that the very end of the novel, when Elizabeth Gaskell applies the "colonial solution" and packs off most of the surviving characters to a new life in Canada, is evasive.

Nevertheless, *Mary Barton* is important for the reasons I have tried to make clear, and for the way in which it establishes the sense of a living provincial culture, different from that of the metropolis but equally to be valued. The use of dialect, for instance, is exceptionally authentic, and is not there simply to give the novel a superficial Mancunian flavor. For these characters, it is the real language of home, in which they can express their feelings, and through which a meaningful and supportive culture can exist. When Margaret Legh sings "The Oldham Weaver" (pp. 72–3) she gives voice to that culture and the painful experience that has gone into its making.

The point is emphasized by the way in which the characters speak about other cities, and particularly London, which to them is the strange and unsatisfactory place. Here, for instance, is John Barton, that "thorough specimen of a Manchester man" (p. 42), describing his visit to the capital with the Chartist petition:

> "How can I tell yo' a' about it, when I never seed one-tenth of it. It's as big as six Manchesters, they told me. One-sixth may be made up o' grand palaces, and three-sixths o' middling kind, and th' rest o' holes o' iniquity and filth, such as Manchester knows nought on, I'm glad to say.
>
> "... we walked on and on through many a street, much the same as Deansgate. We had to walk slowly, slowly, for th' carriages an' cabs as thronged th' streets. I thought by-and-by we should may be get clear on 'em, but as th' streets grew wider they grew worse, and at last we were fairly blocked up at Oxford Street. We getten across at last though, and my eyes! the grand streets we were in then! They're sadly puzzled how to build houses though in London; there'd be an opening for a good steady master-builder there, as know'd his business. For yo see the houses are many on 'em built without any proper shape for a body to live in; some on 'em they've after thought would fall down, so they've stuck great ugly pillars out before 'em."
>
> (pp. 142, 143)

The humor in these words is more at the expense of London than of John Barton. His eye is innocent, but it is not stupid, and his fresh vision is used to punctuate metropolitan pretensions. The real places of Manchester (which Gaskell uses as if they were as familiar to her readers as Oxford Street) are the standard of comparison for John Barton. It is London that may be pitied for its "holes o'iniquity and filth"; problems, that is to say, do not simply lie "out there," in places at a safe remove from the capital – they lie at its very heart.

One final view of Manchester as against other cities is worth noting. Job Legh follows John Barton's story with one of his own, when he tells how he went to London to bring back the infant Margaret after the death of her parents. He hates leaving his daughter and son-in-law in the "big crowded, lonely churchyard in London ... as I thought, when they rose again, they'd feel so strange at first away fra Manchester" (p. 147). On the way home, he stops at Birmingham which, he tells his listeners "is as black a place as Manchester, without looking so like home" (p. 150).

NORTH AND SOUTH

North and South, published in 1855, is a different kind of novel from *Mary Barton*. It is set at a time near contemporary with its publication and deals with a more settled, less painful period in the development of the manufacturing town. Conflict between the classes is present as an issue in the novel, but occupies a less prominent place than in *Mary Barton*. The title (changed from the earlier "Margaret Hale") suggests an abstract interest in regional differences, and by presenting a full-length study of a northern employer, Elizabeth Gaskell attempts to engage with the philosophical standpoint of this class of man; but she was never satisfied with dealing with people as classes and, as in *Mary Barton*, her real interest lies in the individual and particular. The most marked difference between *North and South* and the earlier novel lies less in setting or subject-matter than in the narrative point of view. Although it displays as much knowledge and understanding of Manchester (here called "Milton-Northern") as *Mary Barton*, *North and South* is not written from within the heart of its provincial setting in quite the same way. *Mary Barton* contains hardly any characters for whom Manchester is a strange place; the whole drama of *North and South* depends on the encounter with Milton-Northern of a central character whose sensibility is conceived as entirely southern. It is in the growth to understanding by Margaret Hale of the new conditions in which she lives and the corresponding alterations in outlook of Mr. Thornton, her main "Northern" antagonist, that the real interest of the novel lies. No easy contrasts can be made, however, no simplistic distinctions between "Northern" and "Southern" ways of seeing; the novel is valuable less for any allegorical rendering of different points of view than for its engrossing study of the road to self-discovery and reciprocal understanding on the part of its two main characters.

Like *Mary Barton*, the novel begins with an idyll, or at least a potential one. Margaret Hale has lived for ten years in London with her aunt, but regards her real home as Helstone, the village in the New Forest where her father is vicar, the location of her "bright holidays" (p. 35). She describes the village to Henry Lennox, brother of the man who is to marry her cousin and a potential suitor for Margaret:

> "Is Helstone a village, or a town? ..."
>
> "Oh, only a hamlet; I don't think you could call it a village at all. There is a church and a few houses near it on the green – cottages, rather – with roses growing all over them."
>
> "And flowering all the year round, especially at Christmas – make your picture complete," said he.
>
> "No," replied Margaret, somewhat annoyed, "I am not making a picture. I am trying to describe Helstone as it really is. You should not have said that."
>
> "I am penitent," he answered. "Only it really sounded like a

village in a tale rather than in real life."

"And so it is," replied Margaret eagerly. "All the other places in England that I have seen seem so hard and prosaic-looking, after the New Forest. Helstone is like a village in a poem – in one of Tennyson's poems. But I won't try and describe it any more. You would only laugh at me if I told you what I think of it – what it really is."

(pp. 42–3)

Even as the idyll is established, doubts are entered. There is no doubt of the real value of the village in Margaret's life, but Henry, by identifying her vision of it as literary rather than real, suggests other dimensions; dimensions that Margaret herself unwittingly hints at when she speaks of the prosaic qualities of other places in England.

Mrs. Hale, Margaret's mother, hates living in Helstone. She, oppressed by the trees, thinks her husband's talents are wasted on farmers and laborers, and misses the company of like-minded people of their own class. If that seems snobbish, Margaret can be equally so, in another direction. Of a family mentioned by her mother as suitable company, she replies

"Gormans ... Are those the Gormans who made their fortune in trade at Southampton? Oh! I'm glad we don't visit them. I don't like shoppy people. I think we are far better off, knowing only cottagers and labourers and people without pretence."

(p. 50)

Again, the point is subtly made: there is no objection to liking people "whose occupations have to do with land" (p. 50), but by having Margaret put her argument in such snobbish terms, Gaskell is able to suggest the limitations of her outlook and those areas of English life that the New Forest necessarily excludes.

At the same time, the narrative does not question the positive aspects of Helstone life, and the scope for outer activity and inner growth that it offers Margaret:

The forest trees were all one dark, full, dusky green; the fern below them caught all the slanting sunbeams; the weather was sultry and broodingly still. Margaret used to tramp along by her father's side ... out on the broad commons into the warm scented light, seeing multitudes of wild, free, living creatures, revelling in the sunshine and the herbs and flowers it called forth.... She took a pride in her forest. Its people were her people. She made hearty friends with them; learned and delighted in using their peculiar words; took up her freedom among them; nursed their babies; talked or read with slow distinctness to their old people; carried dainty messes to their sick; resolved before long to teach at the school.

(p. 48)

A kind of Eden then, if a heavily qualified one; and it is interesting that the cause of Margaret's exile comes from within, rather than as the result of external forces. Her father's religious doubts make it impossible for him to continue as a Church of England minister, and a friend with property in Milton-Northern offers him an opportunity to work there as a private tutor. The choice of Milton-Northern is in part determined by the need to earn a living, but Margaret suspects another motive and in framing it to herself articulates all her prejudices about the North.

> There was a secret motive, as Margaret knew from her own feelings. It would be different. Discordant as it was – with almost a detestation for all she had ever heard of the North of England, the manufacturers, the people, the wild and bleak country – there was yet this one recom-mendation – it would be different from Helstone, and could never remind them of that beloved place.
>
> (p. 72)

There is an almost masochistic quality in Margaret's thoughts here: if she cannot have what she believes to be the best, then she will submit to what her prejudices tell her is the worst. Eden will be replaced by Purgatory.

Margaret and her father's early encounters with the North both confirm and confound these expectations and fears. In Heston, the seaside town where they spend a couple of nights before moving on to Milton, life seems more "purposelike":

> The country carts had more iron, and less wood and leather about the horse-gear; the people in the streets, although on pleasure bent, had yet a busy mind. The colours looked grayer – more enduring, not so gay and pretty. There were no smock-frocks, even among the country-folk; they retarded motion, and were apt to catch on machinery, and so the habit of wearing them had died out.
>
> (p. 95)

That last explanatory remark seems to come less from Margaret than from the narrator, emphasizing the nature of the new reality Margaret is entering and pointing out the gap between her observation and her knowledge. Briefly, Heston (a name interestingly close to "Helstone") offers a kind of dreamy repose in which Margaret can contemplate "the great long misty sea-line touching the tender-coloured sky ... in ... a luxury of pensiveness" (p. 96). She knows, though, that "the future must be met, however stern and iron it be," and the expectation of an iron future is confirmed by the first sight of Milton-Northern:

> For several miles before they reached Milton, they saw a deep lead-coloured cloud hanging over the horizon in the direction in which it lay. It was all the darker from contrast with the pale gray-blue of the

wintry sky.... Nearer to the town, the air had a faint taste and smell of smoke; perhaps, after all, more a loss of the fragrance of grass and herbage than any positive taste or smell. Quick they were whirled over long, straight, hopeless streets of regularly-built houses, all small and of brick. Here and there a great oblong many-windowed factory stood up, like a hen among her chickens, puffing out black "unparliamentary" smoke.... As they drove through the larger and wider streets from the station to the hotel, they had to stop constantly; great loaded lurries blocked up the not over-wide thoroughfares. Margaret had now and then been into the city in her drives with her aunt. But there the heavy lumbering vehicles seemed various in their purposes and intent; here every van, every wagon and truck, bore cotton, either in the raw shape in bags, or the woven shape in bales of calico.

(pp. 96–7)

This is the kind of set-piece description that never occurs in *Mary Barton*, where the manufacturing town is accepted as the norm. Here, Margaret experiences only difference: smoke, the absence of "natural" smells, a sense of being crowded and a single industry that everywhere announces itself. In the early part of the novel, various members of the Hale family (and one or two Milton inhabitants) remark on the almost tangible enveloping quality of the staple industry: not only smoke, but also fluff, the residue of the substance on which Milton's prosperity is founded, fills the air and enters the lungs. And for the Hales the enveloping medium is not only physical and external: "all other life seemed shut out from them by as thick a fog of circumstance" (p. 105).

For Margaret the experience of difference in Milton does not only concern appearance and atmosphere; it has also to do with people. The narrative emphasizes the crowded streets and Margaret's fear, in her early days in the town, of the mill-girls who come "rushing along, with bold, fearless faces, and loud laughs and jests, particularly aimed at those who appeared to be above them in rank or station"; and even more of the men, for "[s]he, who had hitherto felt that even the most refined remark on her personal appearance was an impertinence, had to endure undisguised admiration from these out-spoken men" (p. 110). After the enclosed, almost secret, life of the forest, Margaret feels exposed and vulnerable in the streets. Here, as elsewhere in the novel, the narrative voice intervenes to point out the limitations of Margaret's vision: "the very outspokenness," it tells us, "marked their innocence of any intention to hurt her delicacy, as she would have perceived if she had been less frightened by the disorderly tumult" (p. 110). It is only when she comes into personal contact with some of the inhabitants of the "crowded narrow streets," such as the weaver Higgins and his daughter Bessy, that she thinks "how much of interest they had gained by the simple fact of her having learnt care for a dweller in them" (p. 143).

Most of all, though, the experience of Milton for Margaret is one of a completely different philosophy and way of life, which requires a series of painful adjustments. When she undertakes the difficult task of engaging suitable servants, she finds that her father is "no longer looked upon as Vicar of Helstone, but as man who only spent at a certain rate" (p. 109). To the snobbish Margaret, fastidious about connections with "shoppy people," the emphasis on production, getting and spending is a dismaying aspect of life in Milton-Northern. But the problem goes deeper, for she soon begins to see the difficulties of a society ruled by the cash-nexus. Mr. Hale is dazzled by "the energy which conquered immense difficulties with ease.... [it] impressed him with a sense of grandeur, which he yielded to without caring to enquire into the details of its exercise." Margaret is more aware that "in all measures affecting masses of people [there] must be acute sufferers for the good of many." She wonders whether "in the triumph of the crowded procession, have the helpless been trampled on, instead of being gently lifted aside out of the roadway of the conqueror, whom they have no power to accompany on his march?" (p. 108).

One of the conquerors on this roadway, and the man who represents for Margaret "the North," is Mr. Thornton, a prominent Milton manufacturer. As I have already hinted, and as one might expect from Elizabeth Gaskell, the North/South debate in which these two characters engage is not a straight-forward one. There are many respects in which Margaret, with her persistent longing for the dream-village of Helstone, and Thornton, with his pride in the achievements of Milton, do represent "South" and "North." To see them simply in these terms, however, is to ignore the real subtlety and variety of the novel. And, in spite of the prominence of the arguments between Margaret and Thornton, to see the North/South question only in adversarial terms would also be mistaken. After all, the only time the phrase is used in the novel is in the context of reconciliation, when Higgins (who comes from Burnley, forty miles further north than Milton) tells Margaret "North and South has both met and made kind o' friends in this big smoky place" (p. 112). Clustered around the central debates are a number of other views, presented through the characters and deep-set in the narrative. There are, for instance, different reactions to Milton among the southerners – Margaret, her aunt, her parents, Dixon, their servant. There are also different reactions from the inhabitants of the town, not simply represented by the spirit of opposition between the classes that so dismays Margaret, but also by Bessy Higgins, a victim of Milton's poisonous atmosphere, or Fanny, Thornton's sister, whose outlook has been so changed by the family's prosperity that she now longs for a fashionable southern life.

The two mothers, Mrs. Hale and Mrs. Thornton, represent much more extreme statements of the South and North positions than do their children. Mrs. Hale is racked by the thought of what her family has left behind, conveniently forgetting her own dissatisfaction with life in Helstone, and is eventually hastened towards death by the shock of the move. Mrs. Thornton is implacably proud of what her son has achieved to reach his present eminence in the town. Thornton, it is easy to see, occupies a different point on the spectrum from his mother, and this is partly explained by the difference in generation. She is conscious of all the effort that has gone into attaining her jealously guarded material prosperity and respect. Thornton has not forgotten his family's past hardships, but feels more secure in his present position. Without relaxing any of his efforts in business he now feels able to enjoy some of the fruits of his success, not least those previously regarded as the property of the aristocratic or landed elite: he is one of those who reads with Mr. Hale. For his mother, on the other hand, "[c]lassics may do very well for men who loiter away their lives in the country, or in colleges; but Milton men ought to have their thoughts and powers absorbed in the work of to-day" (p. 159).

Margaret has to learn that Thornton is not entirely the man she at first perceives him to be. She is too locked in her own prejudices to observe his real nature; or if she observes it, to make the right connections. After all, we might argue, a busy manufacturer who finds time to read Greek and Latin is not altogether in need of southern "softening." Nor, we might add, is the necessary softening going to be achieved only by an absorption in the classics: they are among the first casualties when things begin to go badly for Thornton.

Their first encounter is full of observations, assumptions and mis-apprehensions on both sides:

> she owed to herself to be a gentlewoman and speak courteously from time to time to this stranger; not over-brushed, nor over-polished, it must be confessed, after his rough encounter with Milton streets and crowds. She wished that he would go ... instead of sitting there, answering with curt sentences all the remarks she made.... He almost said to himself that he did not like her, before their conversation ended; he tried so to compensate himself for the mortified feeling, that while he looked upon her with an admiration he could not repress, she looked at him with proud indifference, taking him, he thought, for what, in his irritation, he told himself he was – a great rough fellow, with not a grace or a refinement about him. Her quiet coldness of demeanour he interpreted into contemptuousness, and resented it in his heart to the pitch of almost inclining him to get up and go away, and have nothing more to do with these Hales, and their superciliousness.
>
> (pp. 100–1)

Later, comparing his face with her father's, Margaret sees in the latter lines that are "soft and waving, with a frequent undulating kind of trembling movement passing over them, showing every fluctuating emotion"; his eyes have "a peculiar languid beauty which was almost feminine." By contrast,

Thornton has "clear, deep-set earnest eyes, which, without being unpleasantly sharp, seemed intent enough to penetrate into the very heart and core of what he was looking at"; while the lines are "few but firm, as if they were carved in marble." His smile can transform his face from "the severe and resolved expression of a man ready to do and dare everything, to the keen honest enjoyment of the moment, which is seldom shown so fearlessly and instantaneously except by children" (p. 121).

The complexities of attraction and doubt are evident here, but they are quickly obscured in the ensuing discussion with her father, when Margaret finds it hard to accept Thornton's strong statements of self-reliance and his Utilitarian view of education. There are, for Margaret, puzzles about Thornton because she cannot reconcile herself to the fact that the qualities she admires in him are "tainted by his position as a Milton manufacturer ... by that testing everything by the standard of wealth" and his apparent contempt for those who lack his "iron nature" (pp. 129–30). Her doubts are increased by her first visit to his house. The house stands close to the mill "whence proceeded the continual clank of machinery and the long groaning roar of the steam-engine, enough to deafen those who lived within the enclosure" (p. 157). Inside, the house is well cared for but colorless, draped in nets and druggets, its ornaments covered with glass shades. Margaret sees "evidence of care and labour, but not care and labour to procure ease, to help on habits of tranquil home employment; solely to ornament, and then to preserve ornament from dirt or destruction" (p. 158). Mrs. Thornton is a tough, unyielding person, guarding the hard-won prosperity that the house embodies. Her vehemence against strikers and her scorn for the South ("'South country people are often frightened by what our Darkshire men and women only call living and struggling,'" p. 163) provides the background to the subsequent argument in which Margaret tries to identify the logical flaw in Thornton's assertions of respect for the independence of his employees and the despotism of his behavior towards them (pp. 169–76).

From these positions both Margaret and Thornton will have shifted by the end of the novel. It would be easy to say that Thornton, the determined iron man of the North, is softened by Margaret's southern, humanizing influence; and that Margaret is toughened by what she learns from Thornton of the sterner realities of the North. To some extent this is true, but it is more accurate to think of the novel in other terms as well, to consider what each releases in the other, and the influence that other people have on them. Mr. Hale, for instance, plays a part in Margaret's increasing acceptance of Milton life, suggesting other perspectives, other ways of seeing what she too readily condemns. Similarly, Higgins plays a part in Thornton's changing view of his role as an employer; though it is only fair to say that Margaret is the agent of their coming together.

The change in Margaret can be felt when she tries to dissuade Higgins from moving south in search of a life less subject to the fluctuations of trade, arguing that existence there would be too stagnant for a man of his nature:

> "You would not bear the dulness [sic] of the life; you don't know what it is; it would eat you away like rust. Those that have lived there all their lives, are used to soaking in the stagnant waters. They labour on, from day to day, in the great solitude of steaming fields – never speaking or lifting up their poor, bent, downcast heads. The hard spadework robs their brain of life; the sameness of their toil deadens their imagination; they don't care to meet to talk over thoughts and speculations, even of the weakest, wildest kind, after their work is done; they go home brutishly tired, poor creatures! caring for nothing but food and rest. You could not stir them up into any companionship, which you get in a town as plentiful, as the air you breathe, whether it be good or bad – and that I don't know."
>
> (p. 382)

Helstone is hardly "a village in a tale" in this speech. There is feeling for the real hardships of agricultural life, and an open-mindedness about the nature of town life. Imagination, thoughts and speculations matter to Margaret and she is beginning to see the ways in which she might find these things in Milton.

Arguing with Mr. Bell or accompanying her aunt on her first visit to Milton, Margaret finds herself in a new situation – that of a resident, defending the town and pointing out the things of which it may be proud. Bell, who enters the novel as a principal character only in its later chapters, functions as a final catalyst for Margaret's developing thoughts about Milton. He performs this function by introducing the idea of change and, albeit unwittingly, bringing Margaret to a new understanding of what it means for communities and individual lives. The effect on Margaret is perhaps other than Bell might wish. A native of Milton, but long entrenched in the fastness of an Oxford college, he is completely unable to reconcile himself to what the town has become. He constantly makes unfavorable comparisons with Oxford and reminds the Hales of the contrast between Milton and Helstone. To Bell, Helstone represents all that is unchanging and happy in his past:

> "It is years since I have been at Helstone – but I'll answer for it, it is standing there yet – every stick and every stone as it has done for the last century, while Milton! I go there every four or five years – and I was born there – yet, I do assure you, I often lose my way – aye, among the very piles of warehouses that are built upon my father's orchard."
>
> (p. 467)

It is inevitable, therefore, that when the narrative, as it must, takes Margaret back to Helstone, it should be Bell who suggests the trip and accompanies her.

They set out from London, where Margaret is staying after the deaths, in quick succession, of both her parents. She already knows that metropolitan society is not for her. When the narrative tells us that she is "surfeited of the eventless ease in which no struggle or endeavour was required" and that "the very servants lived in an underground world of their own, of which she knew neither the hopes nor the fears; they only seemed to start into existence when some want or whim of their master and mistress needed them" (p. 458), it does not have to insist on the implicit contrast with either the visible busy activity of Milton or the daily occupation with which Helstone once provided her. The journey to Helstone offers other contrasts, of a mixed kind: there are "country-towns and hamlets sleeping in the warm light of the pure sun, which gave a yet ruddier colour to their tiled roofs, so different to the cold slates of the north"; and also "few people about at the stations ... as if they were too lazily content to wish to travel; none of the bustle and stir that Margaret had noticed in her two journeys on the London and North-Western line" (p. 471). The scene reminds Margaret of "German Idylls" (p. 472), but makes her think of the changes that have taken place in her own life. What is unchanging in the scene, paradoxically, is the most painful: "It hurt her to see the Helstone road so flooded in the sun-light, and every turn and every familiar tree so precisely the same in its summer glory as it had been in former years" (p. 472).

But any sense of stability in Helstone is quickly shattered:

> Here and there old trees had been felled the autumn before; or a squatter's roughly-built and decaying cottage had disappeared. Margaret missed them each and all, and grieved over them like old friends. They came past the spot where she and Mr Lennox had sketched. The white, lightning-scarred trunk of the venerable beech, among whose roots they had sat down, was there no more; the old man, the inhabitant of the ruinous cottage, was dead; the cottage had been pulled down, and a new one, tidy and respectable, had been built in its stead. There was a small garden on the place where the beech-tree had been.
>
> (p. 475)

In response to Margaret's dismay at these changes, Bell tells her, "'I take changes in all I see as a matter of course. The instability of all human things is familiar to me, to you it is new and oppressive'" (p. 475). At the cottage of an old friend she hears a cruel story of a stolen cat, the pain and shock of which not even the "soft green influence" (p. 478) of the wood can dispel; while a visit to the village school leaves her feeling out of place. The greatest changes, however, are reserved for the parsonage, now occupied by a large and noisy family and structurally altered, not least in her father's study "where the green gloom and delicious quiet of the place had conduced, as he had said, to a habit of meditation, but, perhaps, in some degree to the formation of a character more fitted for thought than action" (pp. 480–1).

Margaret's first reaction to these changes is pure nostalgia, a regret for what has been lost in the name of improvement:

> There was change everywhere; slight, yet pervading all. Households were changed by absence, or death, or marriage, or the natural mutations brought by days and months and years, which carry us on imperceptibly from childhood to youth, and thence through manhood to age, whence we drop like fruit, fully ripe, into the quiet mother earth. Places were changed – a tree gone here, a bough there, bringing in a long ray of light where no light was before – a road was trimmed and narrowed, and the green straggling pathway by its side enclosed and cultivated. A great improvement it was called; but Margaret sighed over the old picturesqueness, the old gloom, and the grassy wayside of former days.
>
> (pp. 481–2)

The modulations of tone in this passage are extremely interesting, for although the paragraph from which it is taken seems to be offered as Margaret's thoughts, there appear to be two voices here. One is clearly Margaret's, heard in the last sentence quoted. The other is an anonymous narrative voice, which quietly preaches an acceptance of the natural inevitability of changes – as in the image of the fruit dropping from the tree – and even hints at the improvements it may bring. Later, after a conversation with Bell about the misunderstandings that have grown up between her and Thornton, she has a much sharper apprehension "of changes, of individual nothingness, or perplexity and disappointment," and is "'so tired of being whirled on through all the phases of my life, in which nothing abides by me, no creature, no place; it is like the circle in which the victims of earthly passion eddy continually'" (p. 488). This Dantesque vision is in turn replaced by an understanding of what the alternative would be:

> "If the world stood still, it would retrograde and become corrupt.... Looking out of myself, and my own painful sense of change, the progress of all around me is right and necessary. I must not think so much of how circumstances affect me myself, but how they affect others, if I wish to have right judgment, or a hopeful trustful heart."
>
> (pp. 488–9)

The breaking of Margaret's self-absorption is a crucial stage in her development. In a final visit to the Vicarage, she finds "the place was reinvested with the old enchanting atmosphere. The common sounds of life were more musical there than anywhere else in the whole world, the light more golden, the life more tranquil and full of dreamy delight" (p. 489). Margaret is irritated by her own changeability – "'now disappointed and peevish because all is not exactly as I had pictured it, and now suddenly discovering that the reality is far more beautiful than I had imagined it'" (p. 489). In fact, this final vision of the Vicarage, couched in language that takes it further towards a dream-like idyll, leads her to find her level:

> she was very glad to have been there, and that she had seen it again, and that to her it would always be the prettiest spot in the world, but that it was so full of associations with former days, and especially with her father and mother, that if it were all to come over again, she would shrink back from such another visit as that which she had paid with Mr Bell.
>
> (pp. 489–90)

Its very associative richness is what causes Margaret to turn her back on Helstone. It has now become "a village in a tale" indeed: the tale of her past life. The way in which this happens in the novel is worth tracing, because it shows another aspect of the subtlety with which Gaskell deals with the North/South issue. Margaret's increasing identification with Milton does not expunge Helstone; she merely places the village in its proper relation to her present life.

Her links with Milton become closer in a different way when Bell dies and Margaret inherits his property in the town. Not that this reconciles her to what she still sees as the disadvantages of life in Milton. When Henry Lennox returns from a visit in connection with her property he is full of admiration for "the character of Milton and its inhabitants. Their energy, their power, their indomitable courage in struggling and fighting; their lurid vividness of existence, captivated and arrested his attention"; Margaret is quick to point out the selfishness and materialism, "the tainting sin in so much that was noble and to be admired." Nevertheless, Henry finds that "an enquiry as to some Darkshire peculiarity of character called back the light into her eye, the glow into her cheek" (p. 508).

The narrative precedes Margaret to Darkshire, and offers a picture of a changed Milton-Northern, which has been affected by a trade depression:

> Meanwhile at Milton the chimneys smoked, the ceaseless roar and mighty beat, and dizzying whirl of machinery, struggled and strove perpetually. Senseless and purposeless were wood and iron and steam in their endless labours; but the persistence of their monotonous work was rivalled in tireless endurance by the strong crowds, who, with sense and with purpose were busy with seeking after – What? In the streets there were few loiterers, – none walking for mere pleasure; every man's face was set in lines of eagerness or anxiety; news was sought for with fierce avidity; and men jostled each other aside in the Mart and in the Exchange, as they did in life, in the deep selfishness of competition.
>
> (p. 510)

At this late stage – only two chapters from the end of the novel – we have its strongest and most outspokenly critical authorial intrusion. Its timing is significant, for it is important that the reader should understand precisely the ways in which Milton has *not* changed. Energy, bustle, noise and smoke are still present, but to little purpose; the cut-throat rapacity of trade has, if anything, been sharpened by the crisis.

The picture of Milton is necessary as background to the downturn in Thornton's fortunes. His dream of a career as a kind of merchant-prince is now threatened, and he begins to understand the ways in which he is of the town and yet detached from it:

> until now, he had never recognized how much and how deep was the interest he had grown of late to feel in his position as manufacturer, simply because it led him into such contact, and gave him the opportunity of so much power, among a race of people strange, shrewd. ignorant; but, above all, full of character and strong human feeling.
>
> (p. 512)

He has realized that he and "his own people ... had led parallel lives – very close, but never touching" (p. 511). The reader is made aware of how much Thornton, a Milton man, has to learn about his home town; as much, in his way, as Margaret.

In some ways this penultimate phase of the novel seems over-schematic. Thornton is ruined in business, showing the positive side of his pride and self-reliance in his insistence on using all the capital available to repay his creditors, and in his refusal to risk that capital in an investment which, ironically, makes his brother-in-law rich. There is perhaps too neat a correspondence between Thornton's fall and Margaret's attainment of fortune and a new sense of independence. Elizabeth Gaskell seems too readily to create the conditions in which the lovers may come together, with Margaret financing Thornton's business and giving him the chance to try some of his experiments in labor relations. In fact, the impressive thing about the end of the novel is its tentativeness, its refusal to admit of easy solutions. Thornton's ideas remain as experiments, still to be tested in real conditions; he hopes that they will be

> "carried on by that sort of common interest which inevitably makes people find ways and means of seeing each other, and becoming acquainted with each other's characters and persons, and even tricks of tempers and modes of speech. We should say understand each other better, and I'll venture to say we should like each other more."
>
> (p. 525)

He does not expect his measures to end strikes, but to "'render them not the bitter, venomous source of hatred they have hitherto been'" (p. 526).

The culmination of the relationship between Margaret and Thornton is achieved with similar restraint: a quiet acceptance on the part of each that they have reached the point where they can be together. When Thornton offers Margaret the roses he gathered on his trip to Helstone "to see the place where Margaret grew to what she is" (p. 530), he is in no sense offering her a return to the past. Helstone is no longer a home for Margaret; more exciting and potentially fruitful is the chance they have to live and work in a town whose conditions of living are still developing and where they can continue to grow and learn.

The quietness of this ending is characteristic of Elizabeth Gaskell's work, just as the picture of the industrial town she presents is so markedly different from those of her contemporaries. We may miss in Gaskell the quicksilver, often grotesque, richly metaphorical treatment of Dickens; the philosophical and historical sweep of Disraeli; the sharper political perceptions of de Tocqueville or Engels. Yet more muted versions of all these ways of seeing are present in her Manchester novels. She understands the extraordinary achievement that the town represents, feels the force of its new reality, perceives the depth of its misery, sympathizes with those who suffer the bewildering effects of change. By writing about these things from the heart of the provincial experience in the mid-nineteenth century she creates above all an image of Manchester, not as a symbol, not as a problem, not as a ghastly blot on the landscape to be kept at a safe distance, but as a place in whose streets there walk living, suffering human beings.

NOTES

1   Benjamin Disraeli, *Coningsby* (1844; ed. Sheila M. Smith, Oxford: Oxford University Press, 1982), p. 134.
2   Quoted in Asa Briggs, *Victorian Cities* (Harmondsworth: Penguin Books, 1968), p. 88.
3   Hugh Miller, *First Impressions of England and Its People* (1874), quoted in Humphrey Jennings (ed.), *Pandaemonium* (London: André Deutsch, 1985), pp. 230–2.
4   Alexis de Tocqueville, *Journeys to England and Ireland* (1835), quoted in Alasdair Clayre (ed.), *Nature and Industrialisation* (Oxford: Oxford University Press), pp. 117–18.
5   ibid., p. 118.
6   ibid.
7   Quoted in Briggs, *Victorian Cities*, p. 114.
8   Friedrich Engels, *The Condition of the Working Class in England in 1844* (1845; London: Allen & Unwin, 1892), pp. 45–6.
9   Briggs, *Victorian Cities*, p. 90.
10 ibid., p. 93.
11 ibid., p. 97.
12 Thomas Carlyle, *Chartism* (1839; London: Chapman & Hall, 1892), p. 6.
13 Angus Easson, *Elizabeth Gaskell* (London: Routledge & Kegan Paul, 1979), p. 38.
14 Elizabeth Gaskell, *Mary Barton* (1848; ed. Stephen Gill, Harmondsworth: Penguin Books, 1970), p. 37 (quoted hereafter by page number only).
15 Charles Dickens, *The Old Curiosity Shop* (1840–1; ed. Angus Easson with an

introduction by Malcolm Andrews, Harmondsworth: Penguin Books, 1970), p. 417.

16 ibid., p. 418.
17 ibid., pp. 423–4.
18 ibid., p. 424.
19 Benjamin Disraeli, *Sybil; or the Two Nations* (1845; ed. Sheila M. Smith, Oxford: Oxford University Press, 1982), p. 139.
20 *Coningsby*, p. 134.
21 Elizabeth Gaskell, *North and South* (1854–5; ed. Dorothy Collin with an introduction by Martin Dodsworth, Harmondsworth: Penguin Books, 1970), quoted hereafter by page number only.
22 This parable is to be found in Luke 16: 19–31. An interesting discussion of its use in *Mary Barton* may be found in Michael Wheeler, *The Art of Allusion in Victorian Fiction* (London: Macmillan, 1979), pp. 44–61. It is also the subject of some English folk songs.

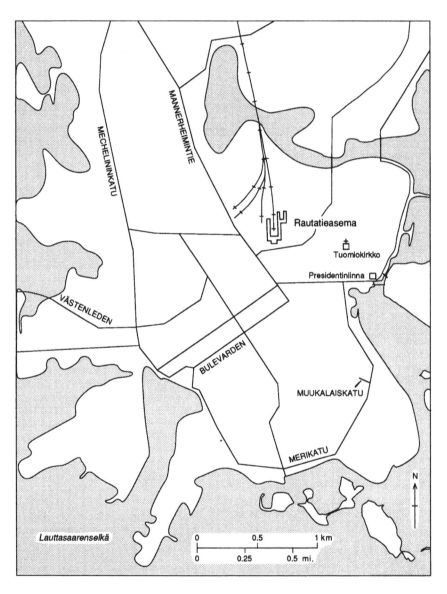

Map 4 Helsinki

4

CONTRASTING THE NATURE OF THE WRITTEN CITY

Helsinki in regionalistic thought and as a dwelling-place

Pauli Tapani Karjalainen and Anssi Paasi

INTRODUCTION

In this chapter we will first give a general view of Helsinki, the capital of Finland, and then examine the ways in which the city has been described in Finnish literature. The main part of the chapter deals with the literary examples of Helsinki as a dwelling-place and as the central point of Finnish regionalism. The city of Helsinki is thus viewed both as a scene of individual experience and as a symbol at the societal level.

Helsinki is a maritime city, sometimes called the Daughter of the Baltic. It is situated on the south coast of the country, so that seen from the sea, Finland spreads out behind Helsinki. But how can one describe the city to a foreigner? The following paragraph by Bo Carpelan (a poet and novelist writing in Swedish; born 1926 in Helsinki) captures something essential about it:

> Is there another city in Europe having such an unperturbed, free and open sea horizon as Helsinki does? For one who approaches Helsinki by ship the city rises up out of the sea. A small row of houses and behind them: nothing, as a traveler coming from afar described the scene a hundred and thirty years ago. Now one finds behind the row of houses bordering the beautiful market-square miles upon miles of life, growing and crowded with traffic: cranes and excavators, tunnels and bypasses, factories and housing, scrap yards, waving grain fields, mixed settlement that you can behold from a train or the airport bus while coming in to the city. No! Helsinki must be approached from the sea; one must see the islands, the rocks, the huge ramparts and walls of Suomenlinna, the busy sailing boats, the bustle of fishing boats and market stalls at the harbor, the glittering of the water around the statue

of Havis Amanda and the flying flags in front of the Town Hall: this
is the classic image of Helsinki, dear both to travelers and to the people
of the city.

(Carpelan 1983: 485–6)

The first part of this quotation gives an image of a mobile city, portraying
Helsinki as growing and expanding. This is to see the city from the point of
view of work and the economy. The second part paints a more stylized
picture: Helsinki is seen from the sea in the light of a bright summer morning,
an aspect of beauty. The images are different, but both of them are correct.

Helsinki was founded by the Swedish king Gustav Vasa in 1550, at the time
when Finland formed a part of Sweden. The king forced burgesses from
various parts of the empire to come to live on a site where there was no prior
settlement. The city is nearly 450 years old, which is not very old in
European terms, and for a long time it was only a paltry village. As Jörn
Donner (a novelist writing in Swedish; born 1937 in Helsinki) puts it, an
image of Helsinki before the 1800s can be obtained by "turning your back
on the city, stopping for a while at the open sea and the cliffs and imagining
that somewhere behind you there is a group of dilapidated wooden hovels,
and muddy paths and rutty pastures between them" (Donner 1987: 21).

At the turn of the eighteenth century Helsinki had some 4,000 inhabitants,
but a hundred years later, under Russian rule, the population was already
close to 100,000. Today the city has half a million inhabitants, which, of
course, is small by comparison with the world's great metropolises –
London, New York, Moscow and Tokyo – but still makes it the only "city"
in Finland, since all the other centers of population are much smaller. If the
urban districts bordering onto Helsinki proper are included, the population
living in the metropolitan area is almost one million. Thus one-fifth of
Finland's total population is now living in only a diminutive fraction of the
total area of the country.

The proportion of the population living in the main city center has been
continually diminishing since the 1950s, and the majority of people in the
greater Helsinki area live in suburbs, in the much criticized impersonal "box
houses," with huge shopping centers, sharp separation between work and
living-space and consequent social problems (Kortteinen 1982). Only a
minority of the people now living in Helsinki were born in the city area, and
still fewer can trace their roots as city dwellers back for several generations
(Donner 1987: 27). Most of the people in the Helsinki metropolitan area have
moved there from the other parts of the country. The "Great Migration" of
the 1960s and 1970s, an outcome of the fundamental restructuring of
Finland's economy, took hundreds of thousands of people to live in the
"south" (Valkonen 1985). The majority of the city's population is thus
composed of "newcomers," so that an authentic city culture of necessity
remains thin on the ground. As Donner (1987: 28–9) says, the city as such has

no intrinsic meaning for the newcomers and this has given rise to social problems, most visible in the suburban areas, where people who lie somewhere "between" city dwellers and people from the country have failed to create an identity of their own.

For the suburban people the reality of the city is not that of the stylized picture quoted at the beginning. Rather, in the morning

> buses and trams roll out of their sheds and the streets become filled with the rumble of wheels. Most people are sitting quiet, many are drowsy, some are sleeping. Now and then eyes are opened for fear of missing the next stop. Those who are awake are reading newspapers. Outside it is snowing heavily.
>
> (Donner 1983: 497)

HELSINKI IN FINNISH LITERATURE

The life of the big city and the triumphant progress of technology so intimately associated with it were highly valued in the Finnish literary movement of the 1920s, mainly among the poets known as the "Torch-bearers." The new liberated mood stemmed partly from the fact that Finland had gained her independence in 1917, and the Torchbearers' generation was turning its back on the past to greet the new age, which in Finland, as perhaps throughout Europe, seemed full of promise (Laitinen 1985: 107). In her first work "Fire and Ash," published in 1930, Iris Uurto (a poet and novelist; born 1905 in Kerimäki) offers an example of the city romanticism of that period:

> I love Helsinki.
> Its streets and cafés.
> Its evenings and mornings.
> Especially those new buildings on the corners.
> Straight, enormous lines.
> A narrow façade that describes
> a longing for heights,
> a longing for the skies,
> just as Gothic churches do.
>
> (Uurto, in Palmgren 1989: 83)

By the 1930s some city literature was taking a more realistic and critical view. After World War II, in the late 1940s and 1950s, there was little to be idealized in Helsinki. The city was suffering from material and spiritual deprivation. This paved the way for a new kind of literature, and especially from the 1960s onwards city life, industry and modern technology became the objects of social and ecological criticism, even condemnation. The enthusiastic descriptions of the city were followed by the rise of rural literature and the historical novel, a protest against the protagonists of internationality and an urban way of life (Palmgren 1989: 356–7; Laitinen 1985: 148). The regional controversies between the developing "south" and the declining eastern and northern parts of the country – bringing with them transformations in regional identity – became an important topic in the "regional novel" (Mäkelä 1986).

The new image of the city was critical. The growth pains in the 1960s – loneliness, longing, environmental deterioration and the problems of identity – are expressed by Claes Andersson (a poet and novelist writing in Swedish; born 1937 in Helsinki) in his poem "The Name of the City is Helsinki" published in 1965:

> I wander around in the city of Helsinki
> among people and their statues
> their dogs and their loneliness
> I search for myself in them
> I feel their loneliness
> how at times it grows into a cry
> which is cut off
> by the scream of the brakes
> I see how their loneliness is growing
> by the growth of the crowd
> The old houses are watching each other
> behind closed curtains
> The old houses which just disappear
> as the new houses reach towards the sooty sky
> ever higher as the crowd grows ...
>
> As the night comes
> when the traffic has fallen asleep under its bonnets
> and the pigeons take over the city
> I see them
> one or two of them
> on the piers
> looking out to the sea
> in the parks
> gazing at the sooty sky.
>
> (Andersson 1987: 41, 43)

As Palmgren (1989: 188) points out, a striking feature in post-war Finnish literature is the scarcity of explicit descriptions of a city. Writers rarely describe the landscape, architecture or physical aspects of a city, which remain largely implicit. Yet the city atmosphere is present, for the city is not only streets and buildings, parks and avenues but also feelings, images and memories (Porteous 1990). The "cityness" of a city is manifested only through feelings: the experience of the environment is always a fusion of the external physical realm and the human being's internal capacities. It is this aspect of the city that is explored in the next section.

HELSINKI AS A DWELLING-PLACE: THREE INSTANCES

As human beings we are always located somewhere, with some particular locality as the surroundings for our lives. The localization of existence is a part of its grounding in fact, which always binds our existential freedom. Nor does the rapid abolition of physical distances, as a consequence of ever more effective means of transport, change this fundamental principle. Dwelling-places are not "pure positions" with equal values, as are the points in cartographic systems; they vary in value with the existential situations of the dwellers (Karjalainen 1991).

The term "dwelling" refers here to the whole spectrum of a person's environmental relations. As Norberg-Schulz (1985: 13) puts it, "to dwell implies the establishment of a meaningful relationship between man and a given environment." To dwell somewhere is to familiarize oneself with one's own environment, to make it a field of care. Here a habitual way of living has a central role to play (Seamon 1979). It is habituality, through familiarity, that gives you "a certain peace of mind" (Donner 1987: 33) but may, on the other hand, make the city self-evident to such a degree that "it almost becomes a barrier, just as if it were too close" (ibid.: 73). Then a long journey, a long absence, a distance of a sort, is required to re-establish the place. For some, as for Claes Andersson, this takes the form of a love/hate relationship:

> I have stood in your ugly railway station longing to get away; and I have traveled away but have always come back, as one comes back to one's first great love, one's mother or one's beloved in whom there is peaceful delight.
>
> (Andersson 1977: 8)

The dimensions of dwelling vary. As we are speaking here of dwelling in a city, the private and public aspects are the most obvious ones (cf. Norberg-Schulz 1985: 13). First, the arena where private dwelling takes place is the home – the place of our own. A house is a material object, but "home" implies a relationship. As Dovey (1985: 34) puts it, home is "an emotionally based and meaningful relationship between dwellers and the dwelling places." In private dwelling, whether it takes place in a rented one-room wooden shack or a privately owned spacious house, we meet with our own personalities. Bordessa (1989: 34) makes a crucial point when he writes that "houses become homes only by some process that culminates in the locking together of self and artifact."

In public dwelling we come outside the private sphere, from homes to houses, streets, the communities of the city quarter and those beyond, institutions of various sorts. Public dwelling is more anonymous; there are apartments, traffic routes and other facilities for people, generally expressed in the form of statistics and regulations of the way homes cannot be. The standards and symbols of living are also a part of public dwelling. The fact that privately owned houses are preferred to rented apartments, in Finnish towns as elsewhere, is an indication of the values prevailing in the sphere of public dwelling.

The following passages illustrate ways of dwelling in Helsinki. The illustrations, drawn from three novels, express the life-worlds and life-styles of different kinds of people in different kinds of city area. The time-span covers the years from the 1950s to the 1980s. The illustrations are only instances; if some other works had been chosen the descriptions would have been different. The focus is on the city as a dwelling-place seen from the angle of the individual, which of course reveals only one aspect of the rich content of the novels cited.

A middle-class housewife looking for her home

Marja-Liisa Vartio (born 1924 in Sääminki; died 1966 in Savonlinna) began her career as a lyric poet writing flourishing dream and visionary poetry related to the folklore and characterized with personal imagination. Her poems are often long and epic, at times virtually a monologue. Had she remained with lyric poetry, there possibly would have emerged in Finnish modernism an expressive line of poetry based on the national myths (Laitinen 1981: 563).

Vartio's move from poetry to prose appeared at first to be a surprising change in form – the wild poetry got a bridle. In her short stories and novels there is, however, much left of the lyric's manner of perception. The work examined here is an example of Vartio's often merciless (albeit with grotesque humor) and impressive depiction of the destinies of human life. The novel *Kaikki naiset näkevät unia* [All Women Have Dreams] (Vartio 1982; orig. 1960) is "a satirical but understanding" (Ahokas 1973: 380) portrait of Mrs. Pyy, a middle-class housewife, and her family, and their difficulties with keeping up with the neighbors in payments for a new house, standard of living and social status. The author does not suggest that the problems of middle-class life are ridiculous, but neither does she idealize or pity her characters. In the end, the author leaves the reader to draw his or her own conclusions, "in an impartial manner typical of the literary style of postwar Finland" (ibid.).

The novel begins with Mrs. Pyy – as the main character is called throughout the book – remembering their move to rented quarters, a terraced house in a suburb of Helsinki, in the 1950s.

When they moved here, it had been raining; drizzly rain that made

64

everything wet. ... She had been standing here, there was a ring at the door, and she had not quite realized that the lorry was carrying their home. The feeling was because of the rain; everything looked so unpleasant in the rain.

(p. 12)

As a person, Mrs. Pyy is neurotic, cautious, repressed, sometimes hysterical, narrow-minded and unsociable: in short, she is not an attractive lady. The point is that she does not like her living quarters, or city life, for that matter. She does not feel comfortable; she is dissatisfied with everything and especially because her home is rented and not owned, an unhappy situation for middle-class people. Mrs. Pyy dreams of a house of her own which, although in Helsinki, should not be a mere house in the city but should be reminiscent of a better life, the forests and her own past. Once the family has bought a building lot and begun to plan the house, her thoughts are revealed:

She had specifically wished that the two pines growing on the rock should be seen as a whole, from the base up to their crowns, just as in a big painting. She had imagined herself sitting in an armchair looking at the trees. They were just the same as the two pines on that cliff from which, as a child, she had plunged into the water. And she could not understand why the window could not be directed the way she had planned. Were it to open toward the yard, it would not be the same window any more. One would have to look at people again, and that would be that. But if the window opened onto the forest and the rock, one could forget that one was living in the city. ... She would not have wanted anything else but this piece of luxury, if it could be called a luxury: that one could see the forest through the window. And the rock.

(pp. 193–4)

For Mrs. Pyy's husband, the position of the window was a matter of indifference. But he was a native city-dweller, whereas his wife was born in the countryside, in a big farmhouse in the middle of fields, forests and lakes. "'You don't understand,' Mrs. Pyy said to her husband, 'because you are a townsman; you don't understand what it means when I say that I'm hungry for soil'" (p. 196). This, of course, is not the whole story. She also wanted the window to be on the rock side because she did not like people. For a human being, she thinks, "it is best to live alone so that in addition to one's own worries one does not have to have trouble with neighbors" (p. 188).

The "house project" is both an economic and a social disaster for the family. They cannot not afford it, and it has to be sold when still unfinished. The family move to a flat in the downtown area which is too small. There, inevitably, Mrs. Pyy's "symptoms" get worse. She cannot stand traffic noise and citizens making a row every night "under the windows of tired people"

(p. 222). She begins to visit a psychiatrist (yet another middle-class maneuver) who finally gives her good advice: "Can't you travel somewhere, just change your environment. It will be worth it. Go somewhere and rethink everything once again right from the very beginning" (p. 268). Mrs. Pyy travels to Italy with her husband. Whether the journey was a success or not does not really come out in the novel. In the Vatican she does not feel anything special, but – so she thinks – "it was because she had become tired," so that she

> was no longer capable of seeing or feeling or understanding, even though she saw all those pictures she had dreamed of so much, in which she had believed – yes, believed that her soul would change, that seeing the pictures would give power and peace to her soul.
>
> (pp. 299–301)

Instead, for this introverted and inhibited woman the Sistine Chapel becomes something like a Finnish "smoke sauna" where "nothing can be seen" (p. 301). Back in Helsinki, she totally drops "out of the game" (p. 104) for a while:

> There were people coming and going, climbing up and down the stairs, walking alone and hand in hand, quietly, talking, laughing. She watched and listened as if she could not remember where she had come from or how she had got here, to the edge of an unknown jungle. She saw the people as if they were strange figures mowing along jungle paths. She only knew that they were the inhabitants of this district; they came and went and they spoke as if they were familiar with the district and its paths. But she was an outsider. She did not feel part of that group and did not know them. She stood still, watching and gazing.
>
> (p. 102)

The scene becomes increasingly surrealistic. The bus that is to take Mrs. Pyy home becomes transformed into an animate creature.

> That was the bus, of course it was; it opened itself and unloaded something from itself, without knowing what it did, still thinking the same thing, as if still going on with the same thought even when it had stopped, and those it carried along were of no consequence to it, something it took inside itself and then unloaded, always thinking the same thing, and the people only thought they were traveling.
>
> (p. 101)

Impressions like these do not belong to an ordinary person going about his or her daily affairs. They cannot be generalized. They were created by Mrs. Pyy, a person with complex inner thoughts of her own. As so often, Mrs. Pyy feels tired. "The whole day in the city, the commotion and the noise and all that rush. They are ingrained in my head" (p. 103). She is annoyed and

66

therefore "I'm sure I won't get to sleep at night" (p. 103). This is the way it always happens. Whose fault is it? Not hers, she thinks, but that of her husband and mother-in-law who have "forced her to live all these years in the way they have had to live" (p. 123). The novel offers no solution on how to manage the situation of this extremely complex woman. She remains an unhappy city-dweller who finds no satisfactory place, for she is unable to handle her life, to control her future.

Lessons about life in a workers' district

Alpo Ruuth, a short-story writer and novelist of many previous trades (born 1943 in Helsinki), has story-telling, joy and morality as the essential features of his broad realism. At first sight his characters seem to be helpless, often even comical, but at bottom thoughtful members of the working-class, who hold on toughly to their desires, rights and truths. Ruuth's novels are often filled with a sense of the warmth and the strength of their everyday existence, much affected by the controversies between the general and the private (Tarkka 1990: 158); in the novel examined here, *Viimeinen syksy* [The Last Autumn], he depicts the consequences the rebuilding of a city block has in the dwellers' lives. Ruuth's novel sympathetically shows how the goodness in life finally overcomes the misfortunes.

> I swish my whisky in the glass: suddenly I get really lonesome for Sörkka, the streets, the smells, and all the rest. Oh damn it, everything is so intimate. I think the memories will be stuck in my mind until the grave. I was a kind of hero there, and we had a good spirit, that is how I remember it. And the people. I've met so many people in my life, but they are not like those in Sörkka, at least they were different there from these unemployed who are always drinking beer and telling jokes.
>
> (Ruuth 1979: 17)

This piece of nostalgia for Helsinki, is filled with the warm sympathies of a young man and his fellows in the old workers' district of Sörnäinen (Sörkka) in 1961. The nostalgic feelings come to mind seventeen years later, in a small locality somewhere further inland in Finland, three hundred kilometres from Helsinki, to which "Flea" – he is short and brisk – moved in autumn 1961 and where he became a successful service-station proprietor. Flea has not visited Helsinki since; and now his wife is going to the capital to manage some business concerning his son, and this kindles old memories.

"After all I sometimes long for my small service-station in Sörkka. Life was like one long Sunday. You could do everything by yourself and there was always some action going on" (p. 6). His own service-station in the country is a splendid one – a real glass palace – but there are so many economic and other commitments and responsibilities to be dealt with. In short, life is so very complicated now.

An old friend of Flea's reminds him of the contradictions between dreams and realities: Helsinki and Sörkka are not the same any more:

> Has it become so golden a place to you? You should go and take a closer look. You can't go anywhere there. The whole city is full of offices and there is not a living soul in the streets. There is no city without people, you know.
>
> (p. 19)

As noted above, the so-called "Great Migration" of the 1960s took people from the countryside to the Helsinki area at an accelerating rate. In a way Flea's situation was contrary to the general trend: he moved to a place people were leaving. The Helsinki city area changed rapidly in this process. New suburbs were built and the old residential areas renewed.

Even in Flea's quarter, with its wooden houses, times were changing rapidly. The key events of the novel are paralleled by the gradual destruction of Flea's old district. "The world is changing and Sörkka is changing, too. It is useless to fight against civilization. It would be foolish" (p. 113), says the real-estate dealer, assuring Flea, an orphan living alone, and adds "you'll find a new dwelling, young careless guy, and if not in any other way just by taking an old widow into your arms, and you'll get central heating, too." New, stately, brick-built blocks, marks of real progress, were to be put up on the site of the wooden shacks. There are already some blocks in the neighborhood and the people living in them think of themselves as much more refined than those living in the wooden slums. Furthermore, the old service-station that Flea worked in is to suffer the same fate, for it is too small and impractical to serve the rapid increase in the number of cars in the city.

> The demolition work is well under way: the yard had changed a lot during the day. The last lilacs had been felled and two excavators stood in their place. At the edge of the stone house – just where we had had our clothes-lines – there was now a deep pit.
>
> (p. 134)

Flea is the last dweller in the wooden house. Finally, in the midst of the remnants, he has "but a tiny piece of house to sleep in" (p. 324). Everybody else has already moved away, mostly to the suburbs, where, as Flea's friend sadly states, people are using country dialects in almost every second flat. "I had never before felt such bare loneliness as I did that night. The only puny light coming from our house was mine" (p. 309). Then:

> In the night I thought for the first time that one could leave this place. I don't think I would have opposed it very much had somebody come to the door and offered me a shack somewhere in Herttoniemi [a suburb] or some other God-forsaken place.
>
> (pp. 250–1)

Flea begins to plan his future. He reads in the newspaper about an old service-station to be hired somewhere in the countryside. He has never visited the locality and knows nothing of it, but he leaves for the new place with his future wife, then the 17-year-old daughter of a neighbor. This is a leap into the unknown: things may go well, but they may go totally wrong. After hearing what Flea has in mind, his former boss remarks that "you're going to a spot with no future. The future is in the city, as a city-guy you should know that. ... Come back after a few months, we'll give you a job" (p. 341). The boss then asks Flea to work the night shifts in a service-station in Helsinki. But Flea has made up his mind. When his future wife wonders if the new place is not too far away, Flea replies: "It all depends on the way you look at it. As I am going there entirely, it will be very near to me" (p. 342).

The young couple arrive at their destination with all their worldly belongings on the back seat of the rickety old car:

> The villages were deserted and the weather a bit rainy but getting colder. The landscape had no other colors than grey: the black of asphalt was totally missing. We drove through the villages, passed a number of milk platforms with notices about dances and the like on their walls. As I drove through the forests and fields my belief in the location of the service-station strengthened, for it seemed that all the roads went through it. As the time will come when there's an automobile in the yard of every single cottage, we will have plenty of butter for our bread.
>
> (p. 353)

Yet Flea realizes that "one could even live here, once one had learned the customs" (p. 354). Moreover, the stories of the shopkeeper in this small village "were in no way unlike those of the stationer or the junk dealer" in Sörkka (ibid.). Things have sorted themselves out. There will be enough butter, and much more besides.

By the end of the book, after seventeen years, in 1978, everything is all right, except that, to Flea's distress, his son does not want to go on with his father's business. He wants to taste the freedom of the city. He wants to move to Helsinki to live on his own.

Young urban dwellers and the rates of exchange of human relations

Pirkko Lindberg (born 1942; writing in Swedish), a freelance radio reporter by profession, has so far published only one novel. She has traveled a lot and worked in various crafts: as a cook in a vegetarian restaurant, a carpenter and a rag-picker.

Lindberg's novel, *Saalis* [The Prey], originally titled *Byte* [Exchanges] in Swedish, is a sketch of life in the capital of Finland at the end of 1980s, a widely and entertainingly written story of young Bohemian artists revelling in their emotional experiences. The novel is, at the same time, a love-story and an analysis of a way of life. Lindberg's work, carefully crafted and spiced with soft irony, reveals that even human relations are somewhat like proprietary and exchange relationships between drifting individuals:

> Three years ... there he went with his belongings, one of the many loads going through the city that day. Life means the moving of goods from one nest to another. ... True. The jolly moving loads are cruising back and forth across the city. And the goods themselves are very much tied up with the tussles between people.
>
> (Lindberg 1990: 438)

The novel is set among young Swedish-speaking people in Helsinki and is a multi-layered characterization of the age, profound and light and mythic all at the same time. The main characters are Inna Lysander, an artist busy with her international project on acid rain, named "Corroded Angels," and Sten Zhurikoff, an actor with successful roles on both stage and television, but also going through periods of heavy drinking followed by moments of extreme asceticism. These two people and their many friends populating the pages of the novel belong to the intelligentsia. Yet it is striking that they spend little time discussing the problems of their work or anything of great current interest. It seems to be more important for both Inna and Sten to understand their personal and subjective life by sifting their complex past via dreams and memories.

Another point worth noting in the novel is the scarcity of descriptions of the city: the inner landscapes are the main thing. We know that the house to which Sten Zhurikoff moves in order to live with Inna after separation from his former companion is "an old 'funkis' house [a house in the functionalist style] with heavy stucco balconies and big windows" (p. 19), a house resembling the one in which Sten himself had grown up. The street is Muukalaistie ("Stranger's Road") in the old quarter of Helsinki. There is also "the photo-chemical pollution cloud floating over the city as it always does when it is already hot in the morning and the air is still" (p. 10). The street is "desolate and empty just as city streets usually are on Sunday mornings" (p. 12). Normally, however, if seen from the window of the flat on the sixth floor, the street abyss "is lost in the blue fog of exhaust" (p. 334). For Sten the scene from the balcony is continually present.

> The scene, to be sure, shows its best qualities on a day like this towards the end of the summer. ... Beyond the mass of roofs the line of the sea is shimmering like gold leaf; it simply seems as if a gigantic finger had drawn that golden stroke.
>
> (p. 346)

As life goes on, streets, shops, restaurants, taxis and parks make up the medium of living. As is normal in everyday life, the city remains implicit; it does not stand out to form the main focus. The places of their early childhood are more important in Inna and Sten's inner landscapes than their present ones. Inna's family has a summer home in Kevätniemi ("Spring Point"), a specific but also very mythical place somewhere in eastern Finland, near the Russian border. Spring Point is like a sanctuary to which one can retire for refreshment or when life becomes hard. It is good to go there when one has almost forgotten "what the world is like outside the city" (p. 103). When he is skiing on the "Lake of Truth" – the name of the lake stems from the fact that it is deep and clear, like truth should be – toward the islands of "Honor" and "Conscience," "the ice glimmers like thousands of billions of diamonds, and Zhurikoff cannot remember the last time he saw anything so beautiful. Nothing ever glimmers like this in the city" (pp. 102–3).

For Inna, but not so much for Sten, Spring Point represents a primal or ideal place which she takes with her to the city. She also takes a symbol of the idea of Spring Point in the form of a stuffed lynx, "grandfather's finest work" (p. 80). Of the keepsakes in her flat, Inna likes the lynx best just because "it comes from the untamed forests of the Lake of Truth. It is as if one had a piece of wild country in the midst of all this urban, artificial stuff" (ibid.).

Once again at Spring Point, Inna looks through the small window and sees a sharp blue star shimmering beyond the hanging branches of the birch tree:

> The birch and the star, Inna says, Spring Point's big birch tree and the star twinkling in its crown. When I see it I know I have come home and nothing evil can happen. Never.
>
> (p. 112)

Here we have a reference to the famous Finnish fairy-tale "The Birch and the Star" written by Topelius, a remarkable figure who promoted the Finnish national identity in the nineteenth century. In the story some children are wandering far away from home and it is the birch and the star that guide them back. For Inna, the modern city-dweller of the 1980s, the very same birch and star come to show her the place where she belongs! This, if anything, is really Finnish.

Inna and Sten even plan to move to Spring Point to live and keep bees. This is, of course, an unrealistic dream of country life, and Sten, after his success in the theatre, quickly comes to realize that he "never in the world could change the passionate pulse of big city life for the tardy trot of the countryside" (p. 367). If she likes, Sten thinks, Inna "may have her dreams but she must have them alone; he is a deep-rooted city-dweller right down to the innermost recesses of his soul" (p. 421). For him, there will always be the call of "the blue glow of the neon light above the door of the local tavern" (p. 480).

Living together begins to fall flat. After three years Sten finally takes his belongings and goes to live with another woman. He is caught again, for "as a public figure he is much sought after" (p. 425). In the end, so the story goes, human beings remain alien to each other and live as strangers among other strangers. "Life means the exchange of goods from one nest to another" (p. 438); life is composed of the rates of exchange between individuals.

REGIONALISTIC THOUGHT

In the sections above we have approached the idea of Helsinki from the insider's point of view, and the meanings of the city have been traced and interpreted mainly through creative literature to illustrate individual experiences of place and identities prevailing among people living there. In this section the written image of Helsinki will be approached from a more general social point of view. The point of departure for the following discussion is the idea put forward by Robbins (1983: 1), who writes that the best regional expression (whether literary fiction, sociological treatise or historical analysis) is grounded in social reality: the world of human aspiration and struggle, of greed and exploitation, of selfless accomplishment and tragedy. Social conditions, Robbins continues, reflect and shape attitudes towards home, the community and work, and these necessarily involve a collective and wider regional consciousness. The basic premiss for the following discussion is that the nature of this collective identity or consciousness is a product of historically contingent social conditions: the collective identity reflects economic, political and cultural conflicts, in the production and reproduction of which individuals are continually involved in their everyday life.

In this section literature and the images it presents will be interpreted in the context of regionalism, a complicated phenomenon which extends into the fields of various social practices, such as politics, administration, art and science. Most of the published work on regionalism is fraught with contradiction and ambiguity, partly due to the fact that it is by definition largely a mental phenomenon (Robbins 1983: 1). However, as Robbins points out, it is also based in part on physical fact, since a region is usually understood as a homogeneous entity and regionalism is usually associated with loyalty to and a sense of a real place. Nevertheless, the emphasis will not be put on the nature or homogeneity of a specific region, as is common in discussions that aim at tracing the expressions of regionalism in specific contexts; here, the nature of socially produced regions is understood as an expression of the production of loyalties and a sense of place, and thus as an expression of social control and power relations prevailing in a society in general. Regionalism is thus understood as one expression of a political principle that aims at certain purposes and the creation of a collective consciousness. The main interest is to interpret the role of Helsinki in regionalistic thought, and not so much the role of Finnish regionalism itself.

Regionalism or provincialism is the artistic "counterpart" of the general political and social movement. Regionalism is expressed in many fields of art, but particularly in literature, where the aim of the author to build up a bond between the text and a specific territorial unit is often very apparent. The connection between the political and literary dimension can be close. Gilbert and Litt (1960), for instance, claim that in many European countries novelists have effectively created regional images and identities and, further, the whole consciousness of a territorial unit other than the country (see also Gilbert 1960). Furthermore, it is typical of regional literature and thought to set images associated with conservative and national connections (Sihvo 1979) against cosmopolitanism and universal cultural values (cf. Robbins 1983: 2; Winks 1983: 16).

Regionalism emerged in Finnish literature during the nineteenth century, in the form of realistic stories of folk life, but it was not until the 1970s that a renaissance in regional literature took place. The social context for this was the rapid change that took place in socio-economic structures, as discussed earlier, and specifically the declining role of the traditional agrarian culture (which was originally one of the basic constituents of Finnish identity), increased urbanization and increased spatial and social mobility (cf. Brown 1983 for the case of America). A number of novelists whose writing has a strong local and regional color and came to prominence during the 1970s reacted strongly to these social processes.

A good example of this trend is Heikki Turunen, the best-known new Finnish writer of the 1970s, whose first three books made an effective contribution to regional thinking. Turunen (born 1945 in Pielisjärvi) comes from the province of Northern Karelia in eastern Finland, and surveys indicate that people in the area think of him as "their" novelist (Ahponen and Järvelä 1983: 179). He is also thought of among Finns generally as a well-known regionalist, especially since the first and best-known of his novels, Simpauttaja [Living it up], has been made into a film.

The illustrations we give show the image of Helsinki as seen from a regional point of view. We will argue that these illustrations, although expressing the feelings of fictitious characters, also represent more general social types, which become illustrations of dialectics between urban and rural, artificial and genuine, modern and traditional – through which the contrasting images prevailing in regionalism are usually constructed. Helsinki has been the main object of regionalistic feelings in Finnish literature.

A completely new regionalistic language was created in the course of the structural change that took place in Finnish society, a language in which it is typical to divide the serious social, economic and political conflicts of society into regionalistic categories and oppositions: the prosperous South versus the impoverished North and East, the cities versus the countryside, and so on (Paasi 1986). In Finnish literary regionalism there appear to be two spatial units with which the authors usually contrast their "own" region: the abstract expression of "South," a greedy evil which in regionalist language exploits the poorer areas of Finland, and Helsinki as a specific territorial unit which is seen as having the power to change people.

Helsinki in a regionalistic frame

In the regional perspective, Helsinki becomes a general cultural manifestation of the decline of what is original and regional, and a callous, oppressive physical framework that controls the actions of its inhabitants and in particular those who have had to leave their native localities and homes to seek work. Heikki Turunen builds up a tension between Helsinki and Northern Karelia within the situations of the fictitious characters he describes. Regional contradictions such as southern Finland versus under-developed areas, educated versus non-educated people, rich versus poor, and so on, are linked together in a critique of cosmopolitanism. The following quotation from Turunen's *Joensuun Elli* [Elli from Joensuu] is indicative of this. It describes the midsummer celebration in a Northern Karelian holiday-camp as experienced by those who have stayed in the countryside and those living in the cities:

> [On the cars] there were number-plates from different provinces and countries. Some people were moving around half-naked as if it was quite natural. Some crouched alone fully clothed and watched with solemn faces the freer and more joyful holiday-making of the others. The foreigners and those from the south spoke loudly and without restraint and were clearly in control of the situation, whereas the people with small cars and tents, middle-aged working-class couples, men and women from the provinces, were mostly quiet and lowered their voice to a whisper when they spoke, as if they were ashamed because of their slow, dialectal mode of speaking.
>
> (Turunen 1975: 121)

The conflict between urban and rural areas is apparent in Turunen's early work and in his first novel *Simpauttaja*, where he describes the emerging basic socio-economic contradiction which derived from the lack of employ-ment in rural areas and the ready supply of work in the cities. The city takes two forms in this novel. First, it is an abstract concept that expresses the pressure of change upon the way of life, and second, it is a spatial expression of Helsinki. Turunen's novels depict Helsinki as a melting-pot: people coming from the countryside can become "real people" there, but also drunks or whores. The dichotomies struck by the author are strong, but in this way they combine very effectively the social content of two spatially different modes of life. For the first, emancipation from the rigorous social control of the rural context is necessary:

> You are right, son. We will not stay here to be watched over by gossips.

74

We will go to south to build roads. We will level the ground and swill wine at the weekends till it oozes from our eyes.

(Turunen 1973: 278)

On the other hand, to identify with Helsinki you have to be able to act like a city resident, to give up your old local or regional identity and your traditional relations with your environment:

Winter brought to Imppa's mind the city, its frosty smoke rising straight upwards, and his stay in Helsinki with Yrjö. You have to cross the streets in a natural manner and walk with an air of importance and without glancing at the shop windows. You have to be free and easy, and hard even if a lousy feeling of strangeness is plucking at your insides.

(Turunen 1973: 259)

These illustrations of regionalism and the ideal identity of Helsinki are not pure fiction since several empirical investigations confirm that the images and identities described really do prevail among people living in the province of Northern Karelia. The specific idea of "stadi" (the city of Helsinki) and its cultural image and cultural codes as they manifest themselves is completely different for those living in the city itself and for outsiders (see Paasi 1984; Ahponen and Järvelä 1983). In fact, the conflict between rural and urban and the stereotypical structures involved within it has been relatively permanent for a long time in Finland (see Eskola 1965).

Recent descriptions of the linguistic behavior of the Finns indicate that people who have migrated to Helsinki from the provinces in fact live in several worlds and have several overlapping sociospatial identities (or social roles) that are expressed differently in different situations and localities (Nuolijärvi 1986: 164–5). Local sociocultural milieux thus have "causal power" over the behavior of individuals and it is obviously difficult for the individuals to resist these "structures of expectations," that is, the cultural codes prevailing in different places (cf. Paasi 1986, 1991). The tension between town and country is perhaps strongest in Turunen's novel *Kiven-pyörittäjän kylä* [The Village of the Stone Roller], which is an impressive description of the death of a village. The principal characters, Pekka and his wife Meeri, who have emigrated from Northern Karelia to Sweden, are beautiful illustrations of the ability of a regional writer to convey in his descriptions the deep social message of the homogenization of space – a process which has been illustrated so effectively by Relph (1976) in his book *Place and Placelessness*. It is not difficult to place the message of the author in the framework of the classical sociologists, above all Émile Durkheim (1964; orig. 1893), who put forward arguments about changing spatial ties during the development of a society: in modern society it is occupation that defines your spatial location and not vice versa.

Pekka lives in a world filled with golden memories of his rural past, and does not accept the "causal" role of his workplace, as noted by Durkheim. Meeri, on the other hand, is a "placeless pessimist," who has been completely socialized into a modern society characterized by the division of labor: her spatial bond is constituted purely mechanistically on the basis of the labor markets and without any scruples – your home is where your bread is:

Oh Lord, what can an average human being do about development? Society will become urbanized in any case. People nowadays simply want to live together. ... I could live anywhere in the world nowadays. I cannot understand why people are mourning over places. They are just places and there exist a lot which are much better. I have accepted the fact that we have been born on wheels.

(Turunen 1976: 52)

The social roles embodied in the characters are made more prominent by stereotyped good and bad attributes: Pekka is kind, introverted and calm – a product of traditional rural life – while Meeri is cold, rational and simply a calculating character – an example of the detrimental aspects of urbanization.

The dialectics of departure and return

The proponents of humanistic geography have emphasized that it is the dialectic between departure and return that makes people discern where they really belong. A prerequisite for a deep sense of place is distance, which makes it possible to distinguish oneself from specific places. This distinction is not purely physical but also experiential (Karjalainen 1986). A classic theme in literature is that of returning home (Sihvo 1969). It is common for regional writers to describe how individuals understand their places of origin only after they have left. Of course, this is only one way of responding to such departures. Often, leaving home is seen as a release from spatial and social bonds.

The idea of leaving and returning is important in Turunen's works, where a tension in one's relation to places emerges from the fact that it is always good to be where you are not. But migration to some other place means change, and as such it is to be resisted. A conservative perspective typical of regional literature arises from the fact that a new situation is often seen as insincere and non-viable, while the past is genuine but lost. The relations between new and old are manifested in the personification of the situations of the principal characters in their own lives. It is from these facts that the oppositions such as city versus rural areas, home versus away, and "own region" versus Helsinki emerge.

EPILOGUE

All cities, not only Helsinki, can be comprehended as places, processes and states of mind (cf. Winks 1983: 18). Furthermore, all cities are parts of larger sociospatial systems that manifest themselves in social, economic and political practices and the power relations emerging from them. These also appear in the cultural and literary realm. We have approached the city of Helsinki here within a dual frame, aiming at illustrating the written image of the city from two essentially interdependent perspectives. First, we have shown the written image of the city from the insider's point of view, as manifested in novels about the daily lives of people living in Helsinki and experiencing its atmosphere. This approach has helped us to understand situations in the lives of the characters, which in turn illustrate more general sociospatial processes. Second, we have tried to contextualize these individual identities and experiences of life by approaching the image of the city in a more general framework as an expression of the regionalistic social and spatial relations prevailing in Finnish society. Helsinki, the capital, has become the most important symbol of social change and urban values in Finnish regional literature. This is especially apparent in the works of Heikki Turunen cited here. Turunen builds his regionalistic perspective on powerful dichotomies which sharply contrast urban and rural values. As a regionalist, Turunen stands firmly on the side of rural people.

This framework has made it possible to demonstrate that the image of a written city is a complicated phenomenon expressing both individual feelings, sorrows and identities and more general sociospatial processes that are produced and reproduced in everyday life.

REFERENCES

Ahokas, J. (1973), *A History of Finnish Literature*, Bloomington: Indiana University.

Ahponen, P.-L. and Järvelä, M. (1983), *Maalta kaupunkiin, pientilalta tehtaaseen: Tehdastyöläisten elämäntavan muutos* [From Countryside to City, Small Farm to Factory: The Changing Way of Life of the Factory Workers], Juva: Werner Söderström.

Andersson, C. (1977), *Armas aika* [Beloved Time], Helsinki: Otava.

——— (1987), *Det som blev ord i mig: Dikter 1962–1987* [That which Became a Word in Me: Poems 1962–1987], n.p.: Alba.

Bordessa, R. (1989), "Between house and home: The ambiguity of sojourn," *Terra*, 101/1, pp. 34–7.

Brown, R. M. (1983), "The new regionalism in America," in W. G. Robbins, R. J. Frank and R. E. Ross (eds.), *Regionalism and the Pacific Northwest*, Corvallis, Oregon: Oregon State University Press.

Carpelan, B. (1983; orig. 1961), "Minun Helsinkini" [My Helsinki], in *Finlandia: Otavan Iso Maammekirja 1: Uusimaa*, Helsinki: Otava.

Donner, J. (1983; orig. 1961), "Helsingfors – Finlands ansikte" [Helsinki – Finland's Face], in *Finlandia: Otavan Iso Maammekirja 1: Uusimaa*, Helsinki: Otava.

——— (1987), *Kaupungin sydämessä: Kahdeksan tapaa kuvailla Helsinkiä* [At the

Core of the City: Eight Ways to Describe Helsinki], Helsinki: Oy Stockmann Ab.

Dovey, K. (1985), "Home and homelessness," in Irwin Altman and Carol M. Werner (eds.), *Home Environments*, New York and London: Plenum Press.

Durkheim, E. (1964; orig. 1893), *The Division of Labor in Society*, New York: Free Press.

Eskola, A. (1965), *Maalaiset ja kaupunkilaiset* [Rural People and Urban Dwellers], Helsinki: Kirjayhtymä.

Gilbert, E. (1960), "The idea of region," *Geography*, 45, pp. 157–75.

Gilbert, E. W. and Litt, B. (1960), "Geography and regionalism," in G. Taylor (ed.), *Geography in the Twentieth Century: A Study of Growth, Field, Techniques, Aims and Trends*, London: Methuen.

Karjalainen, P. T. (1986), *Geodiversity as a Lived World: On the Geography of Existence*, Joensuu: University of Joensuu, Publications in Social Sciences, 7.

—— (1991), "Matters of environmental interest: Thoughts of meaning and place," *The National Geographical Journal of India*, 37/1–2, pp. 9–16.

Kortteinen, M. (1982), *Lähiö: Tutkimus elämäntapojen muutoksesta* [Suburb: A Study of the Changes of the Way of Life], Helsinki: Otava.

Laitinen, K. (1985), *Literature of Finland: An Outline*, Helsinki: Otava.

—— (1981), *Suomalaisen kirjallisuuden historia* [A History of Finnish Literature], Helsinki: Otava.

Lindberg, P. (1990), *Saalis* [The Prey], Helsinki: Otava.

Mäkelä, M. (1986), *Suuri muutto 1960–70 -lukujen suomalaisen proosan kuvaamana* [Great Migration as Depicted in Finnish Prose 1960–1970], Helsinki: Otava.

Norberg-Schulz, C. (1985), *The Concept of Dwelling: On the Way to Figurative Architecture*, New York: Rizzoli International Publications.

Nuolijärvi, P. (1986), *Kieliyhteisön vaihto ja muuttajan identiteetti* [The Change of Linguistic Community and the Migrant's Identity], Jyväskylä: Suomalaisen Kirjallisuuden Seura.

Paasi, A. (1984), *Opiskelijoiden tilapreferenssit aluetietoisuuden indikaattorina* [Spatial Preferences of Students as an Indicator of Regional Consciousness], Suunnittelumaantieteen yhdistyksen julkaisuja, 14.

—— (1986), "The institutionalization of regions: a theoretical framework for understanding the emergence of regions and the constitution of regional identity," *Fennia*, 164/1, pp. 105–46.

—— (1991), "Deconstructing regions: notes on the scales of spatial life," *Environment and Planning* A 23, pp. 239–56.

Palmgren, R. (1989), *Kaupunki ja tekniikka Suomen kirjallisuudessa: Kuvauslinjoja ennen ja jälkeen tulenkantajien* [City and Technology in Finnish Literature: Lines of Description before and after the Torchbearers], Helsinki: Suomalaisen Kirjallisuuden Seura.

Porteous, J. D. (1990), *Landscapes of Mind: Worlds of Sense and Metaphor*, Toronto: University of Toronto Press.

Relph, E. (1976), *Place and Placelessness*, London: Pion.

Robbins, W. G. (1983), "Introduction." In W. G. Robbins, R. J. Frank and R. E. Ross (eds.), *Regionalism and the Pacific Northwest*, Corvallis, Oregon: Oregon State University Press.

Ruuth, A. (1979), *Viimeinen syksy* [The Last Autumn], Helsinki: Tammi.

Seamon, D. (1979), *A Geography of the Lifeworld: Movement, Rest and Encounter*, London: Croom Helm.

Sihvo, H. (1969), "Kotiseutukirjallisuus" [Regionalistic literature], in: *Kansantaide ja perinnepolitiikka* [Folk Art and Traditionalism], Vammala.

Sihvo, H. (1979), "Regionalismi" [Regionalism], in: *Suuri Ensyklopedia* [Great Encyclopedia] 7, Helsinki: Otava.

Tarkka, P. (1990), *Suomalaisia nykykirjailijoita* [Contemporary Finnish Writers], Helsinki: Tammi.

Turunen, H. (1973), *Simpauttaja* [Living it up], Porvoo: Werner Söderström.

—— (1975), *Joensuun Elli* [Elli from Joensuu], Porvoo: Werner Söderström.

—— (1976), *Kivenpyörittäjän kylä* [The Village of the Stone Roller], Porvoo: Werner Söderström.

Valkonen, T. (1985), "Alueelliset erot" [Regional differences], in T. Valkonen, R. Alapuro, M. Alestalo, R. Jallinoja and T. Sondlund (eds.), *Suomalaiset: Yhteiskunnan rakenne teollistumisen aikana* [The Finns: The Structure of Society in the Era of Industrialization], Porvoo: Werner Söderström.

Vartio, M.-L. (1982; orig. 1960), *Kaikki naiset näkevät unia* [All Women Have Dreams], Helsinki: Otava.

Winks, R. W. (1983), "Regionalism in comparative perspective," in W. G. Robbins, R. J. Frank and R. E. Ross (eds.), *Regionalism and the Pacific Northwest*, Corvallis, Oregon: Oregon State University Press.

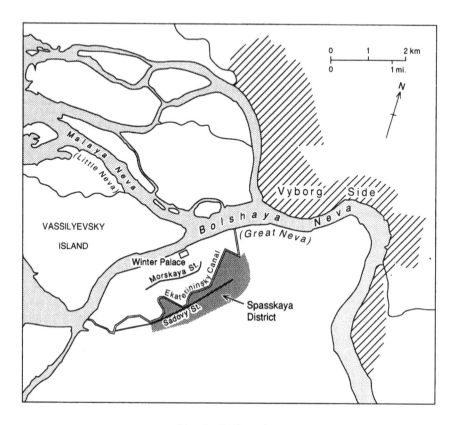

Map 5 St. Petersburg

5

THE ST. PETERSBURG OF *OBLOMOV* AND *CRIME AND PUNISHMENT*

Alec Paul

St. Petersburg was the dominant city of pre-revolutionary Russia, the seat of the tsars and the setting for many works of literature. This chapter examines the treatment of St. Petersburg in two novels by Ivan Goncharov and Fyodor Dostoyevsky, *Oblomov*[1] and *Crime and Punishment*[2] respectively.

The city was begun in 1703, on the isolated and swampy delta at 60° North latitude where the River Neva debouches into the Gulf of Finland. Named after Peter the Great, it was a testament to his determination that Russia would have a splendid capital city, both to symbolize and to lead the "Europeanizing" of his backward Empire. Its port would command the shortest sea routes from Russia towards the capitals and trading cities of northern and western Europe. Peter's selection of this location, remote from the cultural and economic heart of Russia, is reminiscent in some ways of that of Brasilia in the modern era. St. Petersburg's early development was carefully planned and its architecture was imposing, but it was built by the forced labor of serfs imported from many parts of Russia. Thus from its very beginnings it was a place of paradoxes, its outward splendor concealing its harsh realities.

St. Petersburg in the 1850s and 1860s had about half a million people. To many of them, however, the metropolis was a foreign environment, difficult to adapt to. Russia was still predominantly rural. Many migrants to St. Petersburg dreamed of a return to the "old life" in their province of origin. The capital city was truly on the edge of Russia in various ways and it could not satisfy the aspirations of many of its people. Nevertheless, it was new, it was where Russia and Europe met, and for many Russians it was a place of culture, activity and vitality in an otherwise backward nation. St. Petersburg had huge potential, but also huge problems.

In this period and city of contrasts and contradictions were set two of the great novels of Russia. *Oblomov* was published in 1859. This second novel by Ivan Goncharov[3] is considered by many to mark a turning-point in Russian literature, both for its realistic treatment of the characters of ordinary human beings and for its creation of Oblomov himself. *Crime and Punishment* came out in 1866 and is acclaimed as a masterpiece of world-wide significance. Dostoyevsky's[4] almost clinical analysis of Raskolnikov, the central character, is a remarkable achievement. Although Dostoyevsky is far better known internationally than Goncharov, the latter's works remain very popular to the present day in his own country.

Neither author set out to write about the city. Both works discussed here have been commonly described as psychological novels; but the authors put their heroes in St. Petersburg and made them both transplants from rural Russia, as so many St. Petersburg residents were. This says a lot about the importance of St. Petersburg to the life and literature of the country during this period. While Moscow was the obvious center of the Russian empire, St. Petersburg at mid-century had been the national capital for over a hundred years. It was also the critical point of contact with Europe, and Europe was in a ferment. The industrial revolution was spreading. Large manufacturing cities had emerged, radical economic and social changes were abroad, and in Russia St. Petersburg was the first place to be affected by all this.

The theme of ferment, of a society in turmoil, is developed in both novels. Neither Oblomov nor Raskolnikov, who come from traditional backgrounds in the rural provinces of Russia, can adapt to life in this city that seems to be perpetually on the edge. Both are eventually alienated by it. But there the similarity between the two stories ends. The two authors had very different experiences of St. Petersburg, which are reflected in the novels.

Goncharov came from a merchant-class family in Simbirsk (Ulyanovsk), then a sleepy provincial town on the Volga. He was sent to school in Moscow at the age of 10, and eventually graduated from Moscow University in 1834. There followed a conventional career in the civil service in St. Petersburg which lasted thirty-three years and included two periods, each of several years' duration, as an official censor. Goncharov never married and, particularly after the enormous success of *Oblomov*, sought to avoid crowds and the social scene. Indeed throughout the 1860s and 1870s he lived a rather secluded life in his St. Petersburg flat.

Dostoyevsky was born in Moscow but lived most of his life, with the exception of the ten years of exile in Siberia, in St. Petersburg. His life was anything but conventional. His propensity for gambling, his epilepsy, his marriages, his frequent periods of near-insolvency and above all his experi-ence of being sentenced to death and then reprieved at the last moment gave him a veritable treasure-house of experiences to draw upon in his writing. *Crime and Punishment* is regarded by many as one of his finest works, and it is certainly the novel that brought him world-wide fame.

Oblomov and *Crime and Punishment* thus spring from very different sources of creativity, and the plots are poles apart. Raskolnikov, driven to the edge of insanity by his inability to reconcile theories of human existence and traditional moral values, commits murder but is redeemed by the love of

Sonia, a prostitute. Ilya Ilyitch Oblomov, on the other hand, lives a pure but wasted life and even the love of Olga rouses him only temporarily from his lethargy. Again, Oblomov becomes gradually divorced from life and from the city, and eventually dies of lack of interest. Raskolnikov, however, is part and parcel of St. Petersburg. He personifies at once its stench and its beauty, its cruelty and its compassion; but he too is defeated by it. Only exile to Siberia can revive him.

What is it about this city of the tsars that is so overpowering? Strikingly, in both novels the inhuman scale of this human creation is the first impression that comes over. Oblomov is introduced to us "in his flat in Gorohovy Street, in one of the big houses that had almost as many inhabitants as a whole country town" (p. 3). Raskolnikov lodges in a "garret ... under the roof of a high, five-storied house ... more like a cupboard than a room" (p. 1).

At the outset, then, both St. Petersburgs are remarkably alike. The two novels begin by reducing the city to the level of rented rooms – dirty and depressing. Raskolnikov is crushed by real poverty, Oblomov obsessed by the fear that he soon will be.

This theme of poverty is an ironic counterpoint to the original concept of St. Petersburg as the model city of Russia, a planned, visually pleasing capital "embodying all that was new in architecture and design", as Bater[5] puts it. That the initial external grandeur of the "Granite City" cannot avoid being sullied by the scars of human existence is beautifully expressed in the passage in *Crime and Punishment* where a drowning woman is rescued from the Ekaterininsky Canal and then laid on the fine granite pavement of the embankment:

> [Raskolnikov] became aware of someone standing on the right side of him ... a tall woman with a kerchief on her head, with a long, yellow, wasted face, and red sunken eyes. She was looking straight at him, but obviously she saw nothing and recognized no one. Suddenly she ... threw herself into the canal. The filthy water parted and swallowed up its victim for a moment, but an instant later the drowning woman floated to the surface, moving slowly with the current.... "A boat, a boat!" was shouted in the crowd. But there was no need of a boat; a policeman ran down the steps to the canal, threw off his great coat and his boots, and rushed into the water. It was easy to reach her; she floated within a couple of yards from the steps.... They laid her on the granite pavement of the embankment. She soon recovered consciousness ... sat up ... stupidly wiping her wet dress with her hands.
>
> (p. 155)

Poverty is rampant in mid-century St. Petersburg. Raskolnikov's sphere of activity is especially squalid, perhaps reflecting a Dickensian influence in Dostoyevsky's depiction of the urban scene as bleak and unrelenting for the poor.[6] Oblomov's St. Petersburg is considerably different, but his flat too has mice, moths, bugs and fleas (p. 10). This treatment of the seedier side of city life is common to many Russian writers but it is a notable characteristic of *Oblomov* and especially of *Crime and Punishment*.

Raskolnikov's experience of St. Petersburg is a never-ending kaleidoscope of sensual encounters with the city, brief images of its filth and noise and its occasional havens of tranquility, and its sights and smells and people. Confusion reigns. No-one is in control, and certainly not Raskolnikov. After committing the killings, he is on edge, wandering along the bank of the Ekaterininsky Canal seeking to get rid of the things he has taken from the murdered women's rooms ("Fling them into the canal, and all traces hidden in the water, the thing would be at an end," p. 99). But there are rafts at the water's edge, with women washing clothes on them, and people "swarming everywhere." He has to find somewhere else, and he does so entirely by chance, a dirty deserted courtyard strewn with all sorts of rubbish.

A short while later, distractedly walking in the middle of the Nikolaevsky Bridge, he is almost run down by a carriage and only a lash across the back from the coachman's whip brings him to his senses. The incident draws an act of charity from a well-to-do woman walking by, who takes him for a beggar and presses twenty kopecks into his hand. Ten paces later, another aspect of the city impresses itself upon Raskolnikov's chaotic mental state – the "truly magnificent spectacle" (p. 105) of the view along the Neva towards the palace from the bridge. The pain from the whiplash is forgotten in the face of the blue water, the cloudless sky, the pure air and the glittering cupola of the cathedral. Yet the vista "left him strangely cold; this gorgeous picture was for him blank and lifeless," He roams the streets for several more hours before returning home without any recollection of where he has been since the Nikolaevsky Bridge.

This sense of being lost in the city is transmitted by Dostoyevsky through feverish activity and seemingly chaotic movement. Raskolnikov's St. Petersburg at times is like an ant-hill.[7] You can be lost in the great crowd of people "like a grain of sand" (p. 80). He particularly likes the vicinity of the Hay Market. "Here his rags did not attract contemptuous attention, and one could walk about in any attire without scandalizing people" (p. 57). There are crowds of rag-pickers and costermongers and other poor tradespeople. But the anonymity of the ant-hill is illusory. The city is also a complex spider's web of intertwined human relationships, all ultimately connected. Raskolnikov's inability to pay his rent to his landlady results in a summons to appear at the police station the very morning after the murders. His disturbed state is clearly apparent at this interview and arouses the suspicions that lead to the drawn-out cat-and-mouse relationship between himself and Porfiry Petrovich, the head of investigations for the police district. Porfiry is a distant relative of Razumihin, a fellow student of Raskolnikov who eventually marries the latter's sister, Dounia. These far-fetched coincidences serve to emphasize the idea of interconnection in the spider's web.

This idea of lack of control, of events, connections and situations seeming to conspire against the individual, is also brought out in Svidrigailov's suicide. Even this debauched man of the world, who has attempted to seduce and blackmail Dounia and who also hears from a neighboring room Raskolnikov's confession of the murders to Sonia, is overwhelmed in the end by feelings of inadequacy in the face of so many different emotions. The final loss of equilibrium which results in his shooting himself occurs against a backdrop of nightmares and depravity in a cheap hotel on the edge of St. Petersburg. Ironically, the city is not in control either. Svidrigailov's sleep is interrupted not only by his own internal tortures but also by the flood-warning cannon:

> "Ah, the signal! The river is overflowing," he thought. "By morning it will be swirling down the street in the lower parts, flooding the basements and cellars. The cellar rats will swim out, and men will curse in the rain and wind as they drag their rubbish to their upper storeys."
>
> (p. 456)

The city that was to be the Venice of the North itself lives on the edge of disaster.

The flood and his awareness of its impact seem to contribute to Svidrigailov's sense of displacement. Next morning he shoots himself in a short but complex scene which for Richard Peace "symbolizes the nature of [Svidrigailov's] inner dichotomy".[8] But there is no dichotomy between Svidrigailov's mood and the mood of the day, with the St. Petersburg summer climate at its worst:

> A thick mist hung over the town. Svidrigailov walked along the slippery dirty wooden pavement towards the Little Neva. He was picturing the waters of the Little Neva swollen in the night ... the wet paths, the wet grass, the wet trees and bushes.... There was not a cabman or passer-by in the street. The bright yellow, wooden little houses looked dirty and dejected with their closed shutters. The cold and damp penetrated his whole body and he began to shiver.
>
> (p. 459)

The St. Petersburg depicted in *Oblomov* is less of an immediate physical and psychological presence. Oblomov and his manservant, Zahar, are relics of the old feudal Russia. They struggle and ultimately fail to come to terms with life in this new European-style city. As we have already seen, St. Petersburg is given an unflattering introduction in the first few pages of Goncharov's novel. This tone is maintained as Oblomov is visited in the first few chapters by an array of idle St. Petersburg civil servants, socialites and swindlers. They provide a counterbalance to the remarkable chapter called "Oblomov's Dream" which follows. This lyrical piece had in fact been published separately ten years before the appearance of the novel. It takes the reader back to Oblomovka, the hero's childhood home, and satirizes the rural idyll, the old way of life of the minor aristocracy on their country estates.

Oblomovka is the antithesis of St. Petersburg:

> Nothing disturbed the monotony of the life at Oblomovka, and its inhabitants did not resent it, for they could not imagine any other existence ... What did they want with variety, change and adventure that other people seek? ... They went on for years yawning and drowsing, laughing good-humouredly at the country jokes or sitting around and telling their dreams to one another.
>
> (p. 128)

Oblomov's parents lived quiet, predictable lives, well fed and untroubled by any "obscure moral or intellectual problems.... The same things were done in the same way in Ilya Ilyitch's father's time as they had been done for generations" (p. 118). Life at Oblomovka was "like a quiet river, and all that remained for them was to sit on the bank watching." Oblomov dreams of a return to this "blessed spot", this "lovely country" which "promises a calm, long life till the hair turns from white to yellow, and death comes unnoticed like sleep" (pp. 96–7).

Ilya Ilyitch's rude awakening from his dream ushers in Part 2 of the novel. The old ways and the quiet life of the country are things of the past. Oblomov has now inherited Oblomovka and must deal himself with the bailiff on the estate, who complains constantly of poor yields, unpaid debts, unruly peasants and the rising costs of shipping produce to market. His reports may not be entirely realistic, since country bailiffs can be swindlers too, but his picture is very different from the one in Oblomov's dream. Life in St. Petersburg may be bothersome to Oblomov, but so too would be life on the estate. He stays in St. Petersburg.

Oblomov originally came to the capital to take up a post in the civil service, but gave it up after two years. For a time afterwards he maintained an active social life – at the start of the book he still considers the Vyborg side, where he finally settles, to be "dull and dreary" (p. 43) – but this required a certain amount of exertion as well as meeting many people he considers vacuous or abhorrent. By degrees he has withdrawn from St. Petersburg society, and has become "firmly rooted in his flat" (p. 58).

The flat is neglected and untidy. Zahar wears "a grey coat torn under the arm and showing his shirt" (p. 6). He wears this old-fashioned coat solely because it is the only thing around which reminds him of the "peace and plenty in the house of his master [Oblomov's father] in the depths of the country." To him St. Petersburg is a come-down from Oblomovka. He despises the German family who live in the flat opposite. Their tidy, frugal life-style Zahar interprets as parsimonious, even niggardly. Indeed, the house in which they all live is typical of the changing, formerly genteel residential area close to the centre of St. Petersburg. The yard is a hive of activity, as in *Crime and Punishment*. Strolling players, pedlars and hawkers, tradesmen and fund-raisers talk and shout; there is traffic noise from the street and the ringing of axes and the cries of workmen from a house being built next door. There is a drinking-house just outside the yard gate which Zahar often visits. Oblomov's reaction (p. 74) is "How disgusting this town noise is! When will the paradise that I long for come at last? When shall I go to my native fields and woods?"

Paradise is fleeting. Oblomov has a summer romance with Olga Ilyinsky and for a short few months even St. Petersburg becomes tolerable. Like Raskolnikov, he too wanders aimlessly on the banks of the Neva, but for a different reason. The relationship blossoms in the summer-resort area beyond the edge of the city on the Vyborg side, with its peace and quiet and greenery. Olga and her aunt move from their town residence on Morskaya Street and take a summer villa not far from the one Oblomov has rented after giving up his St. Petersburg flat. The move from the city and his love for Olga even give Oblomov the energy to walk by himself across the marsh to the forest three days in a row. During one of their meetings in the huge park nearby, Olga picks a spray of lilac and says how delicious it smells, but Oblomov picks some lilies of the valley and remarks that they "smell better, of fields and woods; there is more nature about them. And lilac always grows close to houses, the branches thrust themselves in at the windows and the smell is too sweet" (p. 211). Lilac symbolizes the city for Oblomov, thrusting itself at him with its smells and noise and frantic activity.

They become unofficially engaged, but the romance peters out with the move back to the city and the cold weather. After a summer in which he has climbed all the hills within three miles of the villa several times and visited "all the places round St. Petersburg" (p. 241) Oblomov relapses into his old idle ways. It is Olga's aunt who decides first on the move back to town, shortly after the end of August when "it began to rain and smoke poured from the chimneys of summer villas that had stoves" (p. 310). The indecisive Oblomov is simply left behind:

> gradually all the summer villas were deserted. Oblomov had not been to town any more. One morning he saw the Ilyinskys' furniture being carted past his windows. Although he no longer thought it a heroic feat to leave his lodgings, to dine in a restaurant and not to lie down all day, he was at a loss where to spend the night.... He walked through [Olga's] empty rooms ... and his heart was oppressed with sadness.
>
> (p. 310)

He tells Zahar and his wife Anissya, to go to the place on the Vyborg side that Tarantyev, an acquaintance, has told him about; he can stay there until he finds another flat in the city. In the evening he visits Olga at Morskaya Street, and things begin to unravel.

"Autumn evenings in town were very different.... In town he could not see her three times a day" (p. 310). Flats near Olga's house are very expensive, and Oblomov is worried about what he sees as major problems on his country estate which will surely result in a reduction of his income. He decides to stay in the flat in the house of Ivan Matveyich's sister on the Vyborg side, as arranged by Tarantyev. It is a long journey by carriage, three hours there and back, from there to Morskaya Street. He visits Olga less and less, and begins to withdraw into the quiet comfort of his lowly new lodging.

Olga makes one last desperate effort to rouse him. She arranges a rendezvous in the Summer Garden around the end of September, but it is too late. Oblomov is already sinking into the lethargy engendered by the pleasant torpor surrounding the establishment of his new landlady, who is an excellent cook and who makes every effort to make life easy for her new "gentleman" tenant. Olga gets him to go for a boat ride with her on the Neva, but the leaves have fallen and one can see right through the tree branches (p. 338). Even the view of the Smolny Church from the river awakens no feelings in Oblomov but impatience. His love for Olga still flickers but the practical problems of a marriage terrify him: "Where is the money? What are we to live on? Even you have to be bought, pure, lawful bliss of love!" (p. 335).

A few weeks later the Neva freezes over and Olga finally realizes that there is no future for her with Oblomov. She breaks off their engagement and Oblomov no longer has any reason to stir from his new nest in Agafya Matveyevna's house. St. Petersburg has nothing left for him, and he dies a few years later in this semi-rural setting on the Vyborg side.

These contrasts of life in the city and in the "country suburbs" are also made in *Crime and Punishment*. Early in the novel Raskolnikov is walking in the inner city and means to go back home, "but the thought of going home suddenly filled him with intense loathing ... that hole ... that awful little cupboard of his" (p. 49). Instead he "walked right across Vassilyevsky Ostrov [Island], came out on to the Lesser Neva, crossed the bridge and turned towards the islands." But although the greenness and freshness of the islands are restful after the dust and the huge houses of the town "that hemmed him in and weighed upon him" (p. 49), he is irritated by the obviously upper-class nature of the district. He meets luxurious carriages and men and women on horseback and passes by the brightly painted summer villas of the well-to-do (p. 50).

This stark contrast with the grim life of the poor quarters of the inner city reminds us that at the time of these two novels St. Petersburg was very much "on the edge." The era of planning and of architectural unity, of the creation of the beautiful capital of the tsars, was over. With the emancipation of the serfs in 1861, the increasing industrialization of the city, and other factors, the rapid population growth became an explosion. Control over the movements of peasants and workers was no longer possible. Housing began to be a critical problem. In the outer boroughs wooden residential structures began once again to proliferate. The stone tenements of the poorer areas of the inner city became increasingly crowded. The murdered moneylender of *Crime and Punishment* lives in a huge house looking onto the canal on one side and into the street on the other, "let out in tiny tenements and ... inhabited by working people of all kinds – tailors, locksmiths, cooks, Germans of sorts, girls picking up a living as best they could, petty clerks, etc." (p. 3). According to Bater, most areas of St. Petersburg at this time had an excess of males, which may help to explain the abundance of taverns, drinking establishments and female prostitution which is incorporated into *Crime and Punishment*.[9] To the traditional military-based male majority of earlier days in St. Petersburg have been added the vast numbers of migrant workers from the rural areas, predominantly men. Unemployment and poverty force many to begging and/or crime to survive.

This is all well known from the works of Dostoyevsky, but *Oblomov* too contains numerous references to fraud and theft, and to bribery and corruption in the civil service. One of the sub-plots in the novel deals with the fraud perpetrated on Oblomov by Tarantyev and Ivan Matveyich, in which they bleed the unsuspecting innocent of extortionate amounts of money for rent and for supposedly setting his rural estate to rights. Meetings of the conspirators take place in what seems to be a typical setting for such events:

> one of the top-floor rooms of a two-storied house that stood in the street where Oblomov lived and faced with its other side on to the quay. It was the so-called restaurant; two or three empty droshki could always be seen outside, while the drivers sat on the ground-floor drinking tea out of their saucers. The top-floor was reserved for the "gentry" of the Vyborg side. Ivan Matveyich and Tarantyev had tea and a bottle of rum before them.
>
> (p. 374)

Such establishments of dubious repute are common in *Crime and Punishment* in which the district police station also figures conspicuously. Raskolnikov's first visit comes on the morning after he has committed the murders. The signs are ominous. He has slept poorly. In the late morning, on his way to the police station, the heat and dust and the stench from the shops and pothouses are insufferable. He has already wondered

> why in all great towns men are not simply driven by necessity, but in some peculiar way inclined to live in those parts of the town where there are no gardens nor fountains; where there are most dirt and smell and all sorts of nastiness.
>
> (p. 68)

The police station is most unprepossessing. It is on the fourth floor of a new house and is approached by a staircase which is

> steep, narrow and all sloppy with dirty water. The kitchens of the flats opened on to the stairs and stood open almost the whole day. So there was a fearful smell and heat. The staircase was crowded with porters going up and down with their books under their arms, policemen, and persons of all sorts and both sexes. The door of the office, too, stood wide open. Peasants stood waiting within. There, too, the heat was stifling and there was a sickening smell of fresh paint and stale oil from the newly-decorated rooms ... [which all] were small and low-pitched.... [The head clerk's] was a small room and packed full of people, rather better dressed than in the outer rooms.
>
> (pp. 87–8)

The location clearly symbolizes the fact that dealings between the police and the public were a very normal part of everyday life in St. Petersburg, at least for Dostoyevsky. The police are busy. These overcrowded lower-class quarters of St. Petersburg (Raskolnikov and the moneylender apparently live in the Spasskaya area, close to the Ekaterininsky Canal) contain all kinds of riff-raff, low taverns and bars and brothels. Typical is the passage about the little street that linked the Hay Market to Sadovy Street. Here

> there is a great block of buildings, entirely let out in dram-shops and eating houses; women ... gathered in groups, on the pavement, especially about the entrances to various festive establishments in the lower storeys.... One beggar was quarrelling with another, and a man dead drunk was lying right across the road. Raskolnikov joined the throng of women, who were talking in husky voices. They were bareheaded and wore cotton dresses and goatskin shoes. There were women of forty and some not more than seventeen; almost all had blackened eyes.
>
> (pp. 143–4)

At the other end of the social scale of the criminal element in Dostoyevsky's St. Petersburg are those corrupt members of the gentry personified by Luzhin and Svidrigailov, both of whom have wronged Raskolnikov's sister Dounia. They are more reminiscent of Tarantyev and Ivan Matveyich in *Oblomov*, symbols of the wolves who come to the Vyborg side in winter (p. 43). The crimes of the middle and upper classes are more subtle and more dastardly. Raskolnikov himself is a peculiar case, guilty of murder inspired by social theory, as it were.

There are good people in St. Petersburg too. Sonia in *Crime and Punishment* has been forced into prostitution to help keep her family, yet she retains an innocent soul and a capacity for redeeming love. Sonia is an almost Christ-like figure. Her plight is equated to that of Dounia, who is planning to prostitute herself in a loveless marriage to Luzhin, in the passages of *Crime and Punishment* (pp. 40–5) where Dostoyevsky shows his sympathy for the roles that many of Russia's women were consigned to play at this period of its history.

Goncharov's Stolz, Oblomov's friend from boyhood days, is an unconvincing creation, too good to be true. The energetic German with the heart of gold is almost a caricature, an inadequate counterpoint to the brilliantly developed character of Oblomov himself. Stolz does not belong in the St. Petersburg of these novels. He has no clearly defined relationship with the city at all. He seems to be an example of the merchants/entrepreneurs who controlled St. Petersburg's economy without being otherwise rooted in the life of the city, who will take off for any part of Europe or America at the drop of a hat to look after their interests. We can understand why Olga finally breaks off with Oblomov, but apart from his boundless energy Stolz hardly seems worthy of her either; yet she marries him.

Oblomov chronicles the gradual withdrawal of the hero from a St. Petersburg that leaves him cold and indifferent. He even loses his dreams of happiness and a return to the country estate of his childhood. St. Petersburg itself does not defeat Oblomov, however; it is his supposedly Russian lethargy that does that. He marries his peasant landlady and settles down into the comfortable obscurity of life on the Vyborg side, where he also dies. Although at times it is written strongly into the story, St. Petersburg is a European intrusion which Oblomov rejects. The city is a largely passive influence, a backdrop to Oblomov's personal tragedy, a tragedy which Goncharov would have us believe is a result of his quintessentially Russian character.

In contrast, in *Crime and Punishment* St. Petersburg is a chief contributor to Raskolnikov's delirium. The place is always in turmoil: new ideas, new theories, new everything. St. Petersburg is part and parcel of his nightmare. It is the ultimate big city, a city which is clearly St. Petersburg yet also a universal, placeless urbanity. In this spider's web of human relationships, individual men and women are minutiae under the crushing weight of what to them is an uncontrollable everyday environment. Their lives are inextricably linked with the city.

Even though a fundamental theme of the novel is the examination of the motivation of Raskolnikov as an individual to commit murder, his thoughts and acts cannot be separated from the circumstances of the society and city in which he lives. Much of the action in *Crime and Punishment* takes place in public places such as squares and streets and hallways. The environment and the rest of society are barely at arm's length. The urban poor in St. Petersburg have no place they can call their own. Raskolnikov cannot even pay the rent for his tiny garret; the impoverished Marmeladov family live in a sort of corridor-cum-room. The walls of Sonia's room are so thin that Svidrigailov can easily hear Raskolnikov's confession from the neighboring suite. People's private lives are at the mercy of the city and they have no retreat from it. Raskolnikov's course of action is crucially influenced by his position in the ant-hill of the inner city and by the coincidences and chance meetings that result.

Dostoyevsky thus views the city as a dominating whole which individuals cannot deny or escape, in which people and events are all interlinked in a system of incredible complexity. This anticipates the view that crystallized in the early twentieth century, of the city as an organism. There is no sense in *Crime and Punishment*, however, of the power of people to influence the direction this organism takes. St. Petersburg, to Dostoyevsky, was on the verge of chaos. The novel is a strong denial of the view of the founders and early designers of St. Petersburg, that the city is an organized entity that can be modified and manipulated by the actions of individuals. In *Crime and Punishment* it is the reverse side of this coin that dominates Dostoyevsky's writing of the city.

NOTES

1   *Oblomov*, by Ivan Goncharov, 1859. Page numbers in this chapter refer to the translation by Natalie Duddington, London: J. M. Dent & Sons Ltd., 1932; repr. 1962.

2   *Crime and Punishment*, by Fyodor Dostoyevsky, 1866. Page numbers in this chapter refer to the translation by Constance Garnett, London: J. M. Dent & Sons Ltd., 1955; repr. 1963.

3   For biographical details of Ivan Goncharov and more traditional literary criticism of his works, in English, see among others Milton Ehre, *Oblomov and his Creator: The Life and Art of Ivan Goncharov* (Princeton: Princeton University Press, 1973); Vsevolod Setchkarov, *Ivan Goncharov: His Life and Works* (Würzburg: Jal-Verlag, 1974); Alexandra Lyngstad and Sverre Lyngstad, *Ivan Goncharov* (New York: Twayne Publishers Inc., 1971); Janko Lavrin, *Goncharov* (New Haven: Yale University Press, 1954).

4   Dostoyevsky's works have spawned a host of literary criticism and biographical studies in many languages. In English, a few of the most useful treatments in the context of the current chapter are those by Richard Peace, *Dostoyevsky: An Examination of the Major Novels* (Cambridge: Cambridge University Press, 1971); Malcolm V. Jones and Garth M. Terry (eds.), *New Essays on Dostoyevsky* (Cambridge: Cambridge University Press, 1983); Harold Bloom (ed.), *Modern Critical Interpretations: Fyodor Dostoevsky's* Crime and Punishment (New York: New Haven and Philadelphia: Chelsea House Publishers, 1988); Malcolm V. Jones, *Dostoyevsky: The Novel of Discord* (London: Paul Elek, 1976); Gary Rosenshield, *Crime and Punishment: The Techniques of the Omniscient Author* (Lisse: The Peter de Ridder Press, 1978).

5   James H. Bater, *St. Petersburg: Industrialisation and Change* (Montreal: McGill-Queen's University Press, 1976), p. 2.

6   For a discussion of the influence of Dickens on Dostoyevsky, see N. M. Lary, *Dostoevsky and Dickens: A Study of Literary Influence* (London & Boston: Routledge & Kegan Paul, 1973).

7   The image of the city as ant-hill is probably widespread, but is certainly used by

the Acadian writer Antonine Maillet in her *Pointe-aux-Coques* (Montreal: Leméac, 1958).

8   Peace, *Dostoyevsky*, p. 51.
9   Bater, *St Petersburg*, p. 169.

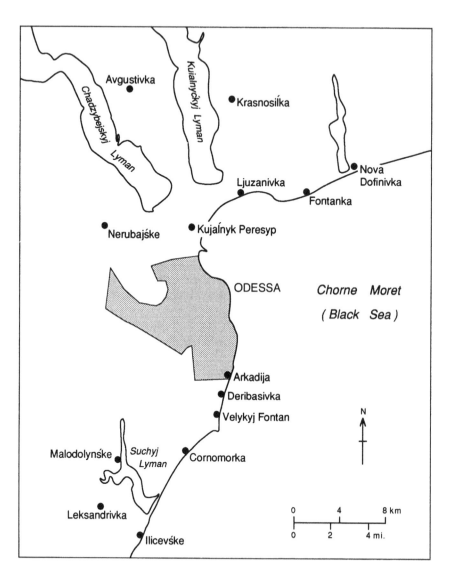

Map 6   Odessa

6

CITY-ICON IN A POETIC GEOGRAPHY

Pushkin's Odessa

Anna Makolkin

The city is situated on an elevation of 100 feet (actually, closer to 200) above the sea; a promenade three quarters of a mile long, terminated at one end by the exchange, and at the other by the palace of the governors, is laid out in front along the margin of the sea, bounded on one side by an abundant precipice, and adorned with trees, shrubs, flowers, statues, and busts, like the garden of Tuilleries, the Borghese Villa of the Villa Recali at Naples. On the other side is a long range of hotels built of stone, some of them with façades after the best models of Italy.

John Stephens, *An American Traveller*, 1838

Oh God! Save this city! Unite all those scattered around who cannot get rid of their own talent and originality in other places. Yes! There is something very special in this neurotic soil giving birth to musicians, artists, singers, crooks and bandits.

Mikhail Zhvanetsky, *One Year for Two*, 1989[1]

PLACE, POEM, POET

The description of the city in literature is a very ancient motif going back to Homer's Troy and the Bible's Holy City with its mythical emblem – the temple.[2] The blossoming of national literatures in nineteenth-century Europe coincided with unprecedented urban growth. The changing faces of European cities were immortalized in poetic landscapes, and the city motif somewhat overshadowed the ancient poetic archetypes of sun, moon and spring.[3] Wordsworth's Venice would compete with his "Daffodils," Byron's "Ode to Nature" with his exalted song to Rome, Lermontov's "Ode to Moscow" with his poem about the Caucasus.[4] The nineteenth century produced numerous impressive images of old cities. Writers and poets rediscovered European capitals and created their verbal monuments to places which were beautiful, central or vital – Dickens's London, Hugo's Paris, Gogol's and Dostoyevsky's St. Petersburg, Byron's Rome, Gaskell's Manchester and Thackeray's Paris.[5] Eventually those literary landscapes immortalized the authors themselves in these urban portraits. The last century also witnessed the birth of new cities which rising literary luminaries put on the cultural map.

One such new city was Odessa, immortalized in verse by Russia's most revered poet, Alexander Sergeevich Pushkin (1799–1837), in his famous novel in verse, *Eugene Onegin* (1825–33). Pushkin, Russia's national poet – and the alleged creator of the modern Russian language – was born in Moscow into an aristocratic family of mixed racial origins. As a child prodigy he was very sensitive about his appearance and suffered from the insulting comments of his classmates at the Lyceum who called him "monkey" (the Gannibals, Pushkin's mother's family, could trace their origins to Ethiopia, which did not prevent them from being accepted in the Court: Peter the Great had once adopted a young black slave). Educated at the privileged Lyceum in Tsarskoe Selo, Pushkin was trained to become a foreign service officer, but his government career was ruined by his rebellious poetry writing. In 1820 he was expelled from St. Petersburg for writing the seditious "Ode to Freedom" and sent to Kishinev (the modern capital of the Moldavan Republic, "Moldova"). His influential friends helped the poet to get a transfer from the less civilized Kishinev to the cosmopolitan and lively Odessa. Pushkin lived in Odessa from July 8, 1823, till August 1, 1824. The Odessa period was very productive for Pushkin, who wrote there more than thirty poems: he began his poem "The Gypsies" and completed two-and-a-half chapters of *Eugene Onegin*. Because of censorship its last chapter dedicated to Odessa was half-destroyed and its fragments appeared as an appendix to the novel. Ultimately, Pushkinists discovered some additional manuscripts and arranged them as the Tenth Chapter of *Eugene Onegin* or the ode to the "Russian Naples" – the port of Odessa on the Black Sea.[6]

The city of Odessa was founded on May 27, 1794, as a part of the Russian expansion in the South and the grand design of bringing Europe to Russia and Russia to Europe. Catherine the Great allegedly decided on the name of the new city, built in the traditional area of the ancient Greek settlements, when she changed the proposed "Odessos" to "Odessa" while dancing at a ball. It is not surprising that she gave preference to a feminine gender. Located by the Black Sea, on the land of the legendary Sarmatians, Scythians, Goths and Greeks, the city was destined to become Russia's "window" on Mediterranean Europe and the second Russian center of culture after St. Petersburg. Patricia Herlihy introduces Odessa in the following way:

> Odessa, as it grew, gradually acquired nicknames. Little Paris, the Southern Palmyra, the second Petersburg, the Golden City, the Southern beauty, the capital of South Russia, Little Venice, Little

Vienna, Naples, Florence, and the Queen of the Black Sea. These honoring titles give testimony not only to the city's physical beauty but also to the vitality of its intellectual and cultural life.[7]

It is not surprising that such a city should become the Russian "cultural Mecca." Pushkin was one of the city's first enchanted pilgrims and it was he who put Odessa on the Russian cultural map. As the "Russian Byron", Alexander Pushkin immortalized Odessa, the original Russian cosmopolitan city, and its unique urban character in his *Eugene Onegin*. As noted above, two-and-a-half chapters of the novel were written in Odessa, and according to Pushkin's correspondence the remainder of the novel was also planned during Pushkin's stay in the city. The poet moved to Odessa in July 1823, but by November he already knew that his hero would be very much like Byron's Don Juan.[8] Like most of his contemporaries, Pushkin was fascinated by Byron, and closely studied his work and was very deeply inspired by him. Byron's Don Juan, a disillusioned individual seeking the purpose and meaning of life, was a very successful role model for Pushkin's Onegin, a man without purpose and goal. Despite the stereotypical view of Pushkin's *Onegin* as the "encyclopedia of Russian life" initiated by Belinsky and the ardent Slavophiles of the last century, one cannot but notice the universal motif of the romantic quest for "self" that unites *Don Juan* and *Eugene Onegin*.[9] Both characters symbolize the unsettled searching side in man which finds its satisfaction in travel. Byron's hero, much like the author himself, travels extensively, visiting various countries and cities: Lisbon, Rome, Athens, Ismail and St. Petersburg.[10] Byron, the aristocrat, and free citizen of Britain, could personally collect urban impressions for his *Don Juan* throughout Europe. Pushkin, having the misfortune to be born in the oppressive Russian Empire, never had the opportunity to visit any foreign country. However, his southern exile in Bessarabia, the Caucasus and Odessa provided the materials for his travel motif in *Eugene Onegin*. Paradoxically, what had been intended as a punishment turned out to be unexpectedly rewarding and happy.

Pushkin's stay in Odessa coincided with the most liberal and progressive period in the life of the city. During 1819–59 Odessa enjoyed the status of free port.[11] This unique status allowed a free exchange of foreign goods, free trade with the Middle East and Europe, and a free flow of ideas, a blossoming in the arts and an influx of foreigners. It is precisely at this time, when Odessa began to play the most vital role in the culture, economics and politics of Russia, that Pushkin happened to be in this exciting city.

THE "RUSSIAN COLUMBUS" DISCOVERS EUROPE

After the stifling gray St. Petersburg, Pushkin was obviously overwhelmed by the colorful urban picture. This is how he summarizes his Odessa impressions and introduces to the reader this young new city, then only twenty-one years old:[12]

> So then I used to live in Odessa, dusty city...
> Where skies are forever bright and pretty,
> Where trade-busy bee bustles and hustles care-free.
> There Europe breathes, flutters.
> There South dazzles and motley crowd glitters.
> Merry street enjoys the tongue of golden Italy
> Where proud Slavs, Frenchmen, Spaniards and Armenians mingle Free
> And Greeks, and heavy Moldavans are
> Next to the son of the Egyptian land,
> The retired corsair, Maurali.

Pushkin, who never left the frontiers of the Russian Empire, fantasizes about a Europe he has never seen. For Pushkin, Odessa was a metaphor for Europe, representing exotic otherness and polyphony. He sees the city with the qualities of a living body where "Europe breathes." This anthropomorphism makes the image of Odessa more exotic for the Russian poet, who vividly captures the new character of a growing city with its intense energy.

Europe, in Pushkin's imagination, is Odessa, a cosmopolitan port with a free flow of goods, people and ideas. Pushkin clearly gives preference to Italian, which he calls "the tongue of golden Italy." Italy, a place of pilgrimage for all Europeans, a mythical country for nineteenth-century Russian intellectuals who knew it only through its art, literature and music. Italy was for Pushkin a golden remote paradise, an impossible dream, the land of Ovid, Dante and Petrarch, which he could only visit through the words of others. He learned about Byron's death while in Odessa, and read *Don Juan* in his southern exile. He would thus relive Byron's Italian impressions while in Odessa, which would then become a substitute for sunny Italy. Byron, who called Rome the "city of the soul" and Italy the "garden of the world," inspired Pushkin, and these Byronic images of Italy were applied to sunny Odessa, a city with 265 days of sunshine, Italian opera and Italian architecture. Thus, the narrator's "I used to live in Odessa, dusty city" is a believable voice; the readers are brought into Onegin's fictional world through the impressions of the author. The city portrait is a life-line of verisimilitude in Pushkin's text.

PHYSICAL GEOGRAPHY IN POETRY

After the first enchanted introduction to Odessa-Europe, Pushkin, quite faithful to reality, describes the climatic and geographic misfortunes that plagued this city located in the midst of the dry steppes. While enjoying its lovely cosmopolitan atmosphere, Pushkin admits that dust and dirt are

Odessa's natural geographical flaws. The traditional metaphor of the "dust of centuries" could have been implied in this description of Odessa, a young city founded upon the dust of the old Greek, Scythian, Gothic, Tartar and Turkish settlements. Italian is not chosen accidentally by Pushkin as the language spoken in his city-paradise, since Odessa's history and prehistory is connected with Italy and Italians.[13] According to one legend, the Italians called the future site of Odessa "Ginestra," after the colorful bloom that welcomed the foreign ships.[14] Allegedly, the first Italian ships came to Odessa in the thirteenth century from Genoa, Pisa and Venice, to the then Turkish harbor "Khadzhibey." Italian also became the lingua franca in Odessa, this Southern Russian belle: the earliest street-signs were in both Italian and Russian, and the first city publication by the local press was a sonnet in Italian, "Al genio musicale della signora Giustina Zamboni."[15] Italian patrons of the arts, artists and merchants were among the founders of this most non-Russian city in the New Russia, "Novorossiia". The city prided itself on its first Italian dwellers and the names of Totti and Rozzi, Florini and Montovani, Zamboni and Pogolotti.

The word "dust" has a double purpose. First, it indicates that although only twenty-one years old, Odessa is a city that reveres the ancient history of the land from the Varangians to the Greeks.[16] The name itself is a tribute to Greek settlements which were prevalent along the Black sea – Khersones, Olvia and Odessos. It is in this ancient "dust" that Odessa is "drowning," and this is also the actual physical dust of the dry steppes. The poetic landscape created through the metaphors thus faithfully brings together in a couple of lines the physical geography and the ancient history of the city and its region.

The discomfort of dust is compounded by the "mud" visited upon the still unpaved young city by the power of the almighty Zeus:

> As I said, I used to live
> In Odessa, dusty city.
> I would not have been a liar
> Had I said "a city full of mud."
> By the will of mighty Zeus,
> Five to six weeks a year,
> Odessa, Southern belle, is in a bed of mud,
> Raped by flood.
> All the grand palazzos soak in mud,
> And the daring Odessa dweller
> Walks on stilts conquering the mud.[17]

The shadow of the ancient Zeus is over the city whose inhabitants struggle against the poor physical conditions. The city lacked stone, which had to be imported. In Pushkin's time the first paved streets were being built and he already saw that mud and dust could ultimately be conquered:

> But one already hears a striking hammer –
> The city's shoe is ready soon to
> safely walk the ever-dusty "rue."

It took Odessa thirty-five more years to pave all its streets (stone had to be brought from Naples, Genoa and Livorno).[18] Apart from the lack of paved streets, this dusty and muddy city by the sea suffered from the lack of fresh water. This natural geographical drawback was a quasi-Biblical urban plague. The new mythical city-garden or city-paradise had no fresh water, and so extra magic was needed to create it. A poetic rearrangement of words evokes concisely the natural obstacles in the way of creating urban magic: "Wet Odessa" is an ironic statement, disclosing the paradox of the landscape and imparting a more fantastic quality to the entire process of its construction amidst arid steppes. Lack of water and lack of stone are the natural obstacles to the creation of an urban paradise, since water is an ancient metaphor of life and its lack is a sign of the insecurity of the city. Odessa's future thus depends upon the creativity and ingenuity of its inhabitants.

The usual Odessa day is full of pleasures, starting with an invigorating early-morning swim in the salty Black Sea. This daily "baptism" is followed by a tasting of the world's coffee-brands from the entire Orient. The poet includes ship- and flag-watching in the harbor as a part of the daily urban routine:

> A merchant goes
> to catch a glimpse of foreign flags,
> to check if friendly sails
> are sent to him from Heaven.
> What goods are locked in quarantine?
> What wines are brought for vino-valentine,
> and about plague? And where is fire?
> Is there famine, war or
> other latest piece-nouveau.[19]

The flag- and ship-watching in those days was the best source of the latest news from around the world. It was a live newspaper, and the morning flag-watching in Odessa simulated the European habit of newspaper-reading over a cup of coffee. Thus Pushkin's Onegin on arrival in Odessa-Europe may taste many imagined European pleasures, as well as enjoy the freedom to communicate with foreigners.

What people in the heart of the country usually receive as news printed in the newspapers the port-dwellers obtain first hand from the sailors, reporters who are more trusted. The sailors are the traditional carriers of culture, the explorers of other lands, and the poet who lands in the blessed port expresses his joy at encountering these messengers of otherness. Pushkin poetically recreates the history of past cultural exchanges with a single glance at the ships, flags and curious Odessa-dwellers. News from journalists is not only unbelievable but frequently late, while friendly conversations in the harbor bring rapidly the most horrible and the most desirable news. Despite the relative freedom of the press in tsarist Russia, censors enjoyed substantial power and news was carefully filtered. Port-dwellers developed the habit of checking the reports by talking to sailors. The most intense political, social and economic debates would first take place in Odessa harbor, by the sea or on board ships. The sailor-merchant thus became the most respected political "expert."

In Pushkin's version, it is only the merchant who is interested in any political news while the typical dweller of Odessa-paradise is mainly concerned about gastronomic pleasures:[20]

> But we, carefree folk,
> cared only about oysters,
> amidst the careful merchant stock.
> Oysters? Arrived? What a joy!
> The glutton-youth flies off
> to swallow alive a plump lady-anchorite,
> A seashell dweller splashed in lemon.

If the "merchant, a child of bills, accounts, business venture" lives a practical life, Pushkin's playboy, Onegin, enjoys Odessa as a city of pleasure. Pushkin creates the aura of a fairy tale around the young city, which needs a rescuer to save it and quench its natural thirst.

STONE AND ITS MYTH

Cities were born in a state of rebellion against Mother Nature and the natural cycles of life. Stone was discovered as a proper construction material for a place that was meant to outlive surrounding nature. Stone possessed a mysterious power for conquering the inevitable, and cities immortalized themselves in stone. Stone thus became a symbol of urban life, and the traces of stone structures could remain forever: the great wall of China, the Wailing Wall in the Holy City, the Parthenon and the Coliseum – all defied Nature. Pushkin must have been aware of ancient history, urban myth and the stone metaphor.

Enchanted with the young city arising from the dust of the ancient past and dust of the Black Sea Steppes, Pushkin tells his readers that stone, the symbol of urban eternity, is already there. He can already hear "the hammer" striking upon the first stone pavements, the same as in the old European cities. In Pushkin's day this lack of stone was another of Odessa's bad omens – a sign of an unpredictable urban future. But this the poet disregarded; he believed in Odessa as a city/paradise, which would conquer the lack of water and stone and survive by means of myth and magic.

101

ODESSA – "LAND OF BLISS"

Pushkin, the magic-maker, is not troubled by Odessa's natural shortcomings and geographical flaws. The poet humorously concludes that lack of water could be largely compensated for by the abundance of wine. Having described "wet" Odessa as actually "dry" and "thirsty," Pushkin comforts the reader:

> So what? It's not a drama
> When tax-free wine is pouring into a salty brine.

If there is no water, one can drink Odessa's wine and savor its sun and sea:

> Friends, what else is here in this land of bliss?

It is a poet's paradise, with wine, sun, sea, coffee, air, cards and music:[21]

> Revived in the Black sea brine,
> As a Moslem in his paradise,
> Drink coffee with the Oriental mist
> And take a promenade
> On a city esplanade.
> One enters the gracious casino,
> Cups jingle in the air.

The author was a gastronome, which may have affected his state of happiness. Pushkin plays upon the erotic susceptibilities of the reader and surprises, as we have seen, with the metaphor "lady-anchorite," who turns out to be an oyster. The "anchorite" could have been an oriental harem-dweller but instead is a seashell recluse. "Lemon" stands for the perfume splashed over the body of the seductive shell-lady. The role of the metaphor is to shock the readers with the substitution of meanings and to amuse since readers are provided with the semantic tools of discovering their own mistake.

The restaurateur Autonne replaces the oriental harem-owner and seduces the inhabitants of the Odessa paradise with edible female sea-creatures:[22]

> Noise, debate – light wine
> from city-cellars – the deeds
> of the obliging host, Autonne;
> Hours fly and bill-giant
> insidiously grows bigger!
> With every sip of wine-divine
> by an inebriated Odessiane.

In the Odessa paradise, the mythical land of Pushkin's bliss, one may endlessly enjoy rivers of wine and colonies of oysters. The mythical day starts in the sea, continues by the sea, consuming the sea/oyster and ends in a theater, another part of Pushkin's paradisal myth. Since Pushkin is a poet and a man of culture, his paradise is a land not only of the traditional plenty but also of the other "plenty." Unlike the Biblical "Land of Milk and Honey," Pushkin's is the "Land of Wine and Oysters" – and Rossini's music:

Biblical	Pushkin's
milk/honey	wine/oysters
reward for work	leisure
worshipping	Godly gift
God	Art
Christ	Rossini, Orpheus
Hebraic garden	Hellenic theater

Pushkin's paradise is unthinkable without music and the theater, which is his sacred temple where Onegin ends his day of pleasure.[23]

> Blue evening grows darker,
> It's time for opera, to rush:
> to hear ravishing Rossini,
> Europe's spoiled brat, to rediscover
> Orpheus always the self-same,
> and ever new, despite the harshness of the critics.
>
> He pours out his magic sounds,
> they boil, flow, burn
> Like flaming youthful kisses.
> All is an ocean of tender caresses
> and Flames of Love.

Pushkin's city is a paradise for art lovers whose passion for art and music is compared to the burning erotic passion. Ultimate pleasure is not, as expected, in the erotic but in the sublime, in an elevated spiritual state. Pushkin likens the state of exaltation during the artistic performance to the catharsis of the physical in the sexual act. The poet's mythical land of bliss is a place of culture, not a natural garden. The concepts of sin, fall and reward associated with paradise are subverted in Pushkin's city-paradise. His city is a place where ultimate existential joys can be obtained through art and music. Odessa, the city-paradise, is a place where, for the man of culture, bliss is totally in the hands of the Hellenic pagan gods (such as Orpheus), the seductive Eve or the Devil, and sin is replaced by Bacchus, God of Wine. Pushkin's poetic formula of pleasure is:

> eau + do-re-mi-sol
> a sip of wine + the sound of music

The blissful day in this city-paradise ends with the last opera tune, when the spectators leave the Opera House and in post-operatic ecstasy face the glittering Venetian lanterns and stars above the garden of Orpheus:[24]

> Roars *finale*; empties the magic Hall,
> Carriages noisily, hurriedly roll,
> The crowd runs towards the square,
> The happy sons of Ausonie
> hum the flirting tune of Italy
> under the glitter of Odessa stars
> and its Venetian lanterns.

After the joys of the passionate communion with Art, blessed by the God of Music, Odessa retires:[25]

> Quietly Odessa sleeps;
> Breathless and warm
> Is its mute night.
> Moon speechless shines above and
> A transparent haze embraces
> the Odessa sky.
> All is silent and only the Black Sea loudly speaks ...

Such is the ending of the day in this mythical city-paradise inhabited by the happy sons and daughters of Ausonia – another name for Pushkin's city. The poet intensifies the basic myth about Odessa – a city of art and a city for artists – through this new name "Ausonia/Ausonie." It moves the Novo-Russian city by the Black Sea to an imagined mythical past by the Aegean Sea. Through the poetic imagery Pushkin rearranges the urban space, combining past and present, ancient Rome and Greece, the cradle of Europe and the new European Russia, the nostalgic dream of the Russian tsars and pro-Western intellectuals and free creative individuals who desire to be a part of the larger human whole – Europe.

It is not accidental that Onegin's quest for self and his forced journey through Russia ends in this artistic Mecca of the new Russia. It is his final stop after Nizhniy-Novgorod, Astrakhan', the Caucasus, Moscow, St. Petersburg and the Crimea. Odessa is Pushkin's cure for Byronic spleen. It is his reply to the later generation of Russian "superfluous men" (a term coined by Ivan Turgenev), disenchanted, cynical and angry. Pushkin seems to have found in Odessa the ultimate answer to the human plight – living in the city, a Temple of Arts.

In his "Hymn to Odessa," Pushkin formulates an entirely new world-view. The Shakespearean "to be or not to be" was replaced by Pushkin's playful to live or not to live in Odessa, Russia's cultural Mecca or the Esperanto city anticipating the future global village. What starts as a topographical passage ends as an urban myth about the city as Savior, saving humanity through art, music and beauty. The poetic icon has become transformed into a religious icon.

NOTES

1 Mikhail Zhvanetsky, *God za Dva* [One Year for Two] (Moscow: Iskusstro, 1989).

2 John Johnston, *The Poet and the City* (Athens: The University of Georgia Press, 1984), p. 1.

3 Doris Kadish, *The Literature of Images* (London: Rutger's University Press, 1986), p. 5.

4 William Wordsworth, "On the Extinction of the Venetian Republic, 1802," *The Oxford Book of English Verse*, ed. Arthur Quiller-Couch (Oxford: Oxford University Press, 27th edn, 1968), p. 616; George Gordon Byron, "The Isles of Greece," ibid., pp. 708–11; Byron, *Childe Harold's Pilgrimage*, IV, 47 (Oxford: Clarendon Press, 1980), p. 415; Mikhail Lermontov, "Tambovskaiia kazna-cheisha" [The Tambov She-Treasurer] in his *Sochineniia* (Works), 4 vols (Moscow: OGIZ, 1948), vol. II, pp. 7–30; Lermontov, "Panorama Moskvy" [Moscow Panorama], ibid., vol. IV, pp. 298–302.

5 The portraits of London in Charles Dickens's *Sketches by Boz* (1833), *The Pickwick Papers* (1836), *Oliver Twist* (1841), *Bleak House* (1852); the image of Paris in Victor Hugo's *Notre Dame de Paris* (1831); St. Petersburg in Nikolay Gogol's "Overcoat" (1842); Fedor Dostoyevsky's *Belyie nochi* [White Nights] (1848); William Makepeace Thackeray "The Ballad of Bouillabaisse," in *The Oxford Book of English Verse*, pp. 861–4; Elizabeth Gaskell's *Mary Barton* (1848) includes a portrait of Manchester.

6 V. V. Kunin, *Zhizn' Pushkina* [Pushkin's Life], 2 vols (Moscow: Pravda, 1987), vol. I, pp. 482–3; N. Ostrovskaya, *Ia zhil togda v Odesse* [So then I Used to Live in Odesse], (Odessa: Maiak, 1987), p. 54. In 1879 Petr Tchaikovsky first presented his opera *Eugene Onegin* based on Pushkin's novel; J. Warrack, *Tchaikovsky* (New York: Charles Scribner & Sons, 1973).

7 Patricia Herlihy, *Odessa* (Cambridge, Mass.: Harvard University Press, 1986), p. 144.

8 Victor Zhirmunsky, *Bairon i Pushkin* [Byron and Pushkin], (Leningrad: Akade-miia Nauk, SSSR, 1924).

9 Vissarion Belinsky, "Eugene Onegin as Encyclopedia of a Russian Life," in *Russian Views on Pushkin's Eugene Onegin*, trans. Sonia Hoisington (Bloomington: Indiana University Press, 1988), pp. 17–43.

10 Byron mentions Odessa's famous city chief, the duc de Richelieu, in his *Don Juan* (Cantos 7 and 8); this legendary figure was a Frenchman who served under the Russian tsar Alexander I and, as a sign of recognition of his devoted service, was appointed governor of Odessa in 1803. The city's rapid and successful urban development was largely attributed to his capable governing. The duc de Richelieu left Odessa in 1822 and returned to France to replace Taleyrand in the capacity of Foreign Minister. Byron's Don Juan is made to serve in the army under Catherine the Great.

11 Herlihy, *Odessa*, p. 7.

12 Pushkin, *Eugene Onegin*, "Excerpts from Onegin's Travels," p. 354. This and subsequent quotations from *Eugene Onegin* are derived from A. S. Pushkin, *Izbrannye Proizvedeniia*, ed. B. Rudakovsky (Minsk: Gosudarstvennoe ucheb-nopedagogocheskoe izdatel'stvo, 1953). The translation was made by the author

of this chapter who felt that existing renderings of Pushkin's account have not mediated the desirable flow of the verse. Vladimir Nabokov even categorically denied any possibility of a poetic version and presented to his English audience a rather precise, but lifeless, prosaic account; his view of the situation may easily be challenged by a contemporary translator.

13 Herlihy, *Odessa*, p. 44.
14 ibid.
15 ibid., p. 321. The Tottis were among Odessa's doctors and lawyers and were the founders of the Odessa Public Library; the Pozzis established the Gymnasium for girls; Fiorini was rewarded by Alexander I for many successful operatic productions in Odessa; Montovani annually brought Italian operas to Odessa; and Zamboni, a famous conductor, was instrumental in shaping Odessa's cultural life.
16 Orest Subteny, *Ukraine* (Toronto: University of Toronto Press, 1988) pp. 1–15.
17 Pushkin, *Eugene Onegin*, p. 354.
18 K. Sarkisian and M. Stavnitser, *Ulitsy rasskazyvaiut* [Streets Tell a Story] (Odessa: Maiak, 1979) p. 9.
19 Pushkin, *Eugene Onegin*, p. 356.
20 ibid.
21 ibid.
22 ibid.
23 ibid.
24 ibid., p. 357.
25 ibid.

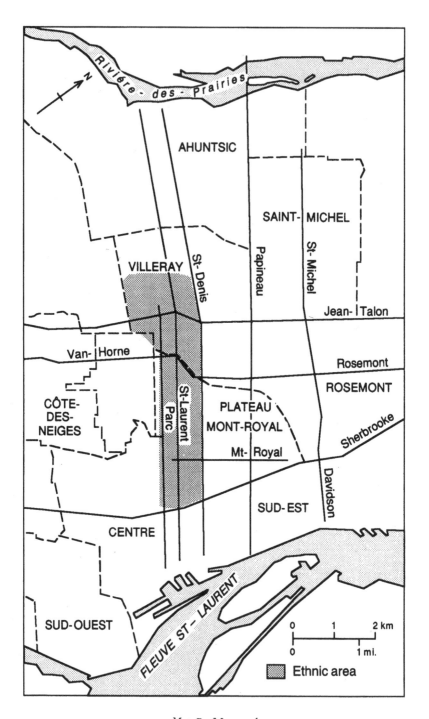

Map 7 Montreal

7

VERY DIFFERENT MONTREALS

Pathways through the city and ethnicity in novels by authors of different origins

Pierre Deslauriers

THE CITY

The city of Montreal is the largest of twenty-eight towns located on the island bearing the same name, in the province of Quebec, Canada. The foundation and initial growth of Montreal were closely associated with its strategic location on the St. Lawrence River, at its imperfect confluence with the Ottawa and Richelieu rivers; the first providing access into the fur-rich interior while the latter led to the British colonies of New England. On the island, and (largely) within the limits of the modern city, Mount Royal rises about 230 metres above the surrounding plain.

The history of Montreal dates from the beginning of European colonization of North America. It was discovered by the French navigator Cartier on his second expedition (1535) and revisited by Champlain (1611). Permanent settlement led by Paul de Chomedey de Maisonneuve followed in 1642. In 1663, the island, up till then the property of the Société Notre-Dame de Montréal, was acquired by the Messieurs de Saint-Sulpice de Paris (Sulpicians) who had an important role in the initial development of the city. Under the French regime Montreal grew, and drawing advantage from its position on the Great Lakes and Mississippi routes, it became an important hub controlling penetration into the interior of North America. Montreal thus was a pre-eminent center in early Canada, and even after being conquered by the English in 1760 its position as the country's main economic and cultural center remained long undisputed.

Following "La Conquête", Montreal began receiving a growing number of immigrants from the British Isles. It was Scottish businessmen who quickly gained control of the lucrative fur trade, and from 1830 there were several consecutive waves of immigration from Scotland and Ireland. This immigration was such that for a short period during the nineteenth century the majority of Montrealers were English-speaking (Levine 1990). The affluent British population settled north-west of the old city in the "Golden Square Mile," "where sumptuous residences surrounded by English-style gardens stood alongside elegant private clubs and hotels" (Rémillard and Merrett 1990: 12). Areas of residence of the anglophone population subsequently expanded westward to the slopes of Mount Royal, and to points further west on the island, although members of an often forgotten anglophone working class settled in the industrial "faubourgs" west of the old core, such as Pointe Saint-Charles. Meanwhile, the predominantly French-speaking working-class population spread to the industrial districts east and west of Old Montreal. Massive settlement of francophone Montrealers in the north-east and east areas of the island was fueled by an exceptionally high birth-rate among this community and by the migration of rural populations attracted by industrial jobs.

Rémillard and Merrett (1990: 13) aptly depicted the setting in which more recent developments occured:

> At the beginning of the twentieth century Montreal had reached maturity. From the top of the mountain, at the geographical center of the city, one can see its long and straight arteries running into the horizon. Several steeples, domes and towers literally spring up from its low dense profile. It is a city dominated by the grays of the abundant local limestone used since the French regime, the silver grays of its tin plate roofs, the blue gray of the slate roofs and the speckled gray of the flat roofs.

In latter years, due to the rectilinear form of expansion of the transportation axes, and the building of high-rise office towers in the business district, Montreal became a city resolutely more American than European.

Today, even after having lost ground to Toronto as the country's metropolis, Montreal continues to fulfill important administrative, financial and commercial functions and maintains a significant influence in the realms of culture and education. Moreover, it maintains significant international visibility through its exchanges and the organization of events of international scope. With almost 65 per cent of its 1,015,420 inhabitants having French as a mother tongue in 1986, Montreal is a leading center among the world's French-speaking countries (*la Francophonie*). For various reasons linked to demography, fluctuations in the economy and politics, the proportion of those having English as a mother tongue has now considerably declined. However, a large proportion of the immigrant population continues to choose English as its first language. The English-speaking people of Montreal retain control of important institutions and maintain a very palpable presence in business and commerce, education and the arts.

The spatial cleavage between the areas of residence of the French- and English-speaking populations is a long-standing and much discussed feature of the city's landscape. It is still common to speak of "the east" as the home of a predominantly working-class French population, while the traditionally more business-oriented English Canadians live in the western area of the city. St. Laurent Boulevard stands as the formal, accepted dividing-line between east and west, as it is at this major north–south artery that street addresses begin to progress in either direction. The central business core, and a south–north corridor along major commercial streets (St. Laurent Boulevard, Avenue du Parc), constitute a buffer zone which has traditionally served as a place both of initial settlement and of transition for immigrant groups.

Among the first to settle in Montreal's "immigrant corridor" were Ashkenazi Jews from Eastern Europe. The Montreal Jewish community is generally recognized as the city's oldest "ethnic minority," religious faith here transcending nationality as the defining criteria of ethnic status (Anctil and Caldwell 1984). At the end of the eighteenth century, there was already noticeable settlement of Jewish people towards the southernmost extremity of St. Laurent Boulevard, close to the old city and the port. There was substantial immigration from Russia and Eastern Europe between 1880 and 1925, and the Jews made their way up (north) the boulevard to the area known today as Montreal's "Old Jewish Quarter," the *shtetl*. Subsequently, they established their religious and social institutions, and economic activities in a "reasonably compact, well identified area, north of Sherbrooke Street" (Waller 1974: 10). That is indeed "Mordecai Richler country" (Baker 1990: 39), the area where he was raised and where he staged several stories: Avenue du Parc or St. Urbain, extending northward from Pine or Duluth, up to Van-Horne, with offshoots (west) in Outremont.

Other waves of Jewish immigration followed (survivors of the Holocaust, Sepharad Jews from francophone North Africa), as immigrants of the "first wave" and their descendants began migrating westward through Outremont, and to suburban municipalities such as Côte-St.-Luc and Hampstead. Montreal's present Jewish community clearly possesses its own institutional infrastructure (schools, hospitals, synagogues, cultural centers). It has been observed that its movement through the city's space has tended to follow progression in socio-economic status, and that Jewish people "are absent from the east end of the island" (Polèse et al. 1978: 37).

Montreal was for a long time the main port of entry for immigrants arriving in the country. Since the middle of the nineteenth century, several people of diverse nationalities have come to settle and have become associated with specific tracts of the city's territory. The city has become an amalgamation of cultural areas, and more than ever today Montreal is a multi-ethnic city. The east–west division between French and English is now seen as an oversimplification (Vincent 1989; for a more accurate and complete picture see Marois 1989) but is still nevertheless recognized as a reality intimately associated with the city's history, and present also in the collective imagination (*imaginaire collectif*) of its inhabitants. As Dupré et al. (1988: 8) have recently noted: "The east–west division along the St. Laurent Boulevard axis remains in the city's imagination, as a linguistic, cultural and social divide." Also, it is obvious that some cleavages are linked not only to matters of language, ethnic origin or religion; socio-economic status also plays a significant role. Status in society, and the perception of others' status may become an important factor affecting mobility. Appraisal of differences, and perceived opportunities, are reflected in the spatial itineraries, by the different pathways chosen by individuals. In Montreal, these choices appear to be closely tied to ethnic identity, and some identifiable patterns have derived from, and have been reinforced by, perceptions nurtured within the various ethnic groups.

THE AUTHORS

Internationally acclaimed playwright and novelist Michel Tremblay, the late English-Canadian author Hugh MacLennan and Jewish Montrealers Mordecai Richler and Leonard Cohen are contemporary writers who have displayed an extensive knowledge of the daily life of different strata of Montreal society. I will try to show how the awareness of belonging, and of being associated with a specific cultural group, may be reflected in their characters' itineraries through the city and in their perception of its space. The titles selected mostly focus on the period spanning from the 1930s to the 1950s, some continuing into more recent times. Although the cleavages between the communities may express themselves in a different fashion today, the stories depict phenomena that remain a part of the city's experience.

Born in Montreal in 1942, Michel Tremblay first made himself known during the 1960s through plays that depict the daily life of the inhabitants of neighborhoods in Montreal's East End. These social and religious satires of the alienation of Quebec (Hamel et al. 1989) mark a milestone in the history of Quebec theater, especially in their introduction to theater of the Montreal working-class "joual." Tremblay's plays have been acclaimed all across Canada, in the United States, Europe and Japan. Since the end of the 1970s, Tremblay has restaged his plays' characters in a series of five novels, "Les Chroniques du Plateau Mont-Royal." One of French-Canada's most prolific writers, he had already produced in 1987, at the age of 45, over twenty plays, two musicals, six novels and several translations and adaptations of plays.

Born in Glace Bay, Nova Scotia, in 1907, Hugh MacLennan was a Rhodes Scholar at Oxford, and held a doctoral degree in classical studies from Princeton University. MacLennan began a teaching career in Montreal in 1935, and retired in 1979, a Professor Emeritus at McGill University. Often labeled "the ultimate" Canadian nationalist among Canadian authors, he published several novels and volumes of essays where he showed mastery and a remarkable understanding of the identity problems of, and difficult relations between, the country's French and English communities. He has illustrated the historical dilemma in two major novels that have the city of Montreal for background: *Two Solitudes* and *The Watch that Ends the Night*. Hugh MacLennan died in 1990 after receiving five Governor General's Awards for his books.

Mordecai Richler was born in Montreal in 1931. He has recorded in autobiographical sketches, and several of his novels (*Oxford Companion* 1983), his experience of growing up in the working-class Jewish neighborhood around St. Urbain Street. After a long stay in England, Richler returned to Montreal in 1972, and has lived there permanently since. He has published several collections of essays, establishing his reputation as a caustic satirist, and attaining international recognition, becoming one of English-Canada's best-known authors abroad.

Born in Montreal in 1934, Leonard Cohen grew up in its affluent Westmount district and graduated from McGill University. Known first as a poet, and recognized by some as the most popular Canadian poet and songwriter of the 1960s, he has subsequently pursued a career as a composer-singer. Cohen has produced two novels in which the action is set in Montreal: *The Favorite Game* (1963) and *Beautiful Losers* (1966). Having traveled extensively, Leonard Cohen has resided in Montreal since the 1970s. He has published several collections of poems.

EAST AND WEST AS FOREIGN LANDS

In the novels chosen, the authors establish a clear distinction between the east and west of Montreal. Their characters infrequently visit, and even less frequently move to, areas other than than those of their own group: when they do so, it is usually for a specific purpose. To the Montreal French, English and Jewish, the areas where people other than "their own" predominate appear as foreign territory, even in the absence of any tangible barriers or boundaries.

In Michel Tremblay's *La Grosse Femme d'à côté est enceinte* the description of a group of French-Canadian women's shopping venture downtown reveals the existence of an invisible, "untold" boundary. At the beginning of the 1940s, shopping in the large department stores was not the sole motivation for a westward expedition. There was also the streetcar ride, an occasion for meeting and gossip for the housekeepers of the Plateau Mont-Royal (north-east of downtown). To these women, the western part of Montreal's main shopping artery, Ste Catherine Street, was a place of temptation, a place for escape at the fringe of wealth. But in their minds, this "escape space" had well-set limits; the Eaton's department store (at University Street) acted as a symbolic landmark: "none went further west ... beyond is the realm of the English, of the rich" (p. 25). St. Henri and La Petite

Bourgogne were French-speaking working-class enclaves which lay south-west of downtown and where some had family, but there is little evidence that they visited them. Life in St. Henri around that time was admirably described by Gabrielle Roy in her classic novel *Bonheur d'occasion* (1947; translated into English as *The Tin Flute*).

Apart from this escape episode to downtown, few passages from Tremblay's "Chroniques du Plateau Mont-Royal" describe movement out of the "quartier" (neighbourhood) to other parts of the city. The quartier appears as little opened to the outside; it is a "place of belonging," a tract of space where small daily dramas and joys are shared by those who inhabit it. It is at once a place for socializing (especially along the commercial Mont Royal Street), recreation (at Lafontaine Park) and settling problems. The tavern is a focal place for men; there, they display their strengths and weaknesses and discuss world issues. Women reign at home where they fulfill their family obligations and exert considerable control on the life of the household; it is often they who decide in the last instance, but this power of decision paradoxically confers to them little mobility since they are tied down by the tasks of keeping their house. It is the children who move around the most, and they acquire intimate knowledge of the quartier's territory through their games. Those such as Edouard who regularly go out to other areas of the city especially for entertainment, or those who once moved away but have now returned, are treated as outcasts (Tremblay 1978: 51). The death of Ti Lou, the degenerate old woman who led a debauched life in Ottawa, epitomizes not only the consequences of having "betrayed" one's origins, but also those of not having chosen the beaten path of a housekeeper's life. She dies alone, outcast and bitter, her only friend and visitor a young woman who has seemingly chosen a path similar to her own.

It is in stories written mostly by non-French authors that we find descriptions of life in the English and in the "ethnic" areas. They reveal perceptions and a sense of identity fueled by different values, and make reference to other landmarks. Hugh MacLennan is one of the few English novelists who has drawn extensively upon the setting of Montreal. The characters of his 1958 book *The Watch that Ends the Night* are English intellectuals, career people or politicians whose lives revolve around the mountain and areas that border it. Their Montreal is very different from that of Tremblay's characters. They are perfectly at ease in the downtown area where they conduct their business: "Montreal is a business centre where English speaking-businessmen dominate" (MacLennan 1967: 70). Sherbrooke Street is a major path leading them from their homes near the mountain to downtown workplaces (offices, hospitals, the old Canadian Broadcasting Corporation (CBC) building). Just like today, those from the upper strata of society meet at the Ritz-Carlton Hotel on Sherbrooke Street West. MacLennan's characters have time for aesthetic considerations (the natural environment of the mountain, architecture) and discuss world issues passionately in the homes surrounding McGill University campus.

The narrator, George Stewart, shows deep attachment to the city but he and the novel's other main characters hardly ever venture eastward into French-speaking areas. During all his years spent in Montreal, George threads the same paths between his home and downtown. One of the few disruptions to this routine occurs when he becomes exceptionally depressed over his situation and by the events occuring in the world in the mid-1930s. Stewart finds himself on Ste. Catherine Street on a Friday night, having lost a girlfriend, unsatisfied with his work, feeling hopeless. His despair leads him to wander aimlessly, walking into taverns and talking with strangers. While Tremblay's characters seek escape from "problems at home" within the closed circle of the neighborhood tavern, MacLennan's wander anonymously. Ste. Catherine Street is where one feels the cosmopolitan character of the city and where George comes closest to its French dimension, without ever really entering the French neighborhoods. Looking east, as the wind blows a page from a newspaper in that direction, he contemplates "the immense empty tunnel of this endless street, and has the impression that it is his own interior life that he is contemplating" (MacLennan 1967: 146).

MacLennan mentions Lafontaine Park as a landmark of French-Canadian life, and a meeting with a French taxi-driver is the occasion to make reference to his rural common sense, expressed by his knowledge of the city's climate and change of seasons. The author frequently refers to Montreal's cultural duality as above all a tacit agreement, a peaceful coexistence that may be difficult for non-Montrealers to understand but gives Montreal its unique character. The infrequent occurrence of effective interactions between the French and English realities can only remind us of the title of MacLennan's major novel *The Two Solitudes*.

The presence of the Jewish community and its significance is apparent in the work of both these authors. In Tremblay's *Grosse Femme* it is exemplified by the merchants of St. Laurent Boulevard to whom the women of the Plateau owe money. This commercial section just north of downtown lies on the streetcar line they use for their shopping trips. As the streetcar runs through the area, the mood stiffens, the women seeking ways not to be recognized by their creditors. The Jewish are perceived as people with strange eating habits, who do not belong to the "right religion." In MacLennan's novel, McGill University is pejoratively perceived by the English old guard as a "Jews' lair" (1967: 84) and the Jewish are associated with communism, suggesting a latent anti-semitism common at the time in the French community as well. The Jewish area of the city is explicitly described as located "between the English and French areas of town" (p. 260); it appears as a kind of no man's land, a twilight zone avoided by members of both these groups.

Jewish writers who grew up in Montreal's immigrant corridor have provided descriptions of life in their neighborhood and of how its inhabitants related to other parts of the city and to the two dominant cultural groups. Among them, Mordecai Richler has certainly become the best known. In his novels set in the St. Urbain Street area, life is quite similar to that described in Tremblay's novels. Family and religion are central to the community's life, socialization and the education of children taking place within the culturally homogeneous area. However, stratification according to wealth is more obvious and becomes a deciding factor in the itinerary followed during one's life. For Montreal's Jewish inhabitants, movement through the city's territory is a long-term process, closely associated to socio-economic ascension. The traditional home area is a place where small business is conducted within a confined network of relatives and friends. Some Jews such as Joshua's parents in Mordecai Richler's novel *Joshua Then and Now* (1980) may spend their whole life there, but for many others the attainment of success calls for movement to upper-class areas nearby such as Outremont and Westmount. In the period under consideration, Jewish communities were already well anchored in those towns; this enhanced the duality, almost a rivalry, between those who "made it" and the others.

In *Joshua Then and Now* Richler depicts the link that existed between the city's space and Jewish aspirations to upward mobility: "St. Urbain women aspire to marry Outremont money" (p. 12). Joshua's mother is one who followed a reverse itinerary: she was from Outremont and married a man of lower status. She moved "down to lower Outremont." In *The Apprenticeship of Duddy Kravitz* (1976) Uncle Benjy's home on the slopes of the mountain in Outremont symbolizes material success. He is one from the *shtetl* who has "made it." This opposition between "lower" and "upper" Outremont is the most obvious sign of cleavages that exist within the community. This internal division within the Jewish community where spatial segregation reflects socio-economic itineraries is just as strong as that between the different ethnic groups.

Among the three groups under discussion, the Jewish appear as those whose movements are the least restricted. They identify with their original area of settlement, but many seek to move to other parts of town. For many, once this is done, the old neighbourhood becomes little more than a "souvenir chest." In their search for wealth and status, the youth of the St. Urbain area have traditionally turned towards the west of the city. In Richler's work the east of Montreal is only infrequently visited or even mentioned. There is little trace of the existence of a vast area mainly inhabited by French speakers. In *The Apprenticeship* the former are depicted more as rural people and associated with the countryside north of Montreal ("les Laurentides") through the character of Duddy's secretary and lover, Yvette. When the urban French-Canadians are mentioned, they are undesirable persons who are chased away from the "ghetto" in streetfights (1976: 12). In the more recent *Joshua Then and Now* the eastern area is presented as the realm of Montreal's leftist and nationalist French intelligensia. Strangely enough, to Richler, St. Denis Street, not St. Laurent, is the city's dividing-line (p. 176). This street of bars and restaurants, east of St. Laurent, is described as "an artificial world where politicians and artists stand as symbols" (p. 75).

West of town, the search for status

In the work of Tremblay, MacLennan and Richler, the western section of the city of Montreal is associated with material and social success and the attainment of status. However, the perceptions of such attainment differ among the communities, as do the paths that lead to it. For the French and Jews, access to the "English section of town" implies adoption of some of its traits and behavior; yet they remain aware of never really becoming fully integrated. When seen through the eyes of MacLennan's characters, Montrealers of British descent, the hierarchic structure of English Montreal's society sometimes appears just as immutable.

Richler's Joshua epitomizes this search for upward mobility and its consequences. In his strive for status he marries a girl from a wealthy Westmount family, a senator's daughter, and after gaining notoriety as a writer he finally settles in Lower Westmount, a more upper-middle-class area of the town. (The rich in Montreal live in hillside properties in Upper Westmount, on the slopes of Mount Royal.) However, Joshua has attained fame more than wealth, this making him more a tolerated outcast (the "artist") than a fully-fledged member of a society where it is money that above all determines the rules. This "special status" does not relieve Joshua from the constraining demands of material subsistence (he has a family) and consequently he adopts the characteristic itineraries of the city's west-end inhabitants. Downtown bars and taverns, the Ritz-Carlton Hotel and the neighborhood of the old Forum hockey arena become the focal points of his search for inspiration – or maybe rather, professional contacts as time passes and his inspiration wanes. He becomes growingly cynical and not only does he despise the old Anglo-Saxon establishment, but he is also contemptuous of the Jewish professionals and financiers who populate Westmount. Actually, even though they are neighbors, these two groups do not mingle. Policeman McMaster's description of the Westmount Jewish community as "a privileged and morally decadent bunch" (p. 19) crudely expresses a widespread perception within the Anglo-Saxon community.

Joshua is deeply rooted in Montreal's cultural reality: he is Jewish, his wife has French blood (through her mother) and he lives in an English neighborhood that corresponds to his aspirations for upward mobility. Even though his marriage to a girl from an affluent Anglo-Saxon family and eventual settling in Westmount represent a successful social rise, Joshua does not, and cannot, cut the ties with his origins. In one of numerous clashes with his wife Pauline over their situation in Montreal society, he articulates a feeling experienced in many different situations by Montrealers: "You and I come from very different Montreals" (p. 135).

In *The Apprenticeship of Duddy Kravitz* Duddy's older brother Lenny personifies this search for higher status. His studies at the McGill school of medicine are being paid for by his uncle Benjy, who "had a mind for business but no education." The pride of having one of the family succeed at university is a constant theme of the story. But not only does Lenny want to have access to the medical profession, he also seeks to become a member of the "in crowd" of the English group on campus. He is ready to do anything to escape his own origins. He socializes with the Anglo-Saxon group and becomes close to a girl from Westmount, the place where "the real rich" live (1980: 263–4). Lenny never finally manages to become one of the bunch, and ultimately becomes marginalized, but not before having been used and humiliated.

The path followed by MacLennan's George Stewart illustrates very similar aspirations: to succeed and attain recognition in the city. However, Stewart's background and itinerary are different. First, he has grown up outside of the city in a West Island suburb. Second, it is not obvious in his case that a McGill education symbolizes the mandatory pathway to success. To George's aunt Agnes, who provides money to pay for his education, it is not certain that university studies will lead to the kind of future she has planned for him. Since she does not believe that he is predisposed to a career in business or liberal professions, she seeks a more practical solution that will show quick results. Thus it is decided that George will not go to McGill, this decision incidentally also putting an end to his first love. Deeply frustrated and in a state of revolt he decides to leave for Toronto, where he will spend ten years. He then follows a circuitous path that eventually leads him back to Montreal. Towards the end of the Great Depression of the 1930s, he meets frequently with a group of Montreal leftist intellectuals among whom one of the book's main characters, Dr. Jerome Martell, is a leading figure. Stewart finally manages to secure a position at the CBC. In later years, we find him comfortably settled in Côte-des-Neiges and, ironically, occupying a teaching position at McGill.

George Stewart's path to success is thus not straightforward: exile out of the province (to learn "the hard way"), an interlude of teaching in a rural area from which he commuted on weekends to meet his Montreal friends, and finally, at a mature age, a slow climb up the ladder, leading him to a modest, but respectable, position in the city. MacLennan's often ironic descriptions of Montreal's English society show the difficulties encountered by one not born "into the right family," possessing neither power nor money.

To the French Canadians of Tremblay's novels the west of the city is also a symbol of material success, although the attainment of material success is a less central theme to this author's stories. The areas west, especially those on the slopes of Mount Royal are also the seat of power, especially religious power. Edouard, a member of the family that gathers the main characters of the "Chroniques du Plateau Mont-Royal" personifies the particularity of the perception of Montreal's French working-class. He leaves the neighborhood to become a salesman in a downtown department store "west of the city, among the rich" (Tremblay 1978: 181). Edouard eventually becomes manager of the shoe department and masters the English language so well that his boss calls him "Eddy" and customers address him in English only. However, these attainments do not bring Edouard any special recognition or prestige in his quartier. They are, rather, seen as a betrayal and even as a "double betrayal." First, he has estranged himself from his own people as a result of spending most of his time away from them. Second, he has betrayed his origins; he has "become English."

For the people of the Plateau Mont-Royal the more advantaged areas of the city are the realm of educated people who pour scorn upon them, and are also in certain cases the seats of oppression. The latter was especially true at a time when Catholic institutions held considerable power in French Quebec. During the 1940s and 50s the elite of Montreal's French community and its leading institutions did not dwell in the more popular areas, among the people. Religious and educational institutions as well as members of a nascent business class had already moved up the slopes of the mountain to Outremont. In the second book of his Chroniques, *Thérèse et Pierrette à l'école des Saints-Anges* (1986: 103–8), Tremblay illustrates this separation in his description of the journey of a teacher who is also a nun from the Plateau to Outremont where she is to face a reprimand from her Superior. As she rides the steetcar west, along the bourgeois St. Joseph Boulevard, her feeling is one of growing enstrangement, even though she knows the area well.

As was the case in Richler's and MacLennan's novels, there is not only a division between the communities, there is one within the community itself, determined not by ethnic belonging but rather by social position, wealth and power. This division is featured once again in a third novel by Tremblay, *La Duchesse et le rôturier*, through the character Lucienne Boileau, a teenage girl from the Plateau whose parents send her to school in Outremont. In a chance encounter in a neighborhood drugstore, she is mocked by her childhood friend Thérèse because of her recently acquired "proper manners" and language (1988: 260). For those who do not foresee any escape from the quartier's boundaries, adopting the behavior of the educated is just as much a betrayal as going to work for the English. The author makes clear reference to the city's space to emphasize the distance between the Outremont rich and the people of the Plateau: Lucienne's new schoolmates "have never set foot east of Parc Avenue." Thus, most ironically, the spatial boundary between the rich and the "ordinary people" is the same for the French of the Plateau Mont-Royal and for the Jewish of the Montreal *shtetl*.

When one looks a little closer, this is not the only point that the two communities have in common: we have already mentioned the importance of religion and family in the communities' life, of a search for material success

119

that makes some individuals turn to the Anglo-Saxon area of the city. Richler's description of the physical setting of the Jewish neighborhood provides other elements of resemblance.

> On each corner a cigar store, a grocery and a fruit man.... Outside staircases everywhere ... winding ones, wooden ones, rusty and risky ones, an endless repetition of precious peeling balconies and waste lots making the occasional gap here and there.... Each street between St. Dominique and Park Avenue represented subtle differences in income.
>
> (1976: 21)

These are features that resemble very much those of Tremblay's quartier.

East of town, a place for escape

In relation to certain characteristics of the city the novels by authors writing in English have several points in common. Most striking is the agreement on the association of life-styles or events with specific landmarks or areas. For example, McGill University is the place to get a good education, the focal point not only of intellectual life, but also of often marginal activities with little to do with academic life. The Ritz-Carlton Hotel is an important place for social exchange, even in the world of Tremblay (1988: 364–70). Downtown, the general area of Peel Street is the seat of a more Bohemian life, a place where the youth rent their first appartments and where there is an active night-life. Westmount, on the slopes of Mount Royal, is the home of a domineering and (often) decadent upper class of Anglo-Saxons or Jewish parvenus. These places all serve as setting for various events of a rather similar nature in the novels of MacLennan, Richler and Leonard Cohen.

Neither English nor Jewish people often visit the area east of St. Laurent Boulevard and there is little interaction, and even less positive exchange, with members of the French community. However, when English Montrealers move towards or visit the French area, they are motivated by strikingly similar considerations. They seek either pleasure, especially with women, or escape from an unpleasant situation. This is most obvious in Leonard Cohen's novel *The Favorite Game* (1971).

In Cohen's novel, the area east of the city is the "place to find real women" (p. 54). Two young Jewish men from rich Westmount families walk or drive in "the narrow streets of the east of town waiting for two beauties to detach themselves from the crowd and take their arms." However, when they do finally succeed in approaching some French girls in a Stanley Street dancehall, the outcome is not so idyllic: they become involved in a brawl with the girls' boyfriends. The scene not only emphasizes the French Montrealers' anti-semitism, but also reveals the two young Jewish friends' contempt for the girls and for the French in general.

When the main character, Breavman, decides to escape from his everyday world with his first lover, Tamara, they rent a room in the east end of town. There they enjoy the pleasures of love, away from their families, going so far as to describe themselves as "refugees from Westmount." From the window of their room they describe the surroundings: "everybody in this house seems to be getting up to go to work ... bulging ashcans sentried the dirty sidewalk"; the chimneys and television antennas on the roofs, the buildings blackened by soot are all striking features of the east's landscape. The east of the city is the workaday world, a place of filth and daily human misery in which the two young lovers go unnoticed, away from the moral pressure of their own.

Among the characters considered in this chapter, Cohen's Breavman is unique; the unbounded poet, the drifter who manages to make a living of his art, his "favorite game" being, of course, women. He seeks to cut the ties with his family and his home area, and eventually with the city itself. Breavman is less bound by geography and by family ties than the others. The city holds too much of the past ("the city is designed to preserve the past," p. 161) and the omnipresence of the past oppresses him. Montreal no longer satisfies his need for change but, paradoxically, when he decides to go to live in New York it is because Montreal is changing too much.

For different reasons, Richler's and MacLennan's main characters also escape eastward. In Joshua's case, to escape the crowd of his usual downtown pub he ends up in a St. Denis Street bar, which he finds distasteful. Overall, the image given by Richler of the French area of the city is mostly negative. In the midst of a sad event, MacLennan's character George Stewart instinctively walked east, along Ste. Catherine Street, to get drunk and forget, losing himself amidst the Friday-night crowd. It is striking that neither George nor any of the other English novel heroes ever gets to the well-known red-light district in the St. Laurent and Ste. Catherine area. Those who know Montreal well know how much "La Main" (Lower St. Laurent Boulevard) and its population of marginals (prostitutes, transvestites, racketeers and mobsters) are part of the city's lore. Tremblay, especially in *La Duchesse et le rôturier*, has provided lively and realistic descriptions of the "Golden Age" of Montreal's night-life in the 1940s.

CONCLUSION

The mixture of French, English and Jewish cultures as a distinct feature of Montreal's character is not just tacit; it is a tangible reality. It is a reality now acknowledged by historians and social scientists that through immigration Montreal is becoming an increasingly multi-cultural city, and it is an aspect of the city's life that arouses growing interest. Work recently undertaken by the Groupe Montréal Imaginaire (1989) is an expression of the desire to recognize literature and the arts as means of acquiring a wider understanding of social reality, and to study the relationship of individuals to social space.

For the purpose of this chapter, closer examination of the work of four Montreal authors born into different communities has been an occasion to explore some long-standing ideas about Montreal. Their characters' values, and their behavior in the city's space, show physical and psychological boundaries that are much less clear-cut than most tend to believe, or at least than what common belief usually articulates. Thus, the divide between east and west of town is in fact blurred, set somewhere between, or at, Avenue du Parc or St. Denis Street, at the fringes of the old immigrant corridor. Moreover, in many instances, even though all authors are aware of division following cultural differences, the cleavage appears predominantly economic. These differences in economic status reveal an aspect of Montreal's reality that may have been overlooked and that provides an alternate view to the old "Two Solitudes." It reveals divisions not only between but also within the communities. Many passages in these books show a city divided along the lines of class rather than language or ethnic origins. Although it can hardly be denied that Montreal is a case where language, ethnicity and economic status have traditionally been closely tied, this cannot fully explain the itineraries of its inhabitants through life and through space. Reality is far more complex, and will increasingly be so as immigrants who are not French, Anglo-Saxon or Jewish contribute more and more to the growth of its population, making the city a cultural and socio-economic mosaic.

Beyond these facts there remains a reality of daily contact and peaceful coexistence which feeds the city's collective imagination. This less palpable side of Montreal is manifest in the work of MacLennan and Cohen, who emerge as more idealistic. Cohen's Breavman proudly considers himself as a cross-breed of the French, the Jewish and the English; to him it is that crossing that makes the greatness of Montreal (p. 146). To Breavman, it is the diversity of languages and origins that make Montreal what it is. Here, the vision of Cohen's character rejoins MacLennan's "tacit agreement that cannot be understood by outsiders" which gives Montreal its unique character. This "underlying magic" of the city is much less apparent in the stories of Tremblay and Richler. More often than not, they cast the image of a city populated by the weak, where many horizons are blocked and only the strong and cunning survive.

REFERENCES

Anctil, P. and Caldwell, G. (1984), *Juifs et réalités juives au Québec*, Quebec: Institut Québecois de Recherche sur la Culture.

Baker, Z. M. (1990), "Montreal of yesterday – A snapshot of Jewish life in Montreal during the era of mass immigration," in I. Robinson, P. Anctil and M. Butovsky (eds.), *An Everyday Miracle – Yiddish Culture in Montreal*, Toronto: Vehicle Press.

Cohen, Leonard (1966), *Beautiful Losers*, Toronto: McClelland & Stewart.

——— (1971; orig. 1963), *The Favorite Game*, Paris: 10/18 Christian Bourgois.

Dupré, L., Roy, B. and Théorêt, F. (1988), *Montréal des écrivains*, Montreal: L'Hexagone.

Groupe Montréal Imaginaire (1989), *Lire Montréal*, Montreal: Université de Montréal, Département d'Etudes Françaises.

Hamel, R., Hare, J. and Wyczynski, P. (1989), *Dictionnaire des auteurs de langue française en Amérique du Nord*, Montreal: Editions Fides.

Levine, M. V. (1990), *The Reconquest of Montreal. Language Policy and Social Change in a Bilingual City*, Philadelphia: Temple University Press.

MacLennan, Hugh (1945), *Two Solitudes*, Toronto: Collins.

—— (1967; orig. 1958), *Le Matin d'une longue nuit* [The Watch that Ends the Night], Montreal: Editions HMH, Collection l'Arbre.

Marois, C. (1989), "Caractéristiques des changements du paysage urbain dans la ville de Montréal," *Annales de Géographie*, 548, pp. 385–402.

The Oxford Companion to Canadian Literature (1983), Toronto: Oxford University Press.

Polèse, M., Hamel, C. and Bailly, A. (1978), *La Géographie résidentielle des immigrants et des groupes ethniques*, Montreal: INRS-Urbanisation, Etudes et Documents, 12.

Rémillard, F. and Merrett, B. (1990), *L'Architecture de Montréal*, Montreal: Editions du Méridien.

Richler, Mordecai (1976), *L'Apprentissage de Duddy Kravitz* [The Apprenticeship of Duddy Kravitz], Montreal: Tisseyre.

—— (1980), *Joshua Then and Now*, Toronto: McLelland & Stewart.

Roy, Gabrielle (1947), *Bonheur d'occasion*, Montreal: Editions Beauchemin.

Tremblay, Michel, "Chroniques du Plateau Mont-Royal":

—— (1978), *La Grosse Femme d'à côté est enceinte*, Montreal: Leméac.

—— (1986), *Thérèse et Pierrette à l'école des Saints-Anges*, Montreal: Leméac, Collection Poche.

—— (1988), *La Duchesse et le rôturier*, Montreal: Leméac, Collection Poche.

Vincent, G. (1989), "Montréal: Un aperçu de sa géographie sociale," *Géographes*, 12, pp. 9–10.

Waller, H. M. (1974), *The Governance of the Jewish Community of Montreal*, Philadelphia: Centre for Jewish Community Studies, Study of Jewish Community Organizations, 5.

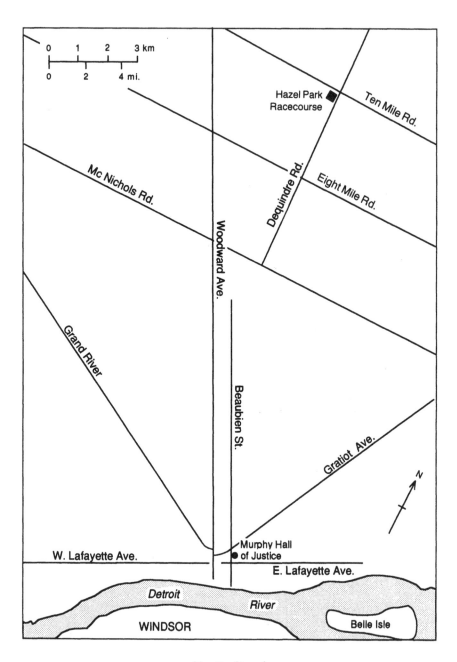

Map 8 Detroit

8

CITY PRIMEVAL
High noon in Elmore Leonard's Detroit
Lorne Foster

Detroit, you can get any kind of piece you want, buy it off a school kid.

(Leonard 1989: 167)

Elmore "Dutch" Leonard was born in New Orleans in 1925 and grew up in Detroit, the setting of several of his more memorable novels. In Detroit he received a Catholic education, a life-long love of baseball and the nickname "Dutch" after the old Washington Senators knuckleballer. No other contemporary author has been so closely associated with the city of Detroit. This is due to the fact that his fictional Detroit strikes a resplendent cord for real citizens, who think of him as their own street-wise story-teller, surveying their terrain and articulating their urban point of view. Thus he has been designated the "Dickens of Detroit." Much has been written about Leonard's alcoholism and subsequent recovery with his second wife, Joan, the acknowledged early influence of Ernest Hemingway and his initial stints as a western script-writer and then an educational movie script-writer for Encyclopaedia Britannica Films. No doubt these varied experiences have enhanced his urban perspective.

INTRODUCTION

Other great cities inspire their own literature, but it is Detroit that inspires Elmore Leonard's signature-form, crime fiction. This chapter explores Leonard's fictional account of Detroit as the unfolding of primeval psychic structure in the midst of technological proliferation.

Detroit is the largest city in the state of Michigan and fifth-largest in the USA. It is the fourth-ranking commercial and industrial city in the nation, following New York, Chicago and Los Angeles. It is headquarters for General Motors, Chrysler and American Motors corporations, and the Ford Motor Company is in nearby Dearborn. Detroit also holds the rather dubious distinction of contesting for the highest homicide rate of any major city in modern history. As a result of these intense dispositions toward industry and homicide, Detroit has come to be known as "The Motor City" ("Motown") and "The Murder City."

This subjective geography or pervasive perception of the environment resonates in Leonard's fictional account of Detroit, punctuating the importance and special relationship between technological rationality and the erosion of social ties in the tenacious and incessant process of urbanization. It further implies that the truly integrated Detroiter – the one who is united with the dominant forces of the environment – is dissociated from the conventional understanding of society as a "sacred object" of reverence and constraint and, therefore, is free from the usual effects of conscience and remorse. Here, the psychic structure is realigned in such a way that negative possibilities can ultimately become real possibilities. Thus technological rationality and social erosion are "counter-sacred" forces and the source of a "negative reality."

Drawing on the collective perception of the environment, Leonard's Detroit is the quintessential "city primeval"; original, primitive and belonging to the first stages of civilization – albeit in the modern age. Against the backdrop of this literary landscape we will explore some of the recurring themes and continuous threads: the Detroit actor, deviance and social control in the city, the good city and the last city, and the sociological imagination.

THE DETROIT ACTOR

There is an old German proverb "City air makes a man free." Today we still believe that an urban environment has corresponding social and psychological consequences. To the modern ear, however, no matter how pithy-sounding, the old adage rings as a trifle too quaint. City air made men free . . . and then came Detroit.

Elmore Leonard's Detroit represents a new parable for our "iron cage" tomorrow, a place where men are so compulsive-obsessively free they are unstable, dissociated and volatile. Compulsive-obsessive "freedom" subsumes the interest of the Detroit actor,[1] and fuels his penchant for social distance and abandonment from his fellows.

Consider this scenario in *City Primeval*: Clement Mansell shoots Judge Alvin Guy in the face five times with a P.38 Walther automatic for cutting him off in a Lincoln Continental Mark VI at Hazel Park Racecourse:

Clement reached down under the front seat, way under, for the brown-paper grocery bag, opened it and drew out a Walther P.38 automatic. He reached above him then to slide open the sunroof and had to twist out from under the steering wheel before he could pull himself upright. Standing on the seat now, the roof opening catching him at the waist, he had a good view of the Mark VI windshield in the flood of light from above. Clement extended the Walther. He shot the chicken-fat jig five times, seeing the man's face, then not seeing it, the windshield taking on a frosted look with the hard, clear hammer of the evenly spaced gunshots, until a chunk fell out of the windshield. He could hear the girl screaming then, giving it all she had.

(Leonard 1982: 17)

Our posture toward the judge is set from the opening of the novel. We become Leonard's "W. W. Beauchamp." Like the biographer Beauchamp in Clint Eastwood's western *The Unforgiven* (USA, 1992), we assume the role of a crucial dramatic foil privy to information that the protagonists do not have. We discover that Judge Alvin Guy is a loathsome bully, who uses his position and the weight of the law to torment people and to sexually harass attractive women. For his part, all Mansell wants to do is keep Sandy and the Albanian in sight. All he knows is that the silver Lincoln Mark VI gets in the way, first in the exit queue, then as Mansell's black Buick Riviera tries to cut into the queue at the gate. The silver Mark becomes the chicken-fat jig with a white girl as Mansell fans his prejudices, the ones we have already adopted from the first chapter. Sandy and the Albanian are lost when the Mark blocks Mansell's Buick at the Eight Mile amber. Judge Guy is set on sending the wise-ass downtown on a bogus charge of assault with a deadly weapon; Mansell is set on sending the chicken-fat jig to jig heaven. The metropolitan equations are balanced. Mansell strikes first. And the reader basks momentarily in a vicarious thrill of urban abandonment and depravity, because we have all raged at the driver in front and wanted to blow him away; we do not do it because we are compelled by communal convention to distinguish between "minor" and "major" in a way that Clement Mansell is not. Our social sanctions against minor violations in street traffic may include hostile stares, fist shaking, finger pointing and horn blowing, in an attempt to bring others into line on public propriety and social graces. But shooting pedestrians in the street is not an act of social control, but an act of social abdication. It is not a conflict of norms, but the embodiment of a moral vacuum. It does not mean "I want you to conform to community standards"; it means "I deny any sense of community with you at all."

The shooting of Judge Guy is more than an eccentric or exaggerated response to a minor incident. Elmore Leonard's bad guys can not only disappear in the crowd; their desire for a principled life disappears in the crowd. It seems the rarefied air of big city life is mixed with dangerous pollutants and hazardous waste by-products that have intensified the actor's central nervous system and heightened his adaptive instincts; while the tattooed skyscrapers, paying homage to technological advancement, have obscured all early views of the universe.

Elmore Leonard's bad guys do not inform the city, but are informed by it. For Clement Mansell, Detroit is both grotesque and captivating. It is a glittering land monster with "Buck Rogers" skyscrapers and a Seven-Eleven gasbar on every corner. As he prowls around Del Weems's twenty-fifth floor apartment on 1300 Lafayette, trying on the man's Brooks Brothers clothes, he is ready to "get acquainted with the finer things in life." He is in awe of the big city – not because of its superordinate presence, but rather because of its seemingly infinite excess and luxury. He embraces its secular geography insofar as it allows him to live with a certain impunity and in the absence of principles. He can hide in the shadows of the towering commercial-industrial complex by day, and commit mayhem and the big score by night.

> In the meantime, cross off the chicken-fat consultant as a score, but use his place to rest up and get acquainted with the finer things in life. Drink the man's Chivas, watch some TV and look out at the twenty-fifth floor view of Motor City. Man oh man.

> The Detroit River looked like any big-city river with worn-out industrial works and warehouses lining the frontage, ore boats and ocean freighters passing by, a view of Windsor across the way that looked about as much fun as Moline, Illinois, except for the giant illuminated Canadian Club sign over the distillery.

> But then all of a sudden – as Clement edged his gaze to the right a little – there were the massive dark-glass tubes of the Renaissance Center, five towers, the tallest one seven hundred feet high, standing like a Buck Rogers monument over downtown. From here on, the riverfront was being purified with plain lines in clean cement, modern structures that reminded Clement a little of Kansas City or Cincinnati – everybody putting their new convention centers and sports arenas out where you could see them. (They had even been building a modernistic new shopping center in Lawton just before the terrible spring twister hit, the same one that picked Clement's mom right out of the yard, running from the house to the storm cellar, and carried her off without leaving a trace.) Clement would swivel his gaze then over downtown and come around north – looking at all the parking lots that were like fallow fields among stands of old 1920s office buildings and patches of new cement – past Greektown tucked in down there – he could almost smell the garlic – past the nine-story Detroit Police headquarters, big and ugly, a glimpse of the top floors of the Wayne County jail beyond the police building, and on to the slender rise of the Frank Murphy Hall of Justice, where they had tried to nail Clement's ass one time and failed. Clement liked views from high places after years in the flatlands of Oklahoma and feeling the sky pressing down on him. It was the same sky when you could see it, when it wasn't thick with dampness, but it seemed a lot higher in Detroit. He would look up there and wonder if his mom was floating around somewhere in space.

(p. 41)

Clement Mansell is a 34-year-old bad guy from Lawton, Oklahoma, with a bright new blue-and-red tattoo of a gravestone on his right forearm, which says "In Memory of Mother." He is the Oklahoma Wildman, nigger-hater, drunk driver and member of the Wrecking Crew. He is the archetypical red-neck cowboy from the dust bowl; but he was made for the negative possibilities of Detroit.

We can imagine that Clement Mansell comes to the big city like any gunslinger out of a 1950s western, ready to make a name for himself, taking on all comers, terrorizing the locals. The substantive difference is that in Motown they wear big cars instead of big hats. Moreover, the analytic difference is that gunning down Judge Alvin Guy goes beyond a simple form of cowboy justice. Clement Mansell is a bona fide "post-modern"[2] sociopath – a pathological killer with an appreciable lack of respect for human dignity and the sanctity of life. The typical 1950s western film cowboy engages in gun-play as an act of courage, or out of the fear of the appearance of cowardice, not out of a lack of respect for his antagonist. Courage or fear justifies the cowboy's deadly sins. Clement Mansell, on the other hand, does not represent any version of justice. There is nothing justified about his sins. He is not like Gregory Peck in *The Gunfighter* (USA, 1950) – referred to in passing in Leonard's text by Detective Raymond Cruz – facing down blustering young toughs with a frontier-man's courage and true grit. Mansell is not a rugged individualist attempting to come to terms with the modern cookie-cutter world, ready to duel to the death to preserve his sense of the right and the good. He does not have one ounce of redemptive tissue. The reader knows that the judge was a slick-headed, beige bully and in Leonard's world, deserved to bite the bullet: he got what was coming to him, cosmic retribution for a downtown slick, as it were. But at the time of the shooting, Mansell was only privy to the traffic encounter, a puny pedestrian skirmish, a breach of public etiquette. In short, he blew the judge's face off, not as an act of courage or fear, but rather because the judge was rude. He violated society's sanctity-of-life clause because the judge breached a mundane norm of politeness. This indicates that Mansell is somehow cut off from the sphere of human values that determine society's elaborate system of rituals and beliefs. For Mansell, society is only arbitrary and pedestrian. He sees no discernable difference between running a stop-sign in his Buick and blowing a man away with his Walther P.38.

Whereas the early migrants and southern share-croppers made their way to the promised land of Detroit's fraternity of auto plants and assembly-lines in pursuit of a better life, the "working gun" is now the high-tech growth industry of the future and an extension of the cityman's soul. Detroit is now a pure dialectic of technology and violence. In an earlier time, Clement Mansell may have been just a poor, gun-toting, white trash, dirt farmer looking for skill-less labor and a grub steak in the factory town; but now "gun-toting, white trash" is a marketable skill. This fact does not elude Clement. He would rather misappropriate capital from the modern support industries than work in the auto industry. He knows, in Leonard's Detroit, he can pick up a piece from any school kid and make more money robbing gasbars than he can making Tempos at the Ford Motor Company. Moreover, there is no compelling spiritual or moral structure inhibiting the actualization of this knowledge, or preserving the integrity of the occasion, save the Seventh Squad of the Detroit Police Homicide Section at 1300 Beaubien.

Clement Mansell shoots Judge Alvin Guy in the face five times because the judge's silver Lincoln Mark VI cut Clement's Buick off at Hazel Park Racecourse. He also shoots Judge Guy because Leonard's techno-city men no longer depend on one another for fundamental survival. Judge Guy is not the "other" for Clement's "self," as it were. Their relation to one another is not as alter-egos. Modern technology is the "other," and has wedged its way into the relations between citymen, as it is increasingly complicit in all their needs. Modern technology purifies modes of thought and action of ethical considerations, as city-men come to impersonate its numbing indifference. Leonard's ideal city-men are villains who relate to each other as technology relates to them. Their "technological rationality" reflects not the breakdown of an ethical situation but the erasure of "ethics" as a category of life.[3] Here, a sense of human evolution, and the perfectibility of man, from a brutish state of nature into a civilized society is transposed; and the original "social contract" of society serving as the fundamental grounds of trust between self and other no longer automatically correlates to the relations between citizens.

Elmore Leonard's bad guys are like the Hobbesian brute, but not entirely. Thomas Hobbes, in *Leviathan* (1651), postulated an imaginary atomic individual in the beginning of time and in the state of nature, alone in a universe not of his own making, surrounded by others in the same predicament. Rational self-interest was his guiding principle, generating the survival motive, and leading him to the prudence of a social contract with others. Here, the first society was motivated by self-survival interest and cemented as a mutually compensatory exchange. Elmore Leonard poses the analytic problem of the last city: what happens when we combine the Hobbesian state of nature with a 9-mm semi-automatic or a Walther P.38 – that is, the technological manifestations of the "will to power"?

In the plot of Detroit, where high technology dominates and subsidizes life, the survival motive is abated and the metropolitan actor is led away from the social contract. The actor is civilized by rational self-interest, but the mutually compensatory exchange of the social contract is written out of the script and replaced with a chilly collective consciousness. Here, metropolitanites eventually only come to share common "feelings" as opposed to a common morality, mutual feelings of reserve and contempt as opposed to feelings of right and wrong. And the cement that binds them is state-of-the-art technology in the war of all against all.

The irony of the Hobbesian brute is that while he comes before society and the social contract, he also necessitates society. His survival depends upon the creation of society, yet the advent of society transforms him from egotistic brute into contracted citizen. In the techno-metaphor of Detroit, however, the city distorts the survival motive and transforms the citizen back into brute again; the post-modern, de-contracted citizen.

Clement Mansell is a Hobbesian brute in a Brooks Brothers suit, without the temperance of a survival motive. But as formidable as Clement is, he may not even be Elmore Leonard's best example of the Detroit actor: in the pages of *Split Images*, for example, we learn that Robbie wants to kill Chichi with either a Baretta 9-mm Parabellum or a German MAC-10 machine pistol with a 32-round clip because Chichi never remembers Robbie's name.

> Angela said, "Does Chichi look like a bad guy to you?"
>
> Bryan said, "Does Robbie? ... When I first mentioned Chichi you got a funny look on your face and you said, 'I just had a thought ...'"
>
> "I wasn't thinking about dope."
>
> "No, it was before we got into that."
>
> "Oh – " Angela came alive. "You said he didn't seem to know Robbie and I thought – right away I thought, what if Chichi's the one he wants to kill?"
>
> "Why would he?"
>
> "That's the trouble, the motive's so flimsy–"
>
> "Because he doesn't like him?"
>
> "Sort of, but more because Chichi never remembers Robbie's name," Angela said. "Does that sound dumb?"
>
> Ruth May Hayes, thirty-seven, was tied by the neck to the rear bumper of her boyfriend's car and dragged in circles over a field until she was dead. Robert Jackson, thirty-four, and James Pope, thirty-five, died in a gunfight that developed when Jackson put his cigarette out in a clean ashtray. Sam James, thirty-five, told his wife their twelve-year-old daughter's shorts were too tight; an argument followed and James was stabbed to death ...
>
> There were approximately six hundred seventy additional homicides, not unlike these, that had taken place in the Detroit area during the past year and were familiar stories to Bryan Hurd.
>
> He said to Angela, "I can tell you how people kill each other and sometimes why and you tell me if any of the motives make sense."
>
> She said, "So it doesn't matter if the reason sounds weird."
>
> He said, "To tell you the truth, killing someone who doesn't remember your name makes more sense than most."
>
> She said, "What do we do now?"
>
> He said, "Let's get out of here and have a Jack Daniels someplace."
>
> (Leonard 1981: 183–5)

A world without a survival motive is a unremitting anomic risk. Leonard's Detroit actor is willing to take the risk.

In conventional society, motives are devices used by ordinary societal members to link particular concrete activities to generally available social rules, thereby clarifying the causal accomplishment of social interaction for the observer (Blum et al. 1974: 21). Yet in Leonard's plot of Detroit, motives take on a quite different form and content. Since modern technology compromises the survival motive, in turn it increasingly compromises all motives for men's actions – any motive at all – until motives for social interaction become characteristically dumbfounding. Here, killing someone who does not remember your name is a motive for murder that – according to Detective Bryan Hurd – "makes more sense than most." Motives look "flimsy" in Detroit. That is, they are intelligible to others as weak or feeble or negligible because they do not have to have a discernible link to a social vocabulary; they do not have to be connected to a common or coherent world-view. Detroit motives do not have to matter in the scheme of things, as they do elsewhere, because the social environment does not require consonance or clarification or grounding. Leonard's fictional Detroit is not only a place where you can be killed for cutting someone off in traffic, or for not remembering someone's name; it is a place where motives tell why but the "why" does not have to make sense. Motives do not need to matter or to clarify. Consequently, one motive is as good as any other in the metaphor of Detroit, because all motives are free from an interest in a coherent and grounded life ("Robert Jackson, thirty-four, and James Pope, thirty-five, died in a gunfight that developed when Jackson put his cigarette out in a clean ashtray..."). Every murder has a motive in the big city, but it seems as if every motive has a murder in the city primeval.

It is important to grasp the fact that in the interplay between self and other, motives are only required instruments for the accomplishment and clarification of action in a world guided by a desire to be coherent and on solid ground – where selfs or egos become alters and not merely others. The attempt to cognitively reconstruct motives is an attempt on the part of ordinary societal members to reconstruct the invisible dimension of social interaction – to make the invisible visible, as it were. Leonard insinuates that this cognitive reconstruction only matters in a conventional world – a post-Hobbesian, pre-Detroit world. Only in a world that has a sense of the perfectibility of man, of reverence or profound wonderment in the social contract, and where actors seek the profound (as opposed to the merely concrete and mundane) in their relations with each other, is there a continual process of clarifying life and things in the pursuit of groundedness. In the conventional world, motives are a pivotal feature of the social vocabulary in a quest for the esoteric – the way things really are, not just the way things look on the surface.

In the conventional world, motives have this vitality and sense of urgency; everybody looks for and wonders about motives, good and bad, because motives affirm life's yearning for the transcendent.

In contrast, there is nothing esoteric about Leonard's Detroit actor and his motives. Detroit motives do not endorse some viable, or misplaced, version of reality. They are not even negative reflections of some esoteric truth beyond utterance; rather, they presence the absence of such a desire.

Not coincidently, it is the professional homicide detective, above all, who is impervious to Detroit motives: Raymond Cruz, Bryan Hurd, Wendell Robinson. In the world of the street-wise, no-bullshit, homicide detective motives do not connote truth; motives are just there. Motives are what they are – which is to say, they are complete in and of themselves, and they make reference to nothing outside the permissiveness of the persons that own them.

The plot twist for the city primeval is that Raymond Cruz, the cop, is the only urban phenomenon standing in the way of Clement Mansell, the robber, turning Detroit into one big convenience store. The unfolding of the relationship between Cruz and Mansell is reminiscent of the 1950s western white hat/black hat mythology of duelling, where good triumphs over the forces of evil; yet Leonard's Detroit is beyond good and evil. As he reminds us "it's not a question that they're going to get their just deserts": redemption is superfluous in the metaphor of Detroit. The professional homicide detective occupies an integral role in a godless environment. He is both a de-contracted citizen and a consummate detached participant observer, following, with precise measurements, Detroit motives as part of a street-calculus that leads not to some profound source, but right back to the definition of the Detroit actor: motive minus the principled pursuit of a grounded life.

COPS AND ROBBERS: DEVIANCE VERSUS SOCIAL CONTROL

The classical sociologist Emile Durkheim (1964; orig. 1893) is credited with the proviso that God is a symbol for society. Durkheim, like Thomas Hobbes before him, was well aware of the fact that the advent of society entails the emergence of a superordinate and prohibitive moral order. It is not only the home of man's material life, but the home of man's spiritual life as well. Therefore, in a Durkheimian sense, God and society are one. Society is man's consecrated Host; it is moral geography in all its contours, to which men are ritually compelled to abide.

This relationship between society and the sacred is assailed by modernity and the metaphor of the great city. The great city does not reflect the divine presence of society, for it refuses to acknowledge there is anything greater than itself, and breeds inhabitants with the same inclinations. The great city seeks to confer freedoms where society only offers constraints and, thereby, delivers men into a liberating state of obliquity, free from the old and the established protocol.

In this regard, Durkheim's society stands to Leonard's Detroit in the way the sacred stands to the profane. Detroit is the antagonist of society's sacred canopy. It is the epitome of the great city that sees nothing greater than itself, and where citizens emulate this conceit: a place where people just might kill others for not remembering their name. Detroit is a woeful city filled with actors in an unholy alliance, like Dante's Lost People in *The Divine Comedy*, blasphemers of all that is venerated, free from the constraints of conventional morality, present to society's rules but more or less absent to its principles, desperately seeking the unauthorized enlargement of their own prerogatives at the expense of honorable motives and high ideals. In Leonard's Detroit, it is the forces of social control – not the sacred – that are the link between the parts and the whole, and the only real adversaries of deviance and profanity.

Picture this station-house scene – Clement Mansell has gunned down the black judge Alvin Guy because Guy cut Mansell off in traffic at Hazel Park Racecourse, and the black police detective Wendell Robinson *knows*:

> "We want to be ugly, we could get you some time over there right now," Wendell said. "Driving after your license was revoked on a D.U.I.L., that's a pretty heavy charge."
>
> "What, the drunk-driving thing? Jesus Christ," Clement said, "you trying to threaten me with a fucking *traffic* violation?"
>
> "No the violation's nothing to a man of your experience," Wendell said. "I was thinking of how you'd be over there with all them niggers."
>
> "Why is that?" Clement said. "Are niggers the only ones fuck up in this town? Or they picking on you? I was a nigger I wouldn't put up with it."
>
> "Yeah, what would you do?"
>
> "Move. All this town is is one big Niggerville with a few whites sprinkled in, some of 'em going with each other. You'd think you'd see more mongrelization, except I guess they're just fucking each other and not making any kids like they did back in the plantation days ... You want to know something?"
>
> "What's that?"
>
> "One of my best friend's a nigger."
>
> "Yeah, what's his name?"
>
> "You don't know him."
>
> "I might. You know us niggers sticks together."
>
> "Bullshit. Saturday night you kill each other."
>
> "I'm curious. What's the man's name?"
>
> "Alvin Guy." Clement grinned.

(p. 82)

Let us say Clement represents deviance and Wendell represents social control. What Clement is to Wendell, Detroit is to Society: the dialectic of deviance and social control. There is a new twist to this dialectic of life, however, that we can only grasp through a closer look.

Wendell, the cop, is not against Clement because Clement is an amoral, red-neck who kills people; Wendell wants Clement because Clement gets away with it. Wendell wants to nail Clement's ass or, better yet, blow his fucking brains out with a Baretta 9-mm Parabellum, because Clement (deviance) is making Wendell (social control) look like a fool.

Wendell, like Raymond Cruz and Bryan Hurd, is the exemplar of every cop in Leonard's Detroit, and cops are the primary agents of social control. As agents of social control, cops once assumed the role of combating the "forgetfulness" of men in regard to the sacred; they served as a "constant reminder" of the portentous gods over the city, intersecting the individual and the collective, as well as society and nature. The new twist here is that social control is no longer guided by the overriding authority of society in its clash with deviance (à la Durkheim's moral society). Or, to put it another way, social control is no longer the agent of society in the city, battling deviance on the moral high ground. In Leonard's Detroit, the agents of social control are men who themselves forget the appropriate legitimations of social reality.

The dialectic of life has been boiled down by technological rationality (that is, the hardware and aptitude to terminate the social contract with extreme prejudice) to the point where social control only cares about its relationship to deviance. Life is an eternal dance between cops and robbers, both vying to own the streets. Wendell the cop only cares that Clement the robber thinks he can live and navigate the city better than Wendell. Wendell does not care that Clement murdered a judge, a representative of the highest ideals of society. Wendell cares that Clement throws the murder of judge Alvin Guy in his (Wendell's) face. The murder does not matter; murders are a dime a dozen; "murder smurder"; it is like Bryan Hurd said – "there were approximately six-hundred and seventy-five other murders in the metropolitan Detroit area last year." Wendell did not like the odious Judge Guy anyway. He was a chicken-fat mark and marks do not matter. What matters is Clement has rubbed a murder in Wendell's face, and that is a challenge. If Clement (deviance) can outmaneuver Wendell (social control) in Wendell's own city, then Wendell knows he might as well give up being a cop and become a robber. Wendell has to outmaneuver Clement in the long run to justify being a cop, for the only justification for social control in the great city is winning the game of metropolitan guile against deviance. Social control has to transcend the bounds of our narrower society and become to some extent as ruthlessly resourceful as deviance itself.

Social control has to beat deviance on its own ground to justify itself and claim home-turf advantage. If social control were in the service of societal goals or values, then social control could afford to resist the challenge of deviance, even lose the day once in a while, because social control would always be morally superior to deviance, having the assurance and endorsement of the right and the good on its side. But, as it is, in lieu of a higher authority, social control in the great city no longer speaks (for) the holy; social control only has its relationship to deviance; its only *raison d'être* is self-preservation and self-determination in the face of deviance. Consequently, social control (Wendell) can only maintain its stake in the city by fending off the blatant challenge of deviance and becoming what sociologist Robert Park (1969: 131–42) once called a "marginal man" – on the move and on the make.

This is the new moral in the story of the city: Clement Mansell (deviance) has challenged detectives Wendell Robinson and Raymond Cruz (social control), and they have to put Clement (deviance) away or blow his fucking brains out; otherwise Wendell and Raymond (social control) lose their place and their standing in the metropolis. Social control no longer stands for society; it stands against deviance.

Deviance, as the classical sociologist Georg Simmel (1950: 118–19) told us, is a relationship, albeit a negative one, to the social. Dialectically speaking, then, deviance and social control need each other (for their own existence and the existence of society). Simmel knew that the relationship between deviance and social control has to be motored by a legitimation process; it depends on the existence of a common version of authority upon which both deviance and social control converge. Without a higher authority, a divine cosmology of the city, social control cannot legitimately interpret the order of society in terms of an all-embracing, sacred order of the universe. Here, social control would lose its standing as the principled agent of the city and guardian of the pure and the holy. In a city where there is no divine, then, social control cannot bind deviance to the whole. Deviance and disorder cannot be related to as the antithesis of civilization, or as a profane affront to the nature of things – deviance merely becomes the antithesis of social control. Deviance and social control face off against one another as differentiated and dissociated members and opponents in the city. In other words, what Simmel did not tell us, and what we can only infer, is that the dialectic between deviance and social control is potentially a new topic in history accompanying the advent of the great city and replacing the former dialectic between deviance and the sacred.

In their elimination of the divine, of God, Mansell (deviance) as well as Robinson and Cruz (social control) struggle to replace the gods over the city. By striving to become god-like they have lost a sense of human civilization's original point of origin and reference and have become their own point of origin. Now the dialectic between deviance and social control presences the need – not for the oneness of the sacred, but to dominate each other and exploit the structured multiplicity of the metropolis.

THE GOOD CITY AND THE LAST CITY

Detroit is Elmore Leonard's literary "ideal type" or "social form" in the Weberian (1964: 61–5) and Simmelian (1950: 22–3) sense, respectively, of what we are becoming. It is our "becoming" – not identical with any particular urban behavior or enterprise, but informing all behavior and enterprise. It is a metaphor or hypothesis for a technological state of nature; a post-modern war of all against all, where the "working gun" is the cottage industry of the future – and (as *Killshot*'s Rickie Nix mused) "you can get any kind of piece you want, buy it off a school kid." This literary landscape is, in some measure, an indication of the historical trajectory of urban life – not because it is a model of the present age; on the contrary, it is like a probe into our paradigmatic future which amplifies our least flattering impulses. In this regard, it offers dramatic contrast to the way the big city ought to be.

Compare, for example, the moving reminiscence and literary landscape of the young Truman Capote's New York:

> The skyline itself was romantic: the first flat-roofed glass skyscraper had yet to be erected, and Manhattan was still an island of grand and ebullient architectural fantasies – minarets, ziggurats, domes, pyramids, and spires. Banks resembled cathedrals, office buildings masqueraded as palaces, and spike-topped towers unabashedly vied for a place in the clouds. Except for a few seamy areas, people walked wherever they wanted, whenever they wanted: street crime was rare. Hustling for news, eight major papers made everything that happened in the five boroughs, no matter how trivial, sound grave and consequential, while a battalion of gossip columnists, like nosy telephone operators in a small town, made the city seem smaller than it was with their breathless chatter about the famous, and those who would like to be famous. Broadway was a never-ending feast; theatergoers, sated with the variety before them, probably expected every year to be as bountiful as 1947, which not only saw the openings of *A Streetcar Named Desire*, *Brigadoon*, and *Finian's Rainbow*, but enjoyed also the continuing runs of *Oklahoma!*, *Annie Get Your Gun*, *Harvey*, and *Born Yesterday* ... [So much exuberance and vitality painted] an unforgettable picture of what a city ought to be: that is, continuously insolent and alive, a place where one can buy a book or meet a friend at any hour of the day or night, where every language is spoken and xenophobia almost unknown, where every purse and appetite is catered for, where every street with every quarter and the people who inhabit them are fulfilling their function, not slipping back into apathy, indifference, decay. If Paris is the setting for romance, New York is the perfect city in which to get over one, to get over anything. Here the lost *douceur de vivre* is forgotten and the intoxication of living takes its place.
>
> (Clarke 1988: 133)

Capote's city coincides with the original sociological hypothesis (beginning with Comte through Tocqueville and Durkheim) that the scope of human sympathy increases with the progress of civilization. In the big city organic solidarity[4] ought to lead to the intoxication of living; rich variety and diversity ought to lead to a never-ending feast; the process of secularization[5] ought to lead to a modern morality that is progressively more humanitarian and tolerant. In short, the big city ought to be the vehicle that preserves the continuity between man and God, the human and the divine.

Elmore Leonard's Detroit, however, is the metropolitan future that defies the classic social-analytic hypothesis, disdains the divine and unhooks men from moral commandments and the progressive order of civilization. The suspended dialectic of the sacred generates the peculiar semblance of the "post-histoire." It is not that the submissions of the original social theorist of urban society were wrong; it is rather that somewhere between Truman Capote's New York and Elmore Leonard's Detroit divine cosmology and the city's connection to it took a nose-dive, and the culmination of civilization began to sprout technological-rational mutants with appetites so various and large that they could not be catered to. The endless feast was nullified by a never-ending appetite. Somewhere between Capote's big city and Leonard's great city a virulent strain of mutant life got a foothold, breeding metropolitanites without an "ought" – mutant angels estranged from the gods of the universe, who no longer see an urban cosmology through the patina of skyscrapers.

The literary landscape of Elmore Leonard's Detroit is filled with technological-rational mutants who are immune to "society's" infinite charms. It is the vortex of a human twister with wars whirling all around, where the once noble life-quest for reason and progress, in the eye of the storm, has given way to an interiorized and solipsistic existence on its fringes. The original social theorists could not have imagined people coming to pride themselves on being able to live on the fringes of society, or pride themselves on being able to move in and out of marginal worlds as comfortably or as fluidly as they move in and out of the conventional social circles. But this is precisely the environment now imaginable and reflected in literature, and "living on the fringe" is precisely the human adaptation that has replaced the pursuit of the good life. Somewhere between Capote's good city and Leonard's last city it ceased to be enough to know merely how to be "civilized." Civilities still exist; but there is stronger emphasis on the idea that for urbanites to possess civility alone is a bit naïve. Mutually compensatory exchanges still exist; but to rely on one's mutuality with others in the great city can be deadly. The civilized man in the great city is a man ripe for the picking. Furthermore, there is a feeling that if you can live marginally, on the fringe of the city, you can go anywhere. To flourish in this metropolitan environment, the Detroit actor must be equipped for every contingency – equipped to bracket conventional rules of conduct when necessary in order to negotiate the vicissitudes of the urban terrain. Relevance structures and provinces of meaning in the streets of the city are interchangeable like differentiated languages in a repertoire; and understanding "the streets" is as enriching, and as handy, as understanding French as a second language.

Ordell Robbie and Louis Gara, the salt-and-pepper kidnapping duo in *The Switch*, depend on their knowledge of "the streets as a second language" in order to carve out a good living. Or as Erving Goffman (1980) might put it, their game is "cooling out the (chicken-fat) mark." To wit:

> Ordell brought out his box of Halloween masks, set it on the coffee table in front of Louis and said, "Now you know how long I've been working on this deal."
>
> They were in Ordell's apartment, Louis stretched out in a La-Z-Boy recliner with the Magic Ottoman up. He'd been sitting here four days on and off, since Ordell had met him at Detroit Metro and told Louis he was coming home with him. Louis had said home where? Some place in Niggerville? Ordell said no, man, nice integrated neighborhood. Ofays, Arabs, Chaldeans, a few colored folks. Ethnic, man. Eyetalian grocery, Armenian party store, Lebanese restaurant, a Greek Coney Island Red Hot where the whores had their coffee, a block of Adult Entertainment, 24-hour dirty movies, a club that locked the doors and showed you some bottomless go-go and a park where you could play 18 holes of golf ...
>
> So Louis had been sitting in the La-Z-Boy and getting up to eat, sitting there and getting up to go out and pick up a lady and get laid, twice – in between listening to Ordell tell him how they were going to make a million dollars and looking at the sights Ordell showed him. Louis did not get excited or ask many questions. He let Ordell make his presentation and dribble out things the way he wanted, taking his time. As Ordell said "Be cool Louis. You ain't got to be anywhere but with me."
>
> (Leonard 1978: 61)

For operators like Ordell and Louis, making it in greater metropolitan Detroit is as straightforward as saying to a mark, "Deposit a million dollars in my off-shore account chicken-fat, or you'll never see your wife alive again."

There are really only two fundamental life-choices or life-categories in Leonard's Detroit: the mark and the operator. You can be a Chivas-sipping, chicken-fat "mark" or a Jack Daniels-guzzling "operator." The mark is a role-player and the operator plays a role. The mark observes strict rules of conduct and social convention, and lives within the bounds of society's institutional role-structure. He is a meticulously compliant status-role bundle: a father, a husband, a business man, a taxpayer. He makes a "reflex-action self" out of society's repertoire of institutionalized roles, as if his self were an involuntary muscle. Safety, of course, is the prime motivation of the mark. It is safe to travel the tested path of convention and tradition. The operator, on the other hand, abandons the sanctuary of a status-role set when it is convenient, or when necessary, using the camouflaging anonymity of the bustling city to mask his self. The operator is a *camoufleur* – continually refining the art of the false front in order to score or "cut a deal in the city." He only looks like a father, a husband, a business man, a taxpayer. His real life-interest is feeding on the marks or protecting them (for a price) from being devoured, which amounts to the same thing.

Ordell and Louis are operators who know how to live in the streets and on the fringes of the city, and weave their way through the cracks separating its irreparably differentiated life-worlds. As operators, they are distinguished by continually acting in ways that revive the fundamental confrontation between light and darkness, nomic security and anomic abandonment, moral order and rational self-interest – confrontations that contemporary society has long since put to rest. They are mutant angels, Hobbesian brutes in business suits, who animate the philosophical and sociological distinction between the good city and the last city.

THE SOCIOLOGICAL IMAGINATION

Formerly, to go against the progressive order of civilization was to risk plunging into a state of anomie. To go against the order of society as normatively and morally legitimated, however, was to make a compact with the primeval forces of darkness. To deny reality as it was societally defined was to risk falling into irreality, because it was impossible in the long run to keep up alone, and without social support, one's own counter-definitions of the world. When the societally defined reality has come to be identified with the ultimate reality of the universe, then its denial takes on the quality of evil as well as madness. The denier then risks moving into what may be called a negative reality – the reality of the devil. This is well expressed, of course, in those archaic mythologies that confront the divine order of the world (such as the ancient *tao* in China, *rta* in India, *ma'at* in Egypt) with an underworld or anti-world that has a reality of its own – negative, chaotic, ultimately destructive of all who inhabit it, the realm of demonic monstrosities.

In the techno-myth of Detroit the analytic distinction between reality and irreality, light and darkness, is for all intents and purposes devoid of all fundamental spiritual or religious overtones. Instead, post-modernism and the Detroit world-view is secularized (even when the topic is religion, as we will see below with the iconoclastic August Murray) and filled with what C. Wright Mills (1970) conceptualized as the "sociological imagination."

In *Swag* Leonard provides the consummate illustration:

"All right," Stick said, "you tell me. What's the best way to make a lot of money fast? Without working, that is."

Frank held up the palm of his hand, his elbow on the edge of the bar. "You ready for this?"

"I'm ready."

"Armed robbery."

"Big fucking deal."

"Say it again," Frank said, "and put it in capital letters and underline it. Say it backwards, robbery comma armed. Yes, it can be a *very* big deal. Listen, last year there were twenty-three thousand and thirty-eight reported robberies in the city of Detroit. Reported. That's everything. B and E, muggings, banks, everything. Most of them pulled by dummies, junkies, and still a high percentage got away with it."

"Going in with a gun," Stick said, "is something else."

"You bet it is." Frank leaned in a little closer.

"Ernie ... Ernest –"

"Stick."

"Stick ... I'm talking about simple everyday armed robbery. Super-markets, bars, liquor stores, gas stations, that kind of place. Statistics show – man, I'm not just saying it, the *statistics* show – armed robbery pays the most for the least amount of risk. Now, you ready for this? I see how two guys who know what they're doing and're business like about it" – he paused, grinning a little – "who're frank with each other and earnest about their work, can pull down three to five grand a week."

(Leonard 1987: 15–16)

Thus begin the promising and lucrative careers of Frank Ryan and Ernest Stickley Jr., equipped with instrumental logic, a Smith & Wesson .38 Chief Special and Colt Python 357 with a ventilated rib over a 6-inch barrel – and, of course, a sociological imagination informing them of their place in society and history. "Statistics" show armed robbery pays in Motown. Armed with social statistics, the hard data, a couple of enterprising guys like Frank and Ernest can end up with a luxury condo in the suburbs and all the pussy they can handle.

C. Wright Mills conceived of "the sociological imagination" as individual thought and consciousness focused on the world historical plane. In his view, it is only by understanding how individual biography intersects society and history that one can grasp what is going on in the world. Through rigorous sociological training, one comes to understand the difference and the relationship between person and structure, private concerns and public issues. This, according to Mills (and as one might suspect), remains the unfulfilled promise of sociology.

Leonard's Detroit is something of an extension on Mill's theme. Whereas Mills would suggest that sociological imagination and literacy is a rare commodity, in Motown it is viewed as standard metropolitan equipment. Here, every street-smart homeboy has sociological imagination oozing out of his pores. He is well versed in the grammar of the substructure and the superstructure of the city. He can refer to sociological statistics and social structures as a matter of routine, and without explicit rule, as easily as he can load a pistol. And as a man on the make, he continually endeavors to convert the otherwise amorphous mass into a potential money making proposition.

In Leonard's Detroit, the sociological imagination not only refers to a method of seeing life, but also to a social psychology of life. Life in interactional and structural context is arbitrary and conventionally forged, it only becomes secure by mutual agreement to take the conventions for granted and to get on with the business of living. Over time even the agreement (even agreeing to agree, or the "precontractual" agreement in society) gets taken for granted, until life appears to be on solid ground and in pursuit of inviolable principles. Sociological imagination makes men city-worthy by reminding them of the social construction of reality. It advises that society's process of sedimentation does not advance people but bogs them down, makes them numb and dumb, noctambulant, "easy pickin's," as they ignore their contingency and surrender their trust to attachments with anonymous others.

Enter August Murray, who makes his appearance in the pages of *Touch*, as perhaps the most driving Leonardesque post-modernist of all – a xenophobic iconoclast with sociological training as well as sociological imagination:

August transferred to the University of Detroit, majored in sociology, minored in philosophy, and graduated cum laude.

He joined the Catholic Laymen's League, which was dedicated to rooting out Communist fronters from the Church and clergy. But the CLL was never active enough for him. He left to found Outrage, along with the Gray Army of the Holy Ghost (membership, a hundred dollars a year), and began writing and distributing pamphlets. The first one he printed, and still his favorite, was the one about the Church needing a good persecution.

August wasn't afraid to be persecuted. He had demonstrated against demonstrators, commie peace makers, political left-wing cowards, lesbian ERAers, fags of all kinds, marijuana creeps in Ann Arbor, equal housing, the right to work, and what's-his-name, the nigger mayor. He'd been arrested a half dozen times and finally convicted of assault, disrupting a Lenten service at Our Lady of Lasalette in Berkley, sentenced to a year probation and ordered to undergo psychiatric examination, which he ducked and they forgot about. Then a disorderly conduct arrest: distributing political literature within three hundred feet of a church, for which he was fined two hundred dollars by a chickenshit colored judge. And right now he was out on a five-hundred dollar bond awaiting his jury trial in the assault on the guinea priest, Father Ravioli ... Navaroli.

(Leonard 1987: 179)

August Murray's sociological imagination carries him well beyond the simple recognition that society is *sui generis* or of its own making. He now knows that a society based on (Durkheim's) organic solidarity and moral tolerance is for weaklings and shame artists.

August Murray also knows:

It was too bad Saint Augustine was not a martyr. He took a concubine and had a son by her. In fact, from what August could find out, Augustine screwed a lot of women in North Africa before deciding to dedicate his life to Christ. Thereafter he was continually engaged in the defense of the catholic faith *against schismatics, heretics, and pagans*. He died August 28, 430, "in a spirit of great courage, humility, and penitence."

The first important paper August Murray ever wrote was entitled "What the Church Needs Today Is a Good Persecution."

The paper wasn't the reason he was kicked out of the seminary; his English teacher, Father Skiffington, agreed with him. But it was an indication of August's problem. He had trouble with rules of conduct, blind discipline, all the no-questions-asked humility shit. He believed the Church needed fighters – anybody could see that – and not the bunch of good little mama's boys the seminary was turning out. It was true that Saint Augustine had died in the spirit of humility (and courage and penitence), but he was seventy-six years old and humility could be a wise move at that age; but not when the Church needed men who weren't afraid to stand up and defend their faith against the cowards within and the Communists without (the sons of bitches).

(p. 178)

August Murray knows *he is* the modern version of St. Augustine. He is St. Augustine without "rules of conduct, blind discipline, and all the no-questions-asked humility shit." Here he carries the so-called "Augustinian solution"[6] to its logical theological conclusion in Detroit: If St. Augustine precipitated the shift in Christological doctrine from God's grace to man's sin, then August Murray's acknowledgment of man's sin and fall from grace is distilled and crystallized into group xenophobia. Detroit city is the end of the fall, the inverse of Eden, the place where an August Murray can despise everyone equally and with impunity. The only source of redemption he can conceive of (as one of the elect) for the barbarians, idolaters, pagans and sons of bitches in the modern thought-tormented world is "his dad's .38 caliber Smith and Wesson Commando." History dictates that St. Augustine's City of God give way to August Murray's City of Infidels – cowards and Communists and "what's-his-name, the nigger mayor."

The belief that the human race is fallen expresses the deep conviction that this is an "odd world," though we know no other. August Murray cannot get a grip on things in the modern world except as failing and weakness; therefore, the things and/or people peculiar to the modern world do not really matter to him except insofar as they might be standing in his way. Of course, he is not subject to the self-same analysis of failing and weakness to which he subjects others, nor is he animated by a strong standard of excellence, since there is no "Eden" to which he aspires. He only knows men are radically truncated and desperately fallen. Consequently he is not at home with anyone or anything. He is a man with only prejudices; and he can only dream of a world and men who do not and cannot exist. In this sense, (even) his xenophobia is without real spirit, or desire, or a real connection to its objects, for he does not hate people for what they are; he hates them for what they are not.

Finally, what August Murray knows – what is hidden or covert within his overt metropolitan prejudices – is that modernity is animated by blunted discrimination. (This, of course, is deeply Simmelian.) His salvation and self-concept depend upon his devaluation of the modern world – the devaluation of all former values. (This, of course, is deeply Nietzschean.) In this prescription, collective xenophobia is the tie that binds and the sign of the true operator. To be a member – a citizen – is to have a little August Murray in you. In analytic terms, this is a profoundly masochistic turn in urban culture. Urban culture "wrongs" the cosmic order and unhooks men from a principled relationship to others. Detroit represents the last city and the final wrong that cannot be righted by the appropriate ritual or moral act. (And this, of course, is deeply Leonardesque.) It is the place where organic solidarity means collective numbness; where variety and diversity means other men could be anything, and so they are not consequential; and where the process of secularization means making a score at the expense of some Chivas-sipping, chicken-fat mark.

NOTES

1   All individuals are social actors insofar as they play specific and reciprocal roles generating a certain complementarity of expectations between them. This complementarity is the basis of group organization and structure. So the sociological premiss here is that the actors depicted in Leonard novels exhibit their own unique complementarity. When I speak of the Detroit actor I am speaking of a prevailing orientation as opposed to a person. This orientation is informed by the environment and gleaned from Leonard villains, bad guys and sociopaths, but is not necessarily identical with them.

2   The conceptual source for "post-modernism" that I use and build on in the

remainder of the chapter is borrowed or extracted from Alan Blum and Peter McHugh's extraordinary analysis of "the estranged modern actor" in their book *Self Reflection in the Arts and Sciences* (1984: 137–9). [Their formulation in turn is a development of Camus, *The Stranger* (1955).] As Blum and McHugh state:

> Although the estranged modern actor is typically thought to be the victim of anomie – of the absence or confusion of convention – we can see here quite another understanding: He is dispirited not because rules are absent but because they are present in the absence of principle.... He does not need another convention, he needs what could matter about convention, unless all is to "come to the same thing in the end."

If the estranged modern actor, portrayed in literature and social theory, is present to rules in the absence of principle, then I view post-modernism as the world-view that unfolds after the fact of this social phenomenon. In short, imagine a world of rules without principle – where members are not only *not* animated by a principled search for a good life but, moreover, are intent on seeing that the issue never arises as a normal feature of membership.

3   This notion of technological rationality may be compared to Jurgen Habermas's (1971: 111) formulation of technocratic consciousness:

> technocratic consciousness is not based on the causality of dissociated symbols and unconscious motives, which generates both false con- sciousness and the power of reflection to which the critique of ideology is indebted. It is less vulnerable to reflection, because it is no longer *only* ideology. For it does not, in the manner of ideology, express a projection of the "good life."

4   The concept of "organic solidarity" was first developed by Durkheim (1964) to refer to modern societies with a complex division of labor, where people are bound together in a highly individualized way and in loose associations based on their economic and political ties.

5   The term "secularization" is used here in the Durkheimian sense to refer to the historical process by which sectors of society and culture are removed from the domination of religious institutions and symbols.

6   St. Augustine is commonly accredited with shifting the question of the justice of God to that of the sinfulness of man. His solution is an affirmation of this shift in Christian theological thought and interpretation that can be expressed as follows: if Christ suffered "not" for man's innocence, but for his "sin," then a prerequisite of man's sharing in the redemptive power of Christ's sacrifice is the "acknowledgment of sin." (Cf. Hicks 1966, Berger 1967.)

REFERENCES

Berger, Peter L. (1967), *The Sacred Canopy: Elements of a Sociological Theory of Religion*, Garden City: Doubleday & Company, Inc.

Blum, Alan, McHugh, Peter, Raffel, Stanley and Foss, Daniel (1974), *On The Beginnings Of Social Inquiry*, London: Routledge & Kegan Paul.

Blum, Alan and McHugh, Peter (1984), *Self Reflection in the Arts and Sciences*, Atlantic Highlands: Humanities Press.

Camus, Albert (1955), *The Stranger*, New York: Vintage Books.

Clarke, Gerald (1988), *Capote*, New York: Ballantine Books.

Durkheim, Emile (1964; orig. 1893), *The Division of Labor in Society*, New York: The Free Press.

Goffman, Erving (1980), "On cooling the mark out: Some special adaptions to failure," in Lewis A. Coser (ed.), *The Pleasures of Sociology*, New York: New American Library.

Habermas, Jurgen (1971), *Toward A Rational Society*, Boston: Beacon Press.

Hicks, John (1966), *Evil and the God of Love*, New York: Harper & Row.

Hobbes, Thomas (1910; orig. 1651), *Leviathan*, London: J. M. Dent & Sons Ltd.

Leonard, Elmore (1976), *Swag*, New York: Dell Publishing Co.

—— (1978), *The Switch*, New York: Bantam Books.

—— (1981), *Split Images*, New York: Avon Books.

—— (1982), *City Primeval*, New York: Avon Books.

—— (1987), *Touch*, New York: Avon Books.

—— (1989), *Killshot*, New York: Warner Books Inc.

Mills, C. Wright (1970), *The Sociological Imagination*, Harmondsworth: Penguin Books.

Park, Robert (1969), "Human migration and the marginal man," in R. Sennett (ed.), *Classic Essays on the Culture of Cities*, New York: Appleton-Century-Crofts.

Simmel, Georg (1950), *The Sociology of Georg Simmel*, ed. Kurt H. Wolff, New York: The Free Press.

Weber, Max (1964), *From Max Weber: Essays in Sociology*, ed. H. H. Gerth and C. Wright Mills, New York: Oxford University Press.

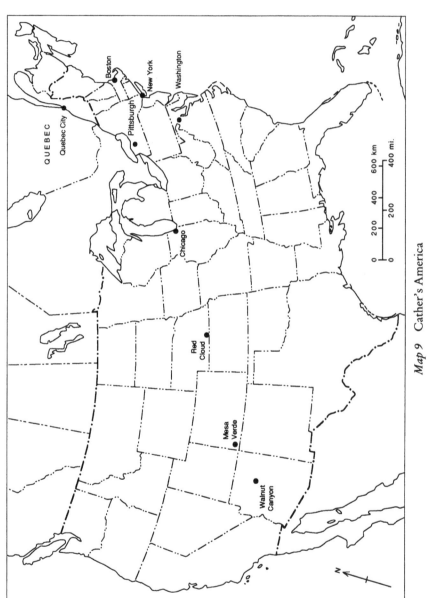

Map 9 Cather's America

9

WILLA CATHER AS A CITY NOVELIST

Susan J. Rosowski

The Willa Cather I propose to write about is not Nebraska's daughter who became America's foremost celebrant of the pioneer period, but a devoted city-dweller who left behind an exceptionally thorough exploration of the urban settings in which she chose to live.[1] Despite our close identification of Cather with the rural town of Red Cloud, the fact is that as soon as she was old enough to leave her parents' home, Cather lived in ever larger cities: Lincoln, Nebraska (where she attended the University), Pittsburgh (where she worked as a teacher and journalist), and New York (where she resided throughout most of her adult life). For Cather, as for many other women, becoming a writer was inextricably tied to leaving the country for the city: it involved confronting and challenging the domestic, female and personal culture so closely identified with nature, to take her place in a public and male culture identified with urban settings (see Squier 1984: 4–5). Not surprisingly – for she was a writer who drew extensively upon her life for her art – Cather made the idea of the city one of the most important and complex subjects of her fiction.

Cather's apprentice writing begins her record of incorporating the city as setting and subject. In her early stories Cather used urban settings conventionally, as backdrops: Chicago for "The Count of Crow's Nest" (1896) and New York for "Nanette: An Aside" (1897) and "A Singer's Romance" (1900). As for her awakening to a "city consciousness," that occurred most dramatically on her 1902 trip to Europe and appeared in essays she wrote while traveling and sent to the Nebraska *State Journal*, where fourteen of them were published. The essays tell of an emerging receptivity to the human story of the city, most often of the working class and particularly of its women. Arriving at Liverpool, Cather contrasted the pageantry of the city's celebration for Edward VII's coronation to its middle-class girls and women, "stoop-shouldered to a painful degree" (1956: 7); in Chester, she admired the self-sufficiency of people living with walled gardens behind their houses; and writing of the canals of England, she imagined the women who carried out housekeeping in barges' cave-like cabins. But it was London that made the greatest impact. Expecting to see the city she had read about as a palimpsest, layer imposed upon historical layer, she was shocked by its human reality of grinding poverty. After securing lodgings in the East End, "to be in the heart of the old city of London, within walking distance of the Tower, Old Bailey, and the Temple," she realized that "the living city and not the dead one has kept us here, and the hard, garish, ugly mask of the immediate present drags one's attention quite away from the long past it covers," for before her the streets seemed "a restless, breathing, malodorous pageant of the seedy of all nations" (1956: 55–6). The experience was so painful that her essay about it is "the least well-balanced, certainly the most compulsive and strongly written article" of her series (Kates in Cather 1956: 50), and she did not draw upon it for her fiction until thirty years later.

Following her European trip Cather continued using the city as a shorthand for culture and art: Boston in "A Wagner Matinee" (1904), London in "The Marriage of Phaedra" (1905) and Paris in "The Namesake" (1907) and "The Profile" (1907); but she began to write of the city also from the perspective of the working class, in "Paul's Case" (1905), for example, contrasting the close quarters and stale odors of his working-class neighborhood in Pittsburgh to the rarified atmosphere in New York's Waldorf-Astoria Hotel. It all seemed a preparation for 1912, when Cather moved the city forward, from background to subject, in "Behind the Singer Tower" and *Alexander's Bridge*.

"Behind the Singer Tower" was Cather's first experiment in "writing the city," an idea she spelled out with uncharacteristic specificity.[2] She selected as her subject "the complete expression of the New York idea in architecture" embodied in the Mont Blanc Hotel, then explained as her theme height and prices that "outscaled everything in the known world," and introduced for her action a fire that destroyed the hotel (1965: 44). Following the fire in which the hotel's famous guests "had met with obliteration, absolute effacement, as when a drop of water falls into the sea," there would at last be an accounting, long overdue, of consequences of the city idea in architecture, not only of deaths of the wealthy but of countless workers, most of them with "names unpronounceable to the American tongue; many of them had no kinsmen, no history, no record anywhere" (p. 45). For this accounting Cather provided a discussion among six men offering varying points of view of the city idea: two newspapermen, the hotel's engineer and one of his draftsmen, a lawyer, and a doctor.

In order to understand Cather's accounting, one must recognize that she casts the city itself as her main character, which she develops by a psychology of place. She describes New York as "enveloped in a tragic self-consciousness," for example, its towers of stone and steel "grouped confusedly together," looking "positively lonely." And she speculates that "one might fancy that the city was protesting, was asserting its helplessness, its irresponsibility for its physical conformation, for the direction it had taken" (p. 44). Expressed by an ideology of success for individuals and of progress for nations, the perpendicular idea creates a landscape of compulsive loneliness. Like the individual, the skyscraper is separated from the ground, where a giant hole is excavated for a foundation, like a mass grave. Like members of a community isolated by ambition, in moving upward sky-scrapers leave behind the connecting network of doors opening onto sidewalks, yards, and streets. Indeed, in this story of an idea, Cather spells out the structure's symbolism: "our whole scheme of life and progress and profit was perpendicular. There was nothing for us but height. We were whipped up the ladder" (p. 46).

For this, her most self-conscious experiment in writing the city, Cather's fiction resembles that of Dickens, whose pen lies behind the city novel as we know it. The "tragic self-consciousness" of her New York echoes the corporate consciousness of Dickens's London (see Williams 1973: 153–64); her depiction of characters' mindless climbing echoes his of aimless wander-ings; and, like him, she looks to society for an accounting of nameless individuals sacrificed to it. Yet even in this most explicit and thematically narrow example of her city fiction, Cather parted ways with him and the other writers most closely identified with city fiction. For whereas Dickens, Woolf, Joyce, Lessing and the rest grounded their imaginations and their sympathies in an urban landscape, Cather grounded hers in the nature that lay outside – or beneath – it. Her description of New York ends, tellingly, with the implicit but no less pained accusation that for the sake of the city, nature was reduced to inert matter, then subjected to instruments of torture. Its skyscraper "was an irregular parallelogram pressed between two hemi-spheres," for example, "and, like any other solid squeezed in a vise, it shot upward" (p. 44). And she contrasts the city idea to an ideal of living close to nature, made personal by recalling Caesarino, a young Italian worker. Caesarino passed his childhood in "sun and happy nakedness" on the island of Ischia, then was drawn to New York, where he fell victim to the chief engineer's maxim "that men are cheaper than machinery" and was killed while helping to dig the great hole prepared for the hotel's foundation (pp. 49, 51).

Cather uses Caesarino's story to introduce the puzzle that is the climax of the story. Why do men, with so "few years to live and ... one little chance for happiness," throw everything they have into New York, "to swell its glare, its noise, its luxury, and its power"? Why did Caesarino and the thousands like him "come, like iron dust to the magnet, like moths to the flame" (p. 53)? For her answer as well as her story's conclusion, Cather refers one last time to the idea with which she began, to assert that "we are all the slaves of" this new idea of progress; "it's the whip that cracks over us till we drop" (p. 54).

Cather never again so blatantly and single-mindedly used the city to structure her fiction. She characteristically approached ideas through indi-vidual lives and human relationships, and once she had got hold of her idea of the city in "Behind the Singer Tower," she began exploring it through the personal lives of its inhabitants. In her first novel, *Alexander's Bridge*, Cather wrote of a world-famous engineer who, realizing that success has failed to satisfy him, searches for his original self; through him Cather tells of the modern individual and the city, each the product of the other. By making Alexander an engineer, Cather identifies him with the technology that builds the city; then by layering allusions, she extends Alexander's search for origins beyond his personal life to the heroic idea that had its genesis in the first cities. Naming her character "Alexander," giving him world renown (even calling him "great" 1977: 84) and setting his memory against "the spoils of conquered cities" (p. 134), Cather links her Bartley Alexander to Alexander the Great, whose urge for glory took the form of conquests of the earliest cities from which classical civilization sprang: Thebes (where he consolidated power), Athens (where he was granted deification) and Babylon (where he died). Finally, by taking Alexander through the great cities of the western world – Boston, New York, Paris, Liverpool and London – she links him to a modern idea of progress.[3] Of these, Boston and London figure most importantly.

Cather begins her novel in America, where Alexander's Boston consists of interiors created by wealth and the fine taste of his wife, Winifred. As if by sleight of hand, the Alexanders make the city vanish, so that its appeal ironically lies in its invisibility. They live on a "quiet, deserted" street from which Mrs. Alexander "disappeared" into their house, where architecture erases the city she has left behind (pp. 2, 3). Rooms are protective and insular, their windows opening upon carefully selected vistas of a civilized nature. The library is a "long brown room [that] breathed the peace of a rich and amply guarded quiet" (p. 5), from which "the wide back windows looked out upon the garden and the sunset" (p. 4); Alexander's study is "a large room over the library, ... [that] looked out upon the black river and the row of white lights along the Cambridge Embankment" (p. 9).

Yet like wombs, such settings eventually become claustrophobic and suffocating, and like the slight hardness Bartley recognizes beneath his wife's beauty, nature glimpsed through a window is remote and slightly cold. The evening sky is pale-colored, the river silver-colored, and an elm, its branches stripped bare, reveals "ragged last year's birds' nests" (p. 4). Inside, too, an impersonal stillness runs beneath the comfortable civility: "The room was not at all what one might expect of an engineer's study," for it offers not a sense of engagement with the world but "the harmony of beautiful things that have lived long together without obtrusions of ugliness or change" (p. 9). The cumulative effect is that Boston, like Bartley's study, comes to seem an unnatural world where, as Cather was to write elsewhere, people live "like the fish in an aquarium, who were probably much more comfortable than they ever were in the sea" but who were cut off from life (1932: 31).

Restlessness propels Alexander to London, where by renewing an affair with the actress Hilda Burgoyne, he finds a Circe-like companion for a journey into the past. In contrast to Boston's interiors, the London scenes are set primarily outside, so that the city's geography is the central fact. With map-like precision Cather traces Alexander's saunterings past the monuments that have become landmarks of history and literature: down the Embankment toward Westminster Abbey, with its view of the Houses of Parliament, Somerset House, and Whitehall; up to the Temple, stopping in the Middle Temple gardens (pp. 35–6); down the Strand past Charing Cross Station, to reach Trafalgar Square, "blazing in the sun, with its fountains playing and its column reaching up into the bright air" (p. 89). For Alexander, walking the city's streets means evoking memories of his youth and, deeper still, echoes of a communal history, and his initial feeling is of comfortable familiarity. Similarly we recognize the city we have inherited with our communal past, for Cather has drawn a psychic map of our collective consciousness – Westminster Abbey a shorthand for religion, the Houses of Parliament for government, the Temple for law and Trafalgar Square for military victory.

The comfort of familiarity is deceptive, however, for Cather's London proves to be a labyrinth leading to a "sullen gray mass" at its heart, the British Museum, which holds Lord Elgin's marbles testifying to "the lastingness of some things" and the Egyptian mummy room, testifying to "the awful brevity of others." By selecting this "ultimate repository of mortality" as its central symbol, Cather radically transforms her characters' (and her readers') experiences in London. Following his memory of the mummy room, for example, Alexander seems less like a conquering hero than like Shelley's Ozymandias gazing upon monuments that will crumble; hereafter, reminders of mortality will cut short his feelings of renewed life. An intimate dinner in London builds to the remark, "Isn't London a tomb on Sunday night" (p. 58); an evening in Boston gives way to a moment of unnatural stillness; and Bartley's rededication to life ends in his death, following which his wife watches over their tomb-like Boston house, his lover continues to visit the British Museum, and his friend resolves he'll remain in London, "even if you have at last to put *me* in the mummy-room with the others" (p. 134).

This contrast between industrial power and human mortality structures not only specific scenes but the London section as a whole. Bartley's and Hilda's widening excursions provide widening perspectives of the city, culminating in one glorious day when they join all London abroad, "a stream of rapidly moving carriages, from which flashed furs and flowers and bright winter costumes.... The parks were full of children and nursemaids and joyful dogs that leaped and yelped and scratched up the brown earth with their paws" (p. 91). Reminders of nature anticipate gender distinctions Cather was to develop in later fiction, and the excursion announces a shift in the pattern of observation in this novel specifically: whereas while walking alone Alexander describes landmarks of culture (as is characteristic of a male way of classifying, Sizmore 1989: 11), with Hilda he observes ways in which districts connect (as is characteristic of female perception, Sizmore 1989: 11–12). Returning means passing through "miles of outlying streets and little gloomy houses," until they reach London, "red and roaring and murky, with a thick dampness coming up from the river, that betokened fog again to-morrow," and finally at London's center a corporate life in which heartbeats keep pace with the machinery driving it:

> There was a blurred rhythm in all the dull city noises – in the clatter of the cab horses and the rumbling of the buses, in the street calls, and in the undulating tramp, tramp of the crowd. It was like the deep vibration of some vast underground machinery, and like the muffled pulsations of millions of human hearts.
>
> (pp. 93–4)

Cather ends the lovers' excursion with her most developed reminder of mortality. "London always makes me want to live more than any other city in the world," Bartley whispers to Hilda; "You remember our priestess mummy over in the mummy-room, and how we used to long to go and bring her out on nights like this?" (p. 94). The scene culminates in the question implicit in the labyrinth analogy: "I was just wondering how people can ever die. Why did you remind me of the mummy?" Hilda responds; "Do you really believe that all those people rushing about down there, going to good dinners and clubs and theatres, will be dead some day, and not care about anything? I don't believe it, and I know I shan't die, ever!" (p. 95).

By questioning "how people can ever die," Hilda poses the psychological horror of the city landscape, where an illusion of corporate power denies the reality of individual mortality. Given this reality, the meaningful question is not whether people will die, but how they may come to terms with death's inevitability. Cather's answer distinguishes her city novel. The immortal soul yearns for assurance of continuity that comes only with nature, with its everlasting cycles. It is a yearning that Cather suggests – but does not develop – in *Alexander's Bridge*, when while riding on a train "tearing" through space to the bridge that will collapse, sending him to his death, Alexander glimpses boys camping beside a fire and wistfully recalls his childhood, when he camped on a sandbar in a western river (pp. 118, 116).

Alexander's Bridge introduced the close identification of the city with male experience that Cather was to maintain throughout her career. Her next novel, *O Pioneers!*, identified the country with female experience. Cather's choice of a name for her central character announces the shift, from Alexander's echoes of Alexander the Great, military conqueror of the city, to Alexandra's echoes of Alexandria, the city from which classical pastoral sprang. As if a new-world shepherd, Alexandra assumes responsibility for her immigrant family upon the death of her father; remaining on the land and living by an ethos based on harmony with nature, she eventually creates a great farm that resembles a communal village, its parts joined in a web-like design. A big white house on a hill resides at its center, sheds and outbuildings are grouped about it, and about them lies "the beauty and fruitfulness of the outlying fields" (1962: 83).

Alexandra's story unfolds against the backdrop of men who are tied to modern cities in one way or another: her father, displaced from Stockholm to wear himself out in fruitless struggle against a land he does not understand; her brothers who, though remaining in Nebraska, comically try to imitate city ways in fashion and thought; and most important, Carl Linstrum, son of the Bergsons' neighbors who, failing to make a living farming, returns to Chicago, whence he goes on to live in New York. When Carl returns as an adult, he describes himself a failure, for he "couldn't buy even one of [Alexandra's] cornfields. I've enjoyed a great many things, but I've got nothing to show for it all" (p. 122). Ironically, his failure qualifies him to return, for instead of becoming "a trim, self-satisfied city man," "there was still something homely and wayward and definitely personal about him" (p. 115). The personal community of the country versus the anonymous individuality of the city – the distinction so basic to city fiction appears in Cather, yet not without qualification. In describing the city to Alexandra, Carl begins a dialog that was to continue through Cather's city fiction:

> Freedom so often means that one isn't needed anywhere. Here you are an individual, you have a background of your own, you would be missed. But off there in the cities there are thousands of rolling stones like me. We are all alike: we have no ties, we know nobody, we own nothing. When one of us dies, they scarcely know where to bury him. Our landlady and the delicatessen man are our mourners, and we leave nothing behind us but a frock-coat and a fiddle, or an easel, or a typewriter, or whatever tool we got our living by. All we have ever managed to do is to pay our rent, the exorbitant rent that one has to pay for a few square feet of space near the heart of things. We have no house, no place, no people of our own. We live in the streets, in the parks, in the theatres. We sit in restaurants and concert halls and look about at the hundreds of our own kind and shudder.
>
> (pp. 122–3)

Yet the knowledge of a world beyond the cornfields is necessary, Alexandra replies, to reconcile one to life in the country, where things seem everlastingly the same and people "grow hard and heavy" (p. 124). Separately, the characters illustrate the gender conventions Cather was to explore through-out her writing, Alexandra linked to the land, Carl to the city; together they anticipate a marriage in which they will live on the land she has settled and from it visit the city-world he has explored. They illustrate also conventions that Cather acknowledged with her narrative strategy – a male narrator who, following a modern script for success, leaves the country for the city, and a female subject who, following a script of domesticity, remains in the country.

Male/city versus female/country – Cather incorporated the familiar gender distinction in subsequent novels. In *My Ántonia* Jim Burden moves to New York while Ántonia Shimerda remains on the land; and in *A Lost Lady* Niel Herbert departs for Boston, leaving Marian Forrester in rural Nebraska. But with each of these novels, women contradict expectations of them. Only in *O Pioneers!* does a man return to find a woman as he remembered her. In *My Ántonia* women defy Jim's script for them, Lena Lingard by moving to cities, where she becomes a successful businesswoman, Ántonia by remaining in Nebraska but defying conventions of pastoral innocence by becoming pregnant while unmarried and refuting the illusion of country ease by having a hard life. And in *A Lost Lady* Marian Forrester, defying Niel's expectation that she preside over her husband's memory in rural Nebraska, takes a lover, then escapes to Buenos Aires.

Thus Cather both adopts and alters the pattern of the classic city novel, "a drama of emotional education whose shape is a spiral journey from country to city and back again". She incorporates the spiral journey as what Raymond Williams has called "the ideology of improvement" (1973: 61). But whereas the classic city novel grants full development to a male protagonist who "learns of life in the city only to live it in the country," Cather divides experience by gender: the man who returns acknowledges the city's shortcomings; the woman who remained articulates the country's limitations (Howe 1971: 62). In doing so Cather writes of country and city as two opposing movements, like the diastolic/systolic rhythm of a heartbeat.

Cather gave only one character the full rhythm of this heartbeat. Twenty years after *Alexander's Bridge* Cather in "Neighbour Rosicky" wrote about a man who in his youth had lived in the great cities of the world, then in middle age settled upon a farm that provided enough to get by, but never to get ahead. The tension between city versus country is made dramatic by Rosicky's concern over his son Rudolph, who had recently married a town girl and is tempted to move to Omaha. Trying "to find what he wanted for his boys," Rosicky examines his life by remembering the cities he had lived in; in doing so, he provides Cather's most mature exploration of a male, working-class experience of the city.

Rosicky first allows his mind to linger over his young manhood in New York, where upon immigrating to the United States he worked in a tailor shop. Living generously – enjoying a good dinner, loaning money to friends, going to the Opera – meant somehow never saving anything; still,

> it was a fine life; for the first five years or so it satisfied him completely. He was never hungry or cold or dirty, and everything amused him: a fire, a dog fight, a parade, a storm, a ferry ride. He thought New York the finest, richest, friendliest city in the world.

(1932: 28)

As in *Alexander's Bridge*, however, a vault-like emptiness lies at the heart of the city. Rosicky's initial satisfaction gives way to a vague uneasiness, until sitting in Park Place in the sun on the fourth of July, he

> found out what was the matter with him. ... The lower part of New York was empty. Wall Street, Liberty Street, Broadway, all empty. So much stone and asphalt with nothing going on, so many empty windows, The emptiness was intense, like the stillness in a great factory when the machinery stops and the belts and bands cease running. It was too great a change, it took all the strength out of one. Those blank buildings, without the stream of life pouring through them, were like empty jails. It struck young Rosicky that this was the trouble with big cities; they built you in from the earth itself, cemented you away from any contact with the ground.
>
> (pp. 30–1)

Eventually this feeling of emptiness led him to reestablish ties with the earth, first hiring out as a farm hand in Nebraska, then – by good fortune he still felt grateful for – purchasing some land of his own.

Were this all to "Neighbour Rosicky," the story might be little more than a celebration of a pastoral retreat, charming and slight. Yet as if in a Chinese box, embedded at the core of Rosicky's recollections lies a city experience for which Cather drew on the conditions that so shocked her during her first trip abroad. In response to Polly's question, "You always lived in the city when you were young, didn't you? ... Don't you ever get lonesome out here?" (p. 37), Rosicky reflects that "big cities is all right fur de rich, but dey is terrible hard fur de poor," then later describes living in London, a time of such wretched poverty that it "had left a sore spot in his mind that wouldn't bear touching" (p. 27): sleeping with eight others in three rooms; carrying water from a pump in a court, four flights down; suffering from dirt, bugs in the place, fleas and constant hunger, while shop windows are "full of good t'ings to eat, an' all de pushcarts in de streets is full, an' you smell 'em all de time, an' you ain't got no money, – not a damn bit" (p. 52). With the brief, powerfully understated memory Cather creates an experience of economic, physical, and psychological entombment – a London which for the poor meant living in corners of rooms hidden in dark recesses of shops, and everywhere "fog gits into your bones and makes you all damp like" (p. 51). Rosicky ends the recollection with a confession: his hunger was so great that he ate half the goose cooked for the family's Christmas dinner; his shame over doing so was so deep that he begged for money to replace it.

The memory of London prepares for Rosicky's most gentle of revelations, as he comes to understand "why it was he so hungered to feel sure [his boys] would be here, working this very land, after he was gone" (p. 58). Kindness

157

and honesty – the most fundamental of values – are at risk in a city's nightmarish landscape of predation. Though on the farm his children would face hardships, he would not have to fear any great poverty or unkindness for them.

> In the country, if you had a mean neighbour, you could keep off his land and make him keep off yours. But in the city, all the foulness and misery and brutality of your neighbours was part of your life. The worst things he had come upon his journey through the world were human, – depraved and poisonous specimens of man. To this day he could recall certain terrible faces in the London streets. There were mean people everywhere, to be sure, even in their own country town here. But they weren't tempered, hardened, sharpened, like the treacherous people in cities who live by grinding or cheating or poisoning their fellow-men....
>
> It seemed to Rosicky that for good, honest boys like his, the worst they could do on the farm was better than the best they would be likely to do in the city.

(pp. 59–60)

Together, *Alexander's Bridge* and "Neighbour Rosicky" present male experience of the city as an economic machine based upon consumption, driven by its own desire and fueled by the energy of ambition and competition. There is no such thing as stasis in the city; when its machinery stops, a hollow emptiness remains. Saying that he has to make the usual effort to succeed, Carl Linstrum speaks for Cather's men more generally, who migrate to cites as a mandatory rite of passage and who return not to demonstrate their success but to pay tribute to the fundamental values of community identified with women.

Still, a return is possible for these men: Carl comes back to marry Alexandra, once he knows she needs him; Jim Burden returns to visit Ántonia and plans to do so again; and by settling in Nebraska Rosicky reestablishes ties with the earth. It is different for Cather's women, for whom a move to the city is irrevocable. Speaking for Cather's women generally, Thea Kronborg says upon leaving her small Nebraska town for Chicago that if she fails, her friends had better forget about her for she will become the worst woman who ever lived (1978: 244). Cather's successful women may return briefly to their country roots, but they do so only to encounter the hostility of provincial suspicion and to feel the suffocation of narrow horizons. Those women who fail or (more often) whose husbands' fortunes fall characteristically break ties with the families and friends they left behind. Even Ántonia, returning from Denver pregnant and unmarried, is not so much readmitted to the rural community as granted the opportunity to establish an alternate community in her own large family – and that only after a prolonged period of atonement.

Yet Cather also grants to her women fullest access to the creative life that the city guards so jealously. Like business success, artistic achievement requires knowledge of the world – for women (and all Cather's functioning artists are women), moving into urban settings that are constructed by men. *The Song of the Lark* and *Lucy Gayheart*, complementary versions of Cather's female city novel, illustrate the experience of an aspiring woman in an urban setting that embodies the male tradition.

The Song of the Lark is Cather's fullest and most positive version of the female city novel. Thea Kronborg passes her childhood in rural Moonstone (so closely based on Red Cloud that a map based upon one provides a guide to the other), until at 17 she moves to Chicago to study first piano, then voice. Chicago is a city where art and manufacture, ideas and business, are made by men; appropriately, Thea's initiation to it is by men: at his death the brakeman Ray Kennedy leaves to her money to study in Chicago; the Moonstone doctor, Dr. Archie, escorts her there and finds lodging for her; her teachers, the generous piano teacher Harsanyi and the cynical voice teacher Madison Bowers, introduce her to the arts; and her mentor-lover, Fred Ottenburg, the wealthy scion of a family of brewers, introduces her to culture.

Lessons Thea learns from others are secondary to her awakening city consciousness, however, and that she learns on her own. For in going to the city, Thea has entered as foreign a place as the empty plains were to Cather's immigrants. Cather structures the Chicago section of her novel by Thea's developing responses to the city, from ignorance to desire, ambition and competition fueled by anger.

At first "Chicago was simply a wilderness through which one had to find one's way," for Thea had as yet "no city consciousness" and

> felt no interest in the general briskness and zest of the crowds. The crash and scramble of that big, rich, appetent Western city she did not take in at all, except to notice that the noise of the drays and street-cars tired her. The brilliant window displays, the splendid furs and stuffs, the gorgeous flower-shops, the gay candy-shops, she scarcely noticed.
>
> (1978: 193)

During those first months Thea only glimpses the city from within enclosures that symbolize her blindness – a street car from which the city seems a blur, a boarding room window looking onto a brick wall, a practice studio into which people come and go.

Awakening begins with awareness of imprisoning ignorance, which Thea feels as "poverty in the richness of the world ... opened to her" (p. 178); it proceeds when Thea, visiting the Art Institute, realizes that a gallery offers "a place of retreat" (p. 196), as nature did in the country. And it climaxes in a two-part revelation of the personal ecstasy of art versus the impersonal hostility of the world. Inside a concert hall, responding to the "reaching and

... yearning" of Dvořák's "From the New World," Thea knew "that she wanted ... exactly that" and felt "the amazement of a new soul in a new world" (p. 199); emerging from the hall, "for almost the first time Thea was conscious of the city itself," newly aware

> of the congestion of life all about her, of the brutality and power of those streams that flowed in the streets, threatening to drive one under. People jostled her, ran into her, poked her aside with their elbows, uttering angry exclamations. She got on the wrong car and was roughly ejected by the conductor at a windy corner, in front of a saloon. She stood there dazed and shivering. The cars passed, screaming as they rounded curves, but either they were full to the doors, or were bound for places where she did not want to go.
>
> (p. 200)

A young man approaches her for sex and an old one whispers something crude, cruelly driving from her the memory of the symphony she has just heard. In that moment, Thea realizes competitive ambition, then assumes an attitude of battle.

> There was some power abroad in the world bent upon taking away from her that feeling with which she had come out of the concert hall. Everything seemed to sweep down on her to tear it out from under her cape. If one had that, the world became one's enemy; people, buildings, wagons, cars, rushed at one to crush it under, to make one let go of it. Thea glared round her at the crowds, the ugly, sprawling streets, the long lines of lights.... All these things and people were no longer remote and negligible; they had to be met.
>
> (p. 201)

Meeting her adversaries means competing in the market-place, and for that Cather provides to Thea instruction by Madison Bowers, a cynical voice teacher. By moving voice lessons offstage in the novel, Cather focuses upon Bowers's temptation of Thea with avarice in art, so that she might prey upon those willing to pay. He is interested in her not because of her voice but because of "her manifest carefulness about money," and his lessons concern not music but business. The "practice" he assigns to her is "the art of making yourself agreeable," and the advice he gives her is crassly commercial:

> When you come to marketing your wares in the world, a little smoothness goes farther than a great deal of talent sometimes. If you happen to be cursed with a real talent, then you've got to be very smooth, indeed, or you'll never get your money back.
>
> (p. 252)

Here Thea's country experience saves her, for she remembers Ray Kennedy's saying "there was money in every profession that you couldn't take" and replies "that's the money I'll have to do without ... the money you have to grin for" (pp. 252-3).

The point, however, is that Cather gives to Thea both the character and the education necessary to make her way in a city structured by a market mentality. From childhood she has a drive toward possession – whether of jewelry ("she wanted every shining stone she saw, ... imagined that they were of enormous value, [and] was always planning how she would have them set" (p. 14)) or of the lesson she was striving "to get." Living in the city means learning to negotiate and discriminate. Seeing tiaras and jewels in a shop window, she determines that these are "things worth coveting," yet understands that buying on credit means leasing one's life to things. When she purchases her first concert dress by placing herself in the hands of a Chicago dressmaker, she candidly admits that the result was "a horror," yet resists going into debt to replace it: "her money was gone, and there was nothing to do but make the best of the dress" (p. 173).

While resisting the seduction of things, Thea is brutally aware of the power that money offers. "To do any of the things one wants to do, one has to have lots and lots of money," she says after her first year in Chicago; money is necessary "to get out" of life what she wants (p. 242). And when financially successful as the artist Kronborg, Thea remains "close about money ... careful"; she retains a standard of the personal, valuing money by its human cost:

> I began the world on six hundred dollars, and it was the price of a man's life. Ray Kennedy had worked hard and been sober and denied himself, and when he died he had six hundred dollars to show for it. I always measure things by that six hundred dollars, just as I measure high buildings by the Moonstone standpipe.
>
> (p. 457)

The question of measurement extends beyond money to a more fundamental issue of aspiration, the "impelling desire" that "has always been the great energizing theme of the American city novel" (Gelfant 1984: 283). Such desire lies behind the staple of the city novel, with its alienated individual versus undifferentiated masses; Cather's variation upon it was to pit her alienated individual against the second rate. Thea suffers for her ideals, and her awakening to Chicago culminates in the discontent and anger necessary to large aspiration. Walking beside Lake Michigan, Thea "believed that what she felt was despair," but Cather clarifies the emotion as "only one of the forms of hope." Leaving Thea's limited vision, Cather offers the wisdom of a writer's omniscience in the world she has created. The result is one of Cather's most condensed and richly poetic statements of a city philosophy for the aspiring artist:

> The rich, noisy, city, fat with food and drink, is a spent thing; its chief

concern is its digestion and its little game of hide-and-seek with the undertaker. Money and office and success are the consolations of impotence. Fortune turns kind to such solid people and lets them suck their bone in peace. She flecks her whip upon flesh that is more alive, upon that stream of hungry boys and girls who tramp the streets of every city, recognizable by their pride and discontent, who are the Future, and who possess the treasure of creative power.

(pp. 264–5)

While Chicago provided necessary lessons of competition, for her most important awakening Cather moved Thea to another, contrasting city – the long deserted cluster of cliff-dwellings in northern Arizona's Panther Canyon. Whereas Chicago was modern and male, the cliff-dwellers's city is ancient and female. Rather than mastering nature, it resides within the earth, within what Ellen Moers has called "the most thoroughly elaborated female landscape in literature" (1977: 258). Thea's awakening to it means moving from exterior to interior, from analysis to intuition, and from judgment to sympathy.

As Thea descends into Panther Canyon, Cather uses precise, analytic detail to describe her perception of the city nestled in its walls: setting (in "a stratum of rock ... hollowed out by the action of time"); building materials ("yellowish stone and mortar"), placement ("the houses stood along in a row, like the buildings in a city block, or like a barracks") and overall design ("the dead city had ... two streets ... facing each other across the ravine") (pp. 297–8). Preparing for the change that will follow by choosing a room and lining it with Navajo blankets, Thea moves into a domestic space, "from her doorstep" describing the city by the "winding path that had been the street of the Ancient People," then describing the interior of her lodge with its flakes of carbon on the rock roof from "the cooking-smoke of the Ancient People" (pp. 298, 302).

In Cliff City, analysis gives way as, climbing the water-trails, Thea "began to have intuitions about the women who had worn the path" and tries "to walk as they must have walked," until she "could feel the weight of an Indian baby hanging to her back as she climbed" (p. 302). Similarly, attitudes of possession and competition fall away, to be replaced by ones of community. Thea feels herself "a guest in these houses," obligated to behave as such: she "visits" the most interesting pottery fragments in the chambers where she found them, for example, rather than removing them to her own lodge. Whereas in Chicago Thea had awakened to city consciousness on the streets, in Panther Canyon she awakens to an alternate city consciousness inside the domestic spaces that she has entered on the most personal and private terms. For in one of them she finds that the women before her have "written" their own idea of a city, leaving behind

half a bowl with a broad band of white cliff-houses painted on a black ground ... just as they stood in the rock before her. It brought her centuries nearer to these people to find that they saw their houses exactly as she saw them.

(p. 305)

In Chicago Thea's awakening of city consciousness had meant that she became fiercely competitive; in Panther Canyon it means that she commits herself to a community that stretches over time. When she realizes that "these potsherds were like fetters that bound one to a long chain of human endeavour," Thea is liberated by "older and higher obligations" than those she left behind in Moonstone and Chicago (pp. 306, 308). By bathing in the stream at the bottom of the canyon, "the only living thing left of the drama that had been played out in the canon centuries ago," she baptizes herself in "a continuity of life that reached back into the old time" (p. 304). In contrast to the metaphors of the marketplace that ran through the Chicago section, ones of reciprocity run through the Cliff City one. Beautifully decorated potsherds "made her heart go out to those ancient potters," as it goes out also to people in her own life: she affirms that Ray Kennedy was "right" about the cliff-dwellers, is "grateful" to Fred Ottenburg for introducing her to them and feels "one ought to do one's best, and help to fulfill some desire of the dust that slept here" (p. 306).

Finally, Cather places Thea in New York (the city she most closely identified with success) to describe what she has gained and to reveal at what cost. By her final section's title, "Kronborg," Cather announces Thea's transformation into a world-renowned diva who has put her personal life into her art. Like the hotel rooms she inhabits, Thea remains a little cold and empty except in performance; only on stage and through art is she fully alive and able to give generously of herself. After a performance, exhausted and perhaps "wondering what was the good of it all," "the only commensurate answer" was the smile of one who, having heard her perform, experienced the ecstasy of her art. The contrast conveys Cather's idea of the city at its best, where the paintings within the galleries and the symphonies within the concert halls are secrets protected by impersonality, as the precious jewel is protected by an unornamented box.

In an essay on Chicago novels written by women, Sidney H. Bremer has noted that *The Song of the Lark* "does not fit the overall pattern of urban imagery in women's novels," for as an isolated newcomer to Chicago rather than one who lives there by "continuities and communal concerns" Thea Kronborg "is anomalous" (1981: 35, 38). Thea is the exception also in Cather's œuvre, where no other character learns so deeply the lessons of the city, modern and ancient, and no other survives so well. As if feeling that her story of a female experience of the city were incomplete, two decades later Cather wrote *Lucy Gayheart*, her final and darkest novel of the modern city. Like Thea, Lucy left the small Nebraska town of her childhood for Chicago, where she also studies music. But unlike Thea, Lucy is "too kind" to survive there. Whereas Thea's was a story of doing battle against the modern city, Lucy's is of tragically falling in love with it.

Chicago provides to Lucy the freedom of breaking out of gender conventions, for its anonymity offers independence, and unlike Thea, Lucy is from the first on her own. Taking rooms above a German bakery, she dismisses the food as "good enough," for what appeals to her is that "everyone had his own little table," where s/he could attend to his own business or read his paper. "For the first time in her life she could come and go like a boy; no one fussing about, no one hovering over her" (1976: 26). As for her room upstairs, she treasures it because it contains "her own things and her own will" (p. 27).

The freedom to follow her own will proves illusory, however. Unlike Thea, who was fiercely committed to protecting her best self, Lucy is fatally pliable. She is content to lie against a man's shoulder and to play as another's accompanist (and then only in practice); she gets "a kind of comfort out of the crowded streets (p. 62), and enjoys feeling herself "a twig ... [or] a leaf swept along on the current" of life in the city (p. 75). Whereas Thea confronted Chicago's harsh reality, Lucy insulates herself against it, embracing the fog that hides the city and looking forward to entering the elevator that lifts her into the rarified privacy of a studio. Most fatally, Lucy Gayheart loves a "city of feeling" that obscures the city of fact.

> Lucy carried in her mind a very individual map of Chicago: a blur of smoke and wind and noise, with flashes of blue water, and certain clear outlines rising from the confusion; a high building on Michigan Avenue where Sebastian had his studio – the stretch of park where he sometimes walked in the afternoon – the Cathedral door out of which she had seen him come one morning – the concert hall where she first heard him sing. This city of feeling rose out of the city of fact like a definite composition, – *beautiful because the rest was blotted out.*
>
> (p. 24; my emphasis)

Again, metaphors of the market-place describe urban relationships, but whereas Thea Kronborg was Cather's ideal consumer, Lucy Gayheart is tragically indiscriminate. She offers her own life in terms of monetary exchange (she loves the city because it provides to her freedom "to spend" her youth as she pleases, p. 86); and embraces the illusion of possession that the market-place invites. When she sees "lovely things in the shop windows," she is content with feeling that they "seemed to belong to her.... Not to have wrapped up and sent home, certainly; where would she put them? But they were hers to live among" (p. 86). Whereas Thea was careful about money, valuing it by a human standard learned at home, Lucy is careless, neither aware of what she spends nor sensitive to those who have helped her. She is oblivious of her family's economies so she could study in Chicago and is indifferent to their increasing debt: though she thinks of sending money home when she begins earning something, she buys clothes instead (pp. 172, 191–2). When Lucy finally realizes that she must have an income if she is to continue living in the city, she is capable only of resolving vaguely that "there must be ways of making money in the world; she had never seriously tried, but now she would" (p. 190).

Lucy is, in other words, keenly vulnerable to the predatory nature of the city, a vulnerability Cather personifies in a love-affair with a middle-aged singer, the attractive but psychologically dangerous Clement Sebastian. "In male discourse the city is a female body," writes Jane Marcus, while in women's fantasy novels the city is a hostile male body" (1984: 138) – the generalization is strikingly apt for Cather's novel, where Clement Sebastian's promise of living more intensely, like the city's, obscures the reality that he, like it, will dominate, even consume, Lucy. For while purporting to encourage her independence, Sebastian (like the city) values her dependency, so that although Lucy believes that in leaving the country she achieves the freedom of a boy, she actually encounters the old gender limitations in another form. As a comparison of her rural and city lovers' expectations of Lucy reveals, beneath the obvious differences of their courtships lies their common hostility against her independence. Whereas Lucy's home-town suitor values her sexual purity, discarding her when he believes (mistakenly) that she has gone "all the way" with another man (p. 111), Sebastian values her because of her purity of motive. Though he knows she is "a young thing with her living to make," he loves her because there is "never the shadow of a claim" in her and because she has "a spirit which disdained advantage" (pp. 80–1). Both men love Lucy for the "innocence" that is so conventionally desirable in a woman, yet that makes her so vulnerable.

Again Cather associates the city with emptiness, but for Lucy it is the emptiness of a woman dependent upon a man. In "the crowded hour in the crowded part of the city," she learns from his valet that Sebastian is leaving Chicago: "She and Guiseppe could scarcely hear each other speak for the clatter of truck wheels on the dirty pavements. Troops of screaming children on roller skates came streaking down the sidewalk, but Lucy hardly noticed them." There is no scene in Cather's writing that makes so dramatic the plight of a woman whose experience of the city has been mediated by men, and who realizes her helplessness against the indifference of its power. Lucy tries to cling to Guiseppe, but "everything around them was blank. An enormous emptiness had opened on all sides of her" (p. 116).

When in *The Song of the Lark* Cather had asked what it meant for a woman to live the life of an artist, she had answered that success provided to Thea Kronborg the insulation of money, the attention of a husband, and the support of a hotel staff – but all at the cost of her personal life. In *Lucy Gayheart* she repeats the question and intensifies the doubt. In asking her about the home-town banker who wished to marry her, Lucy's teacher

165

Auerbach poses the dilemma for the woman striving to leave the country and to make her way in the city:

> "In the musical profession there are many disappointments. A nice house and garden in a little town, with money enough not to worry, a family – that's the best life."
>
> "You think so because you live in a city. Family life in a little town is pretty deadly. It's being planted in the earth, like one of your carrots there. I'd rather be pulled up and thrown away."
>
> Auerbach shook his head. "No, you wouldn't. I've heard young people talk like that before. You will learn that to live is the first thing."
>
> Lucy asked him if there were not more than one way of living.
>
> "Not for a girl like you, Lucy; you are too kind. Even for women with great talent and great ambition – I don't know. Some have good success, but I don't envy them."
>
> (p. 134)

Again, however, there is no going back for the woman in Cather's fiction. After Sebastian's death Lucy returns to her home-town to rest, but she feels trapped within the flimsy walls of her father's house and by the narrow conventions of her rural community. Chicago might save her, she believes; but before she can return to the city, she flees into the country, where she confronts a nature transformed by winter and emotion into an alien landscape. Lucy's "foot kept catching in the walls of the ruts" of the roads, which "had been rutted during the thaw; "mud had frozen in jagged ridges, rough and sharp like mushroom coral"; the light was cold, the country "very dreary," the sun making "a mere glassy white spot in the low grey sky," the plum bushes stripped bare of snow by the wind. Tired, Lucy determines "to beg a ride from anyone who came by" – but rather than finding the generosity conventional of a rural community, she meets her former suitor who refuses her (p. 197). Angry and hurt, she reaches the river bank and begins to skate, seeking to "get away from this frozen country and these frozen people"; not knowing that the river had changed its course, she headed for the center, where she broke through the ice and, groping cautiously for the bottom, "felt herself gripped from underneath" (pp. 198–9).

Lucy Gayheart is far more typical than Thea Kronborg of Cather's women's experiences of the city – dependent upon men generally and their husbands particularly in modern cities defined by male cultures. After Bartley Alexander's death, his wife and mistress seem to stop living too, one presiding over his tomb-like Boston house, the other visiting the mummy room in London's British Museum. In *A Lost Lady* Marian Forrester is cut off from vital life in Denver when her husband's bank collapsed, and in *My Mortal Enemy* Myra Henshawe, after eloping with Oswald to New York, suffers from his economic decline, ending her days in a seedy rooming-house in San Francisco. A similar story is suggested by many of the women we only glimpse, in *The Song of the Lark*, for example, Mrs. Harsanyi, losing her looks early from the strain of maintaining the order her husband needed to work, and in *Lucy Gayheart* Mrs. Auerbach, plaintively remarking that as they grow older, there is little she can do for her husband.

As an alternative to the male/female, city/country division in her account of the modern city, Cather offered another model of cities that are communities living by natural rather than industrial rhythms. She presented the idea first in Alexander Bergson's farm, developed it in Thea Kronborg's awakening to the ancient city of Panther Canyon, and then developed it further in Tom Outland's discovery of another ancient city in *The Professor's House*. Like Bartley Alexander in Cather's first novel, so Professor Godfrey St. Peter is dissatisfied with success in *The Professor's House*; but in the later novel Cather suggests an alternate way of living in the form of a memory of a former student, who while herding cattle in the South West, came upon "a little city of stone," nestled within a large cavern on the face of a cliff. The idea of the sleeping city is the harmony of houses with one another and with the land into which they are built:

> It all hung together, seemed to have a kind of composition: pale little houses of stone nestling close to one another, perched on top of each other, with flat roofs, narrow windows, straight walls, and in the middle of the group, a round tower.... I knew at once that I had come upon the city of some extinct civilization, hidden away in this inaccessible mesa for centuries, preserved in the dry air and almost perpetual sunlight like a fly in amber, guarded by the cliffs and the river and the desert.
>
> (1973: 201–2)

Cather offers an alternative city to Thea and Tom alike, then suggests gender differences in developing their responses to it. Whereas Thea considers herself a guest, then makes herself a receptacle for the desire she inherited from the ancient people, Tom behaves as a modern archaeologist, excavating Cliff City, removing its artifacts and cataloging them, and finally going to Washington, DC, to enlist "Uncle Sam" in sending an expert to "dig out all its secrets" (p. 224), then "revive this civilization in a scholarly work" (p. 222).

Tom's encounter with Washington is a study in disillusionment: newly off the train, he stands silently watching the Capitol's white dome "with a very religious feeling," then begins an unhappy initiation into bureaucracy (p. 225). Again, Cather creates a city as a labyrinth, but this time the maze is composed of the procedures and protocols, shifts and stratagems, of departmental life. The Indian Commission, the War Department, the Smithsonian Institution, the Treasury Building – all filled with people who, like the couple from whom Tom rented a room, were "in office" and struggling to keep up appearances. Walking about the city streets each day, waiting to be granted an appointment with an official, Tom passes his days "until the time when all the clerks streamed out of the Treasury building and the War and Navy. Thousands of them, all more or less like the couple I lived with. They seemed to me like people in slavery" (pp. 233–4). Enslavement to the idea of progress embodied in the size of a city, thousands of people attracted as iron dust to a magnet – it is the idea of a modern city that Cather had spelled out in "The Singer Tower." In *The Professor's House* she continued her "accounting" of the modern idea in architecture, this time indicting it of betrayal.

Cather's ideal cities are, it would seem, dead cities – Panther Canyon and Cliff City, preserved from ancient time as mute testimony that alternatives exist to modern ideas in architecture. The most notable exception appears in *Shadows on the Rock*, where Cather's search for an alternative city intensifies. In telling the story of Quebec in 1697, Cather extends the Native American idea of the city to one of a European culture transplanted to North America. Like the ancient cliff cities, Quebec's communal design results from following its setting: "a mountain rock, cunningly built over with churches, convents, fortifications, gardens, following the natural irregularities of the headland on which they stood" (1977: 5). And as the cliff cities' isolation protected their people against intruders, so Quebec's isolation enabled its colonists to safeguard tradition against change indiscriminately hailed as progress. And, to my mind most interesting, as she had moved Thea Kronborg inside domestic space to describe her awakening to a city idea of community, so Cather placed domestic space – Cecile's kitchen – at the center of her version of a communal Quebec.

From the New York of "Behind the Singer Tower" to the Quebec of *Shadows on the Rock* – the scope has startled me. Though for some years now I have focussed my research upon Cather, reading her recently as a city novelist has meant discovering texts that have become new to me by a writer I had scarcely met. For I began by expecting that the Chicago section of *The Song of the Lark* would provide a paradigm for Cather's city idea as a "rite of passage" (the idea of my original title); I soon recognized how naïve that expectation was. Cather's idea of the city is larger, more experimental and more provocative than anything I anticipated. It includes Chicago, of course, but the Chicago of Lucy Gayheart as well as that of Thea Kronborg, and juxtaposed to Thea's awakening to a "city consciousness" in Chicago is her awakening to an alternate and female city consciousness in Panther Canyon. Alexander's London is part of Cather's idea of the city, a London that embodies the idea of the modern heroic that, in turn, is rooted in the earliest cities. An idea of male possession continues through Tom Outland's attitude toward Cliff City, and, ironically, through Washington, DC's, reception of

Tom. Concluding my paper, I revised its title to the inclusive and, I now believe, more accurate "Willa Cather as a City Novelist."

NOTES

1   I am grateful for comments and suggestions on this paper by members of the 1991 Cather colloquium at the University of Nebraska-Lincoln, and among those groups wish to thank Polly Duryea, Evelyn Funda and Kari Ronning for particular suggestions that helped shape the final draft.

2   Marilyn Arnold interprets "Behind the Singer Tower" in terms of Cather's "finding a fictional stance," then interprets her subsequent short stories as playing upon the theme of "the city in landscape" (1984: chapter title). In discussing the city in Cather's novels, I envision my chapter as complementary to Arnold's treatment of Cather's short stories.

3   On Alexander the Great see Lehan 1987; for Bartley Alexander as Cather's comment on the modern hero see Ammons 1986 and Graf 1991.

REFERENCES

Ammons, Elizabeth (1986), "The engineer as a cultural hero and Willa Cather's first novel, *Alexander's Bridge*," *Arizona Quarterly*, 38: 746–60.

Arnold, Marilyn (1984), "The city as landscape," in *Willa Cather's Short Fiction*, Athens, Ohio: Ohio University Press, pp. 99–132.

Bremer, Sidney H. (1981), "Lost continuities: Alternative urban visions in Chicago Novels, 1890–1915," *Soundings*, 64: 29–51.

Cather, Willa (1932) "Neighbour Rosicky," in *Obscure Destinies*, New York: Alfred A. Knopf.

——— (1956), *Willa Cather in Europe*, intro. and notes by George N. Kates, New York: Alfred A. Knopf.

——— (1962; orig. 1913), *O Pioneers!*, Boston: Houghton Mifflin.

——— (1965; orig. 1912), "Behind the Singer Tower," in *Willa Cather: Collected Short Fiction 1892–1912*. Lincoln: University of Nebraska Press, pp. 43–54; originally published in *Collier's*, 49 (May), pp. 16–17, 41.

——— (1973; orig. 1925), *The Professor's House*, New York: Random House.

——— (1976; orig. 1935), *Lucy Gayheart*, New York: Random House.

——— 1977; orig. 1912), *Alexander's Bridge*, Lincoln: University of Nebraska Press.

——— (1977; orig. 1931), *Shadows on the Rock*, New York: Random House.

——— (1978; orig. 1915), *The Song of the Lark*, Lincoln: University of Nebraska Press.

Gelfant, Blanche (1984), "Sister to Faust: The city's 'hungry' woman as heroine," in Susan Merrill Squier (ed.), *Women Writers and the City: Essays in Feminist Literary Criticism*, Knoxville: University of Tennessee Press, pp. 265–87.

Graf, Nan (1991), "The Evolution of Willa Cather's Judgment of the Machine and the Machine Age in Her Fiction," dissertation, Lincoln: University of Nebraska-Lincoln.

Howe, Irving (1971), "The city in literature," *Commentary*, 51/5, pp. 61–8.

Lehan, Richard D. (1987), "Cities of the living/cities of the dead: Joyce, Eliot, and the origins of the myth in modernism," in Lawrence B. Gamache and Ian S. MacNiven (eds.), *Modernism: Essays in Honor of Harry T. Moore*, Rutherford: Fairleigh Dickinson University Press.

Marcus, Jane (1984), "A wilderness of one's own: Feminist fantasy novels of the twenties," in Susan Merrill Squier (ed.), *Women Writers of the City*, Knoxville: University of Tennessee Press, pp. 134–60.

Moers, Ellen (1977), *Literary Women: The Great Writers*, Garden City: Doubleday, Anchor Books, 1977; orig. 1976.

Sizmore, Christine Wick (1989), *A Female Vision of the City: London in the Novels of Five British Women*, Knoxville: University of Tennessee Press.

Squier, Susan Merrill (1984), "Introduction," in Susan Merrill Squier (ed.), *Women Writers and the City: Essays in Feminist Literary Criticism*, Knoxville: University of Tennessee Press, pp. 3–10.

Williams, Raymond (1973), *The Country and the City*, New York: Oxford University Press.

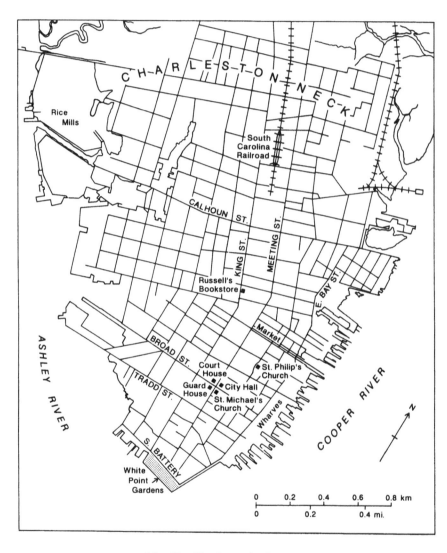

Map 10 Charleston in the 1850s

10

A PLACE OF TOMBS
The Charleston of William Gilmore Simms
John P. Radford

Do not think of Charleston. Whatever your talents,
they will there be poured out like water on the sands.
Charleston! I know it only as a place of tombs.
William Gilmore Simms Sr., 1825[1]

The old man was right. All that I have ever [done]
was poured to waste in Charleston.... And I, too, know
it as a place of tombs.
William Gilmore Simms Jr., 1858[2]

INTRODUCTION

William Gilmore Simms was born in Charleston in 1806, the son of an Irish immigrant father. He maintained an intimate association with the city, and after his death in 1870 his body was interred in Magnolia Cemetery on its outskirts. Ever since the Southern defeat in the Civil War, Simms has remained an obscure figure in American letters, cited chiefly as an eloquent defender of all things Southern, including slavery. He was, indeed, the Old South's major literary champion, but he was also a prolific novelist, a competent poet and an insightful critic, fully conversant with contemporary trends in literature in both Britain and the Northern United States. His fiction withstands comparison with that of all but the best of contemporary New England writers. He founded and edited several magazines, and presided over the Russell's Bookstore Group, by far the most talented community of writers in the antebellum South to identify with the Southern cause and sympathize with what they saw as the region's distinctive *genre de vie*. Apart from occasional travels in the North, he never left Charleston or his rural retreat at Woodlands plantation in its hinterland. Although he married into Charleston society, much has been made of his apparent feelings of rejection at the hands of the city's elite. He died lamenting what was later to be called the Lost Cause of the South, and in despair of his life's mission:

the creation in his time of a Southern literary tradition.

Charleston was founded in 1680 by colonists from England and Barbados, who were soon followed by French Huguenots. After experimenting with a number of forms of agriculture, the colonists began to specialize in the large-scale cultivation of rice and indigo. During the eighteenth century these crops underwrote the development of the fourth largest city in the American colonies as well as the accumulation of some of their greatest per capita wealth. By the turn of the nineteenth century rice and indigo had largely been replaced by cotton. Charleston was the regional focus of plantation agriculture throughout this sequence of staple crops and it acted as the entrepôt through which they were exported. About half the city's population in the antebellum area was African American, mostly slaves. The city's life was bound up in the social relations of planter, merchant and slave, planter interests remaining paramount even throughout the long secular decline in cotton prices that began in 1820. Herein lie the origins both of its resistance to modernization and of the unilateral secession of South Carolina from the Union in 1860 which presaged the Civil War. Although General Sherman did not carry out his threat to raze the "cradle of secession" to the ground following his march through Georgia in 1865, Charleston had already suffered great damage, loss of life and impoverishment. Its steady decline as a major city in the United States continued into the middle of the present century. Members of the white planter-professional elite remained socially dominant inside the city, and much of the built environment was painstakingly restored according to their tastes. The result today is a preserved, anachronistic downtown core (now a tourist haven) near the tip of the peninsula, engulfed by large areas of post-World War II suburban sprawl.

Among the many Southern rejoinders to the publication of *Uncle Tom's Cabin* in 1852 was a book entitled *Uncle Robin's Cabin in Virginia and Tom Without One in Boston*. This title serves to illustrate three important characteristics of the Southern mind in the late antebellum era. First, it was highly defensive. Although Harriet Beecher Stowe's villains were mainly Northerners and her target was not so much slavery as the patriarchal society of America in general,[3] the book was seen by white, literate Southerners as an attack on their way of life. Second, the title illustrates that the Southern mind conceived the United States as divided between North and South. Southerners were not alone in holding this view of a divided culture, but Southern intellectuals and political "fire-eaters" expressed it most intensely. Third, the title hints at the rural–urban dichotomy that formed part of this divided culture. Rural settings predominated in American fiction, but Southerners tended increasingly to equate the North with its cities, over-flowing with immigrant workers. The Southern white elite, in particular, championed its own supposedly noble, caring, rural society as superior to the dangerous, heartless industrial system, exploitative of blacks as well as whites, that was developing in the Northern states. This agrarian ideal, fostered by a planter-professional class whose interests were bound up with land and slaves, was a constant in the nineteenth-century South.

If there exists an *urban* symbol for the antebellum South it is not to be found among the towns of Virginia, although Richmond was an important literary center, nor in the border cities of Baltimore or New Orleans, but in Charleston, South Carolina. In the eighteenth century, Charleston had been the fourth largest city in the thirteen American colonies. In the early decades of the nineteenth, it began to lose ground to other emerging centers, and it never acted as a literary center for the whole South as Boston did for New England. It did become, however, the unrivaled focus of that area commonly known as "the cotton kingdom," wedged between the upper South, the deep South and the frontier. Moreover, it was from Charleston that there emanated after the 1820s the idea of the South as an organic unity. South Carolina assumed the leadership role in the secession movement. From the nullification controversy of 1832–3[4] to South Carolina's unilateral secession from the Union in December 1860, Charleston was the center of the idea of a Southern nation.

By most accounts, this leadership ended abruptly with the Southern defeat; but in reality a New South did not quickly emerge, and when it did Charleston continued for several generations to serve as a powerful symbol of the past, as the epitome of the Old South. From the end of the Civil War through Reconstruction to the Depression of the 1930s, the South experienced and constantly re-visited the un-American experience of defeat. Social and political leaders in other regions in the United States, together with a sizeable portion of the burgeoning middle classes, could isolate themselves from poverty and squalor, and to some extent even from infectious disease, and "look down" on the masses still trapped by these inevitable forces. They were, in a sense, afforded the luxury of feeling themselves above history. Southern history, on the contrary, includes a long experience of guilt, poverty and failure,[5] and its effects are nowhere more vivid than in Charleston.

Between the Civil War and the advent of modern tourism, Charleston was reckoned by many a dismal place. Having suffered fire, bombardment and virtual abandonment, and having buried countless of its dead in graveyards all over the South, it was subjected in the post-war years to the further indignities of military occupation and radical reconstruction. By the end of the century the shift in South Carolina's power-base from the low country to Columbia was complete, and Charleston was cut off from whatever inroads the New South movement was able to make into the Southern economy. In the depression of the 1930s, the South was "the nation's number one economic problem." Charleston, for its part, seemed to be sinking into the coastal swampland which had formerly yielded the bounty of rice, the foundations of its earlier prosperity. Its reputation was widespread as a place utterly devoid of opportunity.

This view of Charleston was not shared – certainly not admitted – by the city's white elite. What remained of the planter-professional class after the Civil War still maintained a tolerable quality of life in large mansions on the tip of the peninsula "south of Broad," or on banks of the Ashley and Cooper rivers. Historically, the planters could claim a reasonably good record in managing to switch from one staple crop to another as the market demanded. Upland cotton had replaced indigo after the Revolution, and when the rice market collapsed in the 1820s, cotton reigned supreme. Cotton production lived on after the Civil War, and, in view of the limited potential for capital accumulation in any staple economy, it brought significant if not bountiful returns. Fortunately, the post-war planting class devoted a portion of its resources to restoring Charleston's built environment; but the elite presided over a largely retrospective culture, intent on evoking former glories. To the majority of its citizens, whether artisan, shopkeeper or freedman, turn-of-the-century Charleston offered little scope in an otherwise entrepreneurial nation. Its society seemed to reject the enterprising and ambitious, and out-migration was rapid. In the view of up-country farmers and many Northerners, Charleston, the so-called cradle of secession, was reaping what it had sown.

It would not be in the least surprising, then, to hear post-Civil War Charleston described as "a place of tombs." The condemnations quoted at the head of this chapter, however, were hurled at the city during its antebellum heyday. Of course, the thirty years between the father's invective and his son's re-affirmation were marked by economic and political frustrations, and many of the intellectual currents already harked back nostalgically to the golden age of the rice plantation. Although the city was past its period of economic pre-eminence, its decline was still relative only to the explosive growth of Northern and western cities. Affluence and leisure remained the hallmarks of the city's planter-professional class. Surprisingly, it was one who counted himself a member of that class, and who is, moreover, universally recognized as the foremost literary mind in the Old South, who echoed the refrain "a place of tombs."

In part this characterization was meant literally. Simms senior, who died in 1830, had long since forsaken Charleston after the death of his wife in childbirth. His son (1806–70), left behind to grow up in Charleston, had by the time of his lament in 1858 already buried a wife and six children, including two sons recently lost to yellow fever on the same day; but Simms was also referring to the demise of his life's mission to gain international recognition for a Southern literary tradition within a new American literature. It was clear by mid-century that American literature would be dominated for at least another generation by New England. Worse still, Simms developed the feeling that his own work and that of his colleagues was little appreciated at home in the South.

Any exploration of antebellum Southern urban literature leads us inexorably to Charleston and to Simms. The younger William Gilmore Simms was a Charlestonian not only by birth but by repeated re-assertion. At the age of 10, given the choice of joining his father on the frontier or staying with his grandmother in Charleston, he chose the latter. On a visit in 1825 his father was unsuccessful in trying to persuade him to begin anew on the frontier. While condemning Charleston as a literary graveyard, he continued to base his life and work there. After his second marriage in 1836 most of his writing was done at "Woodlands," his inherited plantation some 70 miles away on the Edisto river, and the basis of his claim to the title of planter; but he was in close contact with Charleston society throughout and was the leader of a literary group which regularly congregated at Russell's Bookstore. There Simms dominated a group which in addition to several minor authors included the poets Henry Timrod and Paul Hayne, and which between 1857 and 1860 published *Russell's Magazine*, modeled on *Blackwood's*.

As a consequence of this attachment to Charleston, it was Simms's fate to share with his native city its largely bitter antebellum career: economic decline, defensiveness, secession and defeat. When he was born there in 1806, Charleston ranked sixth in the urban hierarchy of the USA. It was the inheritor of perhaps the greatest per capita wealth of any region in the United States[6] accumulated on the basis of slave labor. When he died there in 1870 Charleston was a mere cipher among American urban places, 26th in the US urban hierarchy and declining rapidly. More significantly, it was a city in defeat, the symbol of a vanquished culture.

As antebellum Charleston's undisputed literary leader, Simms has often been represented as epitomizing its culture. Certainly he was imbued with its distinctive *genre de vie*, its finely drawn distinctions of class and race. In the words of a leading Simms scholar, "If ever a city can be said to have shaped a man, Charleston shaped Simms."[7] Since Charleston itself in many ways epitomized the antebellum South, Simms has acquired the status of literary representative for the whole region. His desire to contribute to a new American tradition of letters was underlain, and eventually overtaken, by a fervid support for a Southern literature. In his early career he came to the forefront of political debate in South Carolina and held public office. Frustrated in politics, he redoubled his literary efforts. His goal became to live as a "man of letters."

Quantitatively, his contribution matched this grand design. His output was prolific by any standards. One enumeration lists 82 books, including 34 full-length works of fiction and 19 volumes of poetry, the vast majority of it produced between 1833 and 1859.[8] Over the years he acted as editor, and often chief contributor, to half-a-dozen literary magazines. These included the *Southern Quarterly Review*, which he edited between 1849 and 1854. His literary criticism is voluminous and shows intimate familiarity with major British and American authors.[9] He traveled frequently in the Northern states and enjoyed the acquaintanceship, and occasionally the friendship, of several prominent literary figures.

Over the years, Simms has been best known for his novels, or "romances" as he preferred to characterize them. These he saw as harking back to medieval traditions, less constrained than the novel, and of epic proportions. They allowed him to be rather undisciplined in aspects of both structure and style. His topics were drawn from the Southern past, especially from South Carolina between its founding in 1670 and his own time. There are three main groupings. Two romances are set in the colonial period, and include *The Yemassee* (1835), undoubtedly the most popular of all. Seven "Revolutionary" romances seek to interpret the American struggle for freedom from a Southern point of view, and a number of "border" romances are set in various states of the South. In addition, Simms is now credited by recent scholarship with authorship of more than 1900 poems.

Important literary figure and prolific author as Simms was, however, his own creative work has been largely forgotten. Though much of his fiction sold well in his own time, even it has subsequently been of interest mainly to political and literary historians. The totality of his creative work has come to occupy a marginal, even obscure, position in American literature. He is much better known as a spokesman for the Southern cause than as an artist. Simms's literary reputation has rested on a tiny fraction of his work, notably a handful of romances. Perhaps only *The Yemassee* which has remained consistently on Southern curricula, can be said to have "survived." A few of the other romances have been in and out of print over the years, and occupy shelf-space in any adequate library of American literature.[10] The remaining novels, like his literary criticism and until recently his poetry, are far from accessible.

Several factors have contributed to this obscurity. The romances, inspired to a degree by the Waverley novels, share some of the worst features of Scott's writings. The narrative is often discursive to the point of tedium and the dialogue stilted and affected. The plots are sometimes undisciplined in construction, and the characterization is of variable quality, favoring arrogant young men and rather colorless heroines. Simms criticized Scott for writing too much too quickly, and went on to emulate him in this respect, for exactly the same financial reasons: failures in publishing ventures and the need to keep a rural estate from bankruptcy. Simms's attempts at realism opened him to charges of indelicacy in his own day, but the sensibilities of most modern readers are more likely be jarred by the subtle racism and sexism that permeate parts of the narrative of his fiction. With respect to the verse, a major reason for the neglect is that much of it was published in obscure outlets, largely anonymously. As for his works of literary criticism, the occasionally re-iterated verdict of one specialist that Simms was "a good but not a great critic"[11] provides one reason for its neglect.

Despite all these negative factors, the past decade has seen an extraordinary revival of interest in Simms as a writer. This has been marked by the publication of several critical essays on his work, including those assembled under the title *Long Years of Neglect*.[12] It has also produced an important re-appraisal of his major fiction.[13] Two new biographies are reported to be in progress. Perhaps the most significant development, however, has been the publication of a new selection of Simms's poetry.[14] Although he published most of his verse anonymously, Simms liked to think of himself mainly as a poet. Kibler, the foremost authority on his verse, has counted 186 proven pseudonyms used by Simms for his poems.[15] Others are possibilities, and another 32 are proven for his prose. The recent collection, while representing less than one-tenth of Simms's output, presents new evidence that the obscurity of the poetry is largely unwarranted. Kibler argues that the reason for neglect is not the quality of Simms's writing but the established conventions of American anthologies which continue to include poems by minor New England poets to which much of Simms's verse compares favorably. These new departures enhance Simms's stature not just as a barometer of society, culture and politics in the antebellum Southern states, but to an unprecedented degree as a writer of verse and fiction.

There lies at the core of Simms's career a discrepancy between boundless energy, enthusiasm and talent on one hand and a sense of failure on the other. Simms, by most accounts, counted himself in the end a literary failure, and for this he was inclined, in his correspondence and private musings, to blame Charleston. The epitaph which he wrote for himself (though it was never used) reads thus: "Here lies one who, after a reasonably long life, distinguished chiefly by unceasing labors, has left all his better work undone." Well before the Civil War, while still in early middle age, he was placing the blame for his failure on the society of the city he refused to leave.

The traditional critical view reflects this perspective: it was the stolidity of his social environment that prevented Simms from fulfilling the promise of his talents as a writer. This judgment, which may be called the "environmental view," was scarcely challenged for a century after his death. As manifested in the secondary sources it is expressed as a combination of five recurrent themes, closely interrelated and partly contradictory. The most salient theme holds that Simms's work did not receive due recognition because Charleston society favored other literary models, more genteel and amateurish than Simms was wont to pursue. Second, there is the resentment, justified or not by any real rejection, that Simms is regarded as having felt towards Charleston, particularly its planters. A third theme is that as a Southerner who chose to build his work around mainly Southern themes Simms would inevitably be excluded from the front rank of American letters. Fourth, the suggestion is made that Simms's identification with the South was so intense that he felt obliged to devote increasing amounts of his energies to its intellectual defence, thereby sapping his creative energy, or at least preoccupying his time and attention. Finally, there is the idea that Simms was so enamored of Charleston and the South that he was blind to its faults, and in the end became little more than an apologist for a society morally bankrupted by slavery.

Modern scholars have placed heavy responsibility for entrenching the "environmental view" on William Trent's 1892 biography of Simms, still the only full-length biography to appear.[16] "Perhaps there has never been a man," Trent concluded, "whose development was so sadly hampered by his environment."[17] Trent's perspective as a reconstructed Southerner of the 1890s would predispose him to see rigidities in the antiquated society of the Old South as frustrating Simms's genius. In addition, the biography emanated from a special set of editorial circumstances which would further encourage such a line of argument.[18] At the same time, any counter to Trent's views must reckon with Simms's own privately expressed bursts of resentment towards Charleston society. Moreover, several of Simms's contemporaries present a picture of a somewhat unsympathetic planter aristocracy.[19] Later critics have added their corroboration. Parrington's assessment is the most direct and the most widely influential of these.[20] He describes Simms as hampered by a "desperate environment" created by a society that considered his work ignoble. His characters (in the romances for which he was best known) were often seen as "coarse" and unworthy of the attention of a cultured society. Simms was unable to detach himself from the foibles of the very society that seemed to treat him so "shabbily." Although inherently a realist, he drank of the cup of Southern romanticism which inured him to its follies. Charleston, says Parrington, must take the blame for "befuddling the generous mind of Gilmore Simms," thereby impoverishing not only itself but the whole of American literature. Similar sentiments have been echoed by Southern historians as diverse as Eaton and Cash.[21] Simms's conversion to sectional priorities, for example, fits the broad thesis of the closure of the Southern mind during the antebellum period.

Recent critics have tended to dismiss the various components of the environmental thesis.[22] Michael O'Brien, for example, not only dismisses Simms's feelings of rejection as unimportant, but, more significantly, argues that antebellum Charleston society remained intellectually open and cosmopolitan.[23] One weakness of the arguments placing Simms's career at odds with a closed Southern mind is that if Charleston and the South were closed-minded about anything it was slavery, an institution to which Simms lent his unequivocal support. This was, indeed, one reason for the failure of his work to gain greater acceptance outside the South. Two greater "Southern" talents than Simms, Twain and Poe, rose above sectionalism; while Poe supported Southern institutions, his perspective is invariably cosmopolitan. Twain's post-war literary sectionalism is little more than a gloss, readily translated into the universal. Simms's work, by contrast, is permeated by the South, and most of it has limited meaning outside that context. Moreover, in the 1850s Simms made much of the necessity of using slavery to define an identity for a united South. There was indeed no more eloquent supporter of slavery, nor one more intent on using the issue to strengthen his state's resolve to proceed towards secession.

For every denunciation of Charleston or the South one can find in Simms's writings a quotation extolling their virtues. It would be difficult to imagine a more lavish and extravagant paean to one's native region than the poem "Song of the South":

> Oh, the South! the sunny, sunny South!
>   Land of true feeling, land forever mine;
> I drink the kisses of her rosy mouth,
>   And my heart swells as with a draught of wine!
> She brings me blessings of maternal love –
>   I have her praise which sweetens all my toil;
> Her voice persuades, her loving smiles approve –
>   She sings me from the sky and from the soil![24]

If this euphoria demonstrates anything, it is the futility of seeking consistency in Simms's writings. Equally, it makes little sense to perpetuate the discussion about Simms's relationship with the South in general and Charleston in particular by arguing for or against the environmental view. Each side has its own legitimacy, and rather than searching for a "correct" view, we should explore the contradictions inherent within Simms's own writings as they are reflected in secondary sources. When approaching the question of his view of the Southern elite, for example, one finds a clear ambivalence which has sometimes been seen as expressed in one of his best-known fictional characters, Lieutenant Porgy.[25] That Simms idolized the planter, but took issue with those who attacked the merchant from the planter perspective, highlights just the most obvious of several inconsistencies in his writings.

Such contradictions can be satisfactorily accounted for, though in no sense resolved, by ceasing to view them as competing alternative explanations, and treating them instead as expressions of genuine paradox. As some recent Simms scholarship has pointed out, Simms's life and works were encompassed by a set of dualisms. These, I suggest, were reflective of the inherent inconsistencies both in the Old South's self-definition and in the social organization of Charleston, the region's urban epitome. The main point to be emphasized is the interdependence between the South's own definition of its identity as reflected in a divided American culture, the tensions within Charleston society and Simms's complex personality. The dualisms, manifested at these three different scales, serve to contextualize Simms's writings. His canon, imbued as it is with the predicament of the South and of Charleston, is full of contradictions that must be addressed rather than dismissed.

We may begin at the level of the South as a "section," a separate part of the nation. The conception of the South as a distinctive society is often said to date from the writings of Thomas Jefferson. In a letter to the Marquis de Chastellux, Jefferson described a continuum of personalities as one traveled southward through the states of the new nation. In the North the people were cool, sober, laborious, independent, interested and in their religion superstitious and hypocritical. Southerners, by contrast, could be described in opposite terms: fiery, voluptuous, indolent, generous and without attachment to any religion but that of the heart. The choicest blend was to be found in the middle colonies, especially Pennsylvania. In the nineteenth century such views took on a marked sectionalist character so that North and South began to be represented as polar opposites. The caricatures reach their most dichotomous, perhaps, in the work of Marshall McLuhan, one of Poe's most appreciative critics. In seeking to contextualize Poe's work, McLuhan developed the thesis that the key intellectual fact about America was the way in which it gave geographic expression to two radically different traditions of European thought. Locked together in an age-old struggle in the Old World, the logical/dialectical/Cartesian tradition was taken by emigrants to the Northern colonies. Those in the South, by contrast, were heirs to the Ciceronian ideal and the philosophy of the forensic encyclopedists. In essence, "the schoolmen went to New England, the quasi-humanist gentry to Virginia."[26] One of McLuhan's biographers, Jonathan Miller, has tabulated these dichotomies so as to reveal a set of extreme sectional caricatures.[27] In his enthusiasm to establish Poe's place in the literary world, McLuhan clearly allowed himself to become a captive of a highly romanticized view of American culture.

This is but one illustration of the force of the Romantic movement in the creation of a Southern identity and its influence on subsequent criticism. Other critics have extended the discussion of Southern Romanticism using greater sophistication and plausibility, and without themselves falling victim to its embraces.[28] The archetypical Southerner was a Romantic invention, as indeed to some extent was the whole notion of a "divided culture." The force of Romanticism has been grossly underestimated in the South.[29] It permeated much mediocre writing, and even Poe and Twain were not above using it, though both transcended it; not so Gilmore Simms. No understanding of either Simms's life or his writing is attainable without a realization that both were deeply imbued with Romantic traditions. Twain and Poe, despite their Southern connections, had little difficulty in maintaining their distance from the Southern romantic legend. They defy neat classification on the basis of a divided culture and are not overtly sectionalist in their writing. Indeed, the universality of their work is part of their greater stature as authors. Simms, on the other hand, provides us with a literary guide to the mind of the Old South, with all its defensiveness, fears and preoccupations.[30]

It is important to recognize that Southern Romanticism is not synonymous with the "moonlight and magnolias" version of the Old South. This superficial view is a late-nineteenth-century construction, popularized in the North as well as the South, and later taken up by Hollywood. Beneath this surface two deeper levels may be discerned. Immediately below lies the

Romanticism of Simms's novels, which represents a revival of the genre of Scott, and his chief American follower James Fennimore Cooper. At this level we have a fiction of coincidences, mistaken identities, disguises, duels, secret passages and so on – hardly great literature but laid on with less syrup than its post-war equivalent. This is the level at which Simms's art has conventionally been analyzed. Recent Simms scholarship, however, suggests the presence of a still deeper stratum, only rarely evident in his prose but present in some of the verse. It owes its inspiration to the English Romantic poets. It is to this level that one must penetrate in order to understand Simms's view of Charleston as a place.

Within Simms's immediate environment of Charleston we find a further set of polar tensions, for which there exists considerable historical evidence. The tension between planter and merchant, scarcely detectible in the colonial era, assumed an importance in the decades before the Civil War as the class interests of the two groups appeared to deviate. There were other tensions: between slavery and the use of free labor, between the commercial port and the social resort and between those sections of society intent on production and those who saw in Charleston an instrument through which to enhance their own conspicuous consumption.[31]

Although the Southern population was largely rural, its intellectual life was urban, and much of the social round of the upper class was also based on cities. Charleston was a vital focus, holding sway over a wide area. While most of the action in Simms's Revolutionary romances is rural, Charleston is the constant intellectual center. When the action does enter the city, the narrative reflects the ease of the transition made by the elite from one milieu to the other. This is seen, for example, in *Katherine Walton* (1851), where the scene passes neatly from plantation to city. Perhaps the nicest demonstration is *The Golden Christmas*, a short novel published in 1852, in which the action begins in Charleston, including a memorable shopping scene on King Street, but is easily transferred to Major Bulmer's plantation "barony" in St. John's Berkeley. This conforms with contemporary accounts of the elite's seasonal migration between city and countryside, though the widely acknowledged ease of this transition may reflect a largely male perspective, since there is evidence that many women found the plantation environment lonely and threatening.[32]

The planters' authority stemmed from the ownership of rural land, but was exercised in the city as well as throughout its hinterland. Mercantile interests were important to the port function, but the rural world of the plantation was never far away. Half the population of the city over which the planter elite presided was composed of black slaves, many of them living in slave quarters on the same plot as the owner in an "urban equivalent of the plantation."[33] In contrast to the urban culture of the North, a rural ethos dominated the Southern city. Objectively, the interests of the planters lay in supporting the expansion of the commercial-industrial economy. Yet their prestige was derived from the slave system, and their perceived interests therefore lay in preserving the status quo.

Simms gave resonance to these dichotomies in his own life and work. As Charleston's economic predicament worsened, the struggle between romanticism and realism in his writing, and indeed within his own personality, intensified. As early as 1844 he was championing a past era, Charleston's "golden age," while condemning modern values:

> Would we recal our virtues and our peace?
> The ancient teraphim we must restore;
> Bring back the household gods we loved of yore,
> And bid our yearning for strange idols cease.
> Our worship still is in the public way, –
> Our altars are the market-place; ...
>   we have lost
> The sweet humility of our home desires,
> And flaunt in foreign fashions at rare cost.[34]

By 1848 Simms had fully accepted the notion of the United States as a divided culture. An article in the *Southern Quarterly Review* in that year paints a picture of Cromwell, the "Puritan fanatic," as "the source of our modern Yankee," espousing a set of values antithetical to those of the South.[35] Fully a decade before the Civil War he was characterizing the North as invader:

> Once more the cry of Freedom peals,
> From broad Potomac's wave to ours,
> The invader's cunning footstep steals,
> Usurping fast our rights and powers.[36]

The South, Simms argued, was being forced to defend itself from the assaults of a North that had failed to remain true to the spirit of the Revolution. One of his early tasks, begun with *The Partisan*, was now intensified, that of fixing in the American mind what he saw as the continuity between the ideals of the Revolution and antebellum Southern realities. Simms was unable to accept as fully as did some champions of planter culture that the means lay in intensifying the ideology of the landed elite as the major bulwark against Northern greed. He continued to favor the expansion of trade and industry, which commentators more closely identified with the planter interests saw as tending to erode the South's distinctive identity. No clear solution presented itself to this dilemma, either in Charleston society or within Simms's own psyche. In McCardell's view, such conflicts were expressed in Simms's personality as a struggle between Charleston, representing maternal forces, and the paternal forces of the frontier.[37]

In the 1840s, Simms still retained the hope of a resolution. Speaking in the guise of "Father Abbot" conducting a discourse on Charleston, he confronted the growing struggle between Charleston's two dominant groups, one of "toil and business," the other of "taste." In Charleston's heyday, he insisted, there had been no such dichotomy; merchants and planters had maintained common interests and their social lives had interlocked. While Charleston must maintain its genteel environment, he argued "the danger is always that in the perfection of our tastes, we lose some of our necessary energies. The secret is to refine our manners without forfeiting our strength."[38] It was a forlorn hope, and during the 1850s Simms's work continued to resonate with contradictions. One moment he was condemning mercenary progress, the next he was criticizing those who opposed Southern industrial and commercial development. It is not surprising, perhaps, to witness in his writings a growing preoccupation with the character of Hamlet. The version of Hamlet that appears is less its Shakesperian form than the "empathic and transcendental hero produced by Romanticism," filtered mainly through Wordsworth, and more particularly through the criticism of Coleridge.[39] Simms came to cast the South in the role of an indecisive Hamlet, while recognizing many of the same tensions within his own personality.[40] This period of brooding was in marked contrast to his earlier active life as a politician and interventionist, as an opponent of nullification and champion of national rather than sectional causes. The whole perplexing set of dichotomies seemed to offer only one escape: Southern secession. Nothing short of political independence would leave Southerners free to develop their own culture, untainted by Northern heresies.

The irreconcilable contradictions from which secession seemed the South's only escape extend into Simms's interrelationship with Charleston, just as they permeate his view of the South within the nation. Equally, they infuse both his complex personality and his canon. An important trend in recent Simms scholarship, and one that lends support to the approach outlined above, is the attempt to penetrate the contradictions by recognizing an inherent dialectic in his writing. In the remainder of this chapter, I attempt to show the implications of some of this recent criticism for our understanding of Simms's Charleston by focusing on his philosophy of place.

SIMMS AND THE GENIUS OF PLACE

A sense of place, indeed the necessity of place, is a vital part of Simms's writings. While almost all the action in his romances is place-specific, however, there is less literal description of landscape and morphology than one might expect. One reason for this may be that when Simms sets an action in "a clearing in the pine barrens," for example, he anticipates instant recognition of such a setting by his readers, whom he expects to be mainly Southerners. A more fundamental explanation is that Simms's deepest feelings on the nature of art led him to subjugate the literal attributes of place in favor of the moral.

Simms's major serial contribution is generally regarded as the seven Revolutionary romances, written over a period of 25 years beginning with

The Partisan in 1835 and ending with *Eutaw*. Charleston is the emotional and intellectual center of this series, and he often refers to it as "the metropolis." Yet the only romance which contains any significant description of the city is *Katherine Walton*, in which the action revolves around episodes during the British occupation of the city in 1780–1. It begins on the plantation, where the central tone is established, but quickly shifts to Charleston as the heroine is taken by the British commander into a loose sort of custody. Where he deems it necessary, Simms includes some strikingly detailed descriptions of the city's morphology. As he describes the movement of groups on various sorts of missions in and out of Charleston, for example, Simms goes to some lengths to compare the northern edge of the city of 1780 with that of his own time. In setting the scene for a lavish party in "a fine, airy mansion ... in Queen street opposite Friend," he goes as far as to add the footnote, "now in the possession of Mr. William Easton" (p. 290).

This is exceptional. Simms's major preoccupation is with the subtleties of Charleston society and the interrelationships between the British forces, patriots and loyalists. He claims considerable accuracy for the historical fiction in this series of romances. As he writes at the close of *Eutaw*, the last Revolutionary romance, the "parties and events of authentic history have framed this truthful chronicle" (p. 582). Wimsatt has recently added immeasurably to our appreciation of Simms's account of society in colonial Charleston by analyzing its inherent dialectic.[41] Briefly stated, she finds in *Katherine Walton* two parallel axes. The first extends through descriptions of loyalist society, in which politics are shown to promote fashion and frivolity, as epitomized in the characters of the widow Mrs. Rivington and the young Moll Harvey. This axis is treated as a novel of manners. Patriot society, on the other hand, rejects frivolity and is interested in manners and fashion only to promote partisan politics, to the extent that the corresponding widow, Mrs. Brewton, uses social occasions to gain information of use to American patriots in the rural hinterland. The foil to Moll Harvey is the heroine Katherine herself, who epitomizes the steadfast qualities of the patriot cause and is the focus of action treated not as a novel of manners but according to the traditions of the romance.

Simms also deals with Charleston in his final romance *The Cassique of Kiawah*. Here, however, the tone is different, for by 1859 Simms's resentment against the city seems to have intensified, even as his Southern partisanship has strengthened. His increasing impatience with the foibles of the Charleston elite comes through in acerbic comments from the narrator, quite unlike anything in *Katherine Walton*.[42] Wimsatt traces the technique directly to the influence of Thackeray. Simms, who had frequently reviewed Thackeray's work in Southern periodicals, met the author of *Vanity Fair* when he visited Charleston on a lecture tour as Simms was writing his final romance. Despite a certain disdain for the work of those American authors still anchored to Old World trends, Simms apparently found himself under the influence of yet another established British author. Whatever his model, it is clear that in the later romances Simms readily adopts the posture of a distanced, and on occasion slightly jaundiced, observer of Charleston society.

For a greater understanding of Simms's approach to place, especially Charleston, we must turn to his verse. It is true that there are a few places in his fiction where a deeper philosophy of place briefly emerges. The most prominent, perhaps, is in the observation of Major Procter in *Katherine Walton* made as he accompanies the disguised patriot Singleton on horseback through the forest: "the genius of place is born always in the soul of the occupant" (p. 127). This statement provides one of the rare clues in Simms's prose to a view of place which is thoroughly Romantic. Building on recent analyses of Simms's philosophy,[43] I suggest that his view of Charleston penetrated more deeply than anything to be found in the romances, reaching down to a view of landscape derived from the intense individualism of the English Romantic poets.

Simms adopted with enthusiasm the notion that the attributes of place were endowed by the interaction between humans and nature. In Kibler's view, Simms went beyond a recognition of the emblematic role of nature, which draws the sort of analogies between nature and humankind common in nineteenth-century American verse. He espoused a belief in a symbiotic relationship between nature and the observer, in which the landscape exists only as perceived and becomes a metaphor for the observer's state of mind. This shows that Simms was conversant with English Romanticism to a degree unrivaled by any other American writer, with the exception of Emerson. The influence of Coleridge on both writers is evident. Further, Simms seems to have been fully persuaded that the poet has a duty to see beyond the literal into the spiritual significance of landscape. It is the artist alone who can see its underlying meaning.[44]

When Simms turned to Charleston as a place, therefore, he was mainly concerned with interpreting its moral or spiritual aspects. Indeed, he often seemed to mistrust the literal. While a description of particular streets and buildings might be useful on occasion in framing the action in a romance, the essence of Charleston as a place lay in its moral, not physical, landscape. Perhaps like the young Wordsworth approaching London armed with a broad, pervasive view of humankind existing in a symbiotic relationship with nature, Simms viewed Charleston through his own romantic lenses. McCardell suggests that in Simms's early work Charleston represents "the maternal embodiment" of his conception of poetry.[45]

One of Simms's best poems, the sonnet entitled "Harbor by Moonlight," provides a poignant combination of the elements of Charleston as place. First published in the *Southern Literary Messenger* in 1844, the poem is on one level a description of the city as viewed from Sullivan's Island on the other side of Charleston harbor. People pursue mundane evening activities on and around Fort Moultrie, a Revolutionary defensive site on the island, used in the antebellum years as a resort for fresh air and recreation. Portents of doom are felt by the poet as "normal" life goes on:

> The open sea before me, bathed in light,
>   As if it knew no tempest; the near shore
> Crown'd with its fortresses, all green and bright,
>   As if 'twere safe from carnage ever more;
> And woman on the ramparts; while below
>   Girlhood, and thoughtless children bound and play
> As if their hearts, in one long holiday,
> Had sweet assurance 'gainst tomorrow's wo: –
> Afar, the queenly city, with her spires,
>   Articulate in the moonlight, – that above,
> Seems to look downward with intenser fires,
>   As wrapt in fancies near akin to love;
> One star attends her which she cannot chide,
> Meek as the virgin by the matron's side.[46]

Charleston is compared to a star on the evening horizon, protected by the moon in the way that a mother shelters her daughter. Yet the city's sense of security is as groundless as that enjoyed by the "thoughtless children" on the beach below. The poem is prophetic in its underlying unease for the future of this seemingly peaceful city, from whose battery the first shots of the Civil War were fired sixteen years later.[47]

CONCLUSION

Writing as "Father Abbot" in the 1840s, Simms evoked the vision of Charleston as America's Venice. It was an unusual comparison. Antebellum Southerners seeking a European urban ideal commonly inclined towards ancient Greece. Charlestonians in particular aspired to the image of the Greek city-state, pointing to the intimacy with Charleston and its tidewater tributary area, and its existence within a slave society. Simms expressed regret at the hold maintained by the Greek model and suggested the Venetian in its place. He called attention to the superficial resemblances in climate and site, where Charleston's ocean "sweeps up to her very doors."[48] Above all, he admired the blend of Venetian culture and commerce. According to "Father Abbot," Charlestonians could continue to take pride in "our moral qualities: our graces of society, our frank hospitality, the elegance of our women, the high character of our gentlemen," without spurning commerce, or even railways and factories.

> You will see this promise realized. You will behold the city, and all its tributary islets and shores … linked together by industrial bands of

iron. You will see this goodly city, seated by the sea, sending forth her messengers, winged by steam and sail, to the remote and mighty cities of the old world.

(p. 211)

As Charleston's economic vulnerability became increasingly apparent, this optimistic, almost booster, view of the city was replaced by a renewed sense of desperation. This was manifested in a growing need to defend city and section from outside criticism. Long before the Civil War Simms had staked his literary reputation irrevocably on a defense of the South. Moreover, as if in sheer bravado, he seized on the issue of slavery.

I began with the suggestion that one could immediately discern something significant about the mind of the antebellum South by observing its responses to *Uncle Tom's Cabin*. Simms's own reaction to the publication of the book was complicated. Rather than immediately attacking it in the pages of his *Southern Quarterly Review*, he assigned that task to Louisa Cheves McCord, presumably because he felt that it could be countered more forcefully by a Southern woman. When he did discuss the book almost two years later, he praised its literary merit. Meanwhile, he had long since completed a romance, originally entitled *The Sword and the Distaff*, later called *Woodcraft* (1856), which he claimed was "as good an answer to Mrs Stowe as has been published." This broad-ranging story of plantation life in South Carolina immediately after the Revolution may not originally have been conceived as a direct rejoinder,[49] but he clearly felt that it would serve the purpose. His most direct response, however, was political; he acted as one of the editors and contributors to a volume of essays setting out an extended justification for the slave system.[50] This book reiterated old arguments. Simms's essay was a re-working of a review published fifteen years earlier, but he now seized on the opportunity to use slavery as a banner under which to unite a disparate South. In various outlets during the next several years Simms published extensive reviews defending slavery, attacking the abolitionists and re-visiting the argument that slavery acted as a civilizing and uplifting institution, sanctioned by the private-property provisions in the Declaration of Independence. The old stand-by of the Greek city-state with its generous allowance for slavery made yet another appearance, eclipsing the Venetian vision for Charleston which had recently seemed so plausible.

There is a double tragedy here: the uniting of the Old South around the least justifiable of all its characteristics and the pre-emption of a powerful intellect in defense of the indefensible. Simms puposefully identified himself with the Old South, and specifically with its most morally repugnant feature. Earlier in the antebellum period the South had been much divided on slavery, as on everything else. Whatever solidarity was achieved in the 1850s was given a powerful measure of intellectual respectability and reinforcement by Simms's writings. The North needed no other issue to justify the most

Figure 10.1 Bust of Simms in White Point Gardens, Battery, Charleston

extreme measures to preserve the Union.

Simms survived the War, although he lost his house and most of his library to fire. Despite deprivation, further bereavement and a long illness, he continued writing in the post-war years. He died, so the legend has it, with his pen gripped tightly between his fingers, and was buried in Magnolia cemetery, whose opening he had heralded with a 29-stanza poem some twenty years earlier. To a degree, his city accepted him. Within eight years of his death, sufficient money had been raised from an impoverished community to erect a plinth and bust in White Point Gardens on the Battery, where it stands today (Figure 10.1). Simms was one of the very few people to be honored in this way by Charlestonians; theirs is not a city of statues.

Even the most vociferous critic of Trent's biography cannot deny that Simms was attached to the South to an extent that affected to a greater or lesser degree all of his art. His diminished reputation as an author stems from his devotion to a lost cause. It is not so much that he defended slavery (much literature has survived despite its author's inhumane social or political views) but that the vanquishing North asserted its culture with such righteous indignation.

Iconographically, only a fraction of Simms's work is about Charleston. Emotionally, this urban symbol of the Old South pervades his entire canon. Simms was attached to a place and an ethos. Dualisms within the Romantic view of the divided culture flowed directly into his soul and are manifest in his writings. The intimacy of the emotional and intellectual connections between the writer and his city are clear. Simms made Charleston's dualisms his own. Kibler has claimed that in his poetry Simms succeeded in reconciling opposites. If so, he achieved reconciliation only at the most personal level and through the most intimate of media. There is little evidence elsewhere of a broader reconciliation between Simms and what has generally been referred to as his environment. In the end the contradictions remained unresolved.

NOTES

1 William P. Trent, *William Gilmore Simms* (1892; New York: Haskell House Publishers, 1968), p. 17.
2 ibid., p. 237.
3 William R. Taylor, *Cavalier and Yankee: The Old South and American National Character* (New York: George Braziller, 1961), p. 309.
4 South Carolinians were so incensed by the imposition of federal import tariffs, fearing among other things British retaliation, that they declared the tariff acts invalid in South Carolina, thus opening up the whole question of states' rights. A compromise was achieved which postponed the inevitable conflict.
5 C. Vann Woodward, "The search for Southern identity," in *The Burden of Southern History* (Baton Rouge: Louisiana State University Press, 1960), pp. 3–25.
6 A recent re-assessment of the enormity of this wealth is provided in Peter A. Coclanis, *The Shadow of a Dream: Economic Life and Death in the South Carolina Low Country 1670–1920* (New York and Oxford: Oxford University Press, 1989), pp. 48–110.
7 John McCardell, "Poetry and the practical: William Gilmore Simms," in Michael O'Brien and David Moltke-Hansen (eds.), *Intellectual Life in Antebellum Charleston* (Knoxville: University of Tennessee Press, 1986), pp. 186–210; quotation on p. 188.
8 A. S. Salley, *Catalogue of the Salley Collection of the Works of Wm. Gilmore Simms* (New York: Burt Franklin, 1943); also Salley's essay in Mary C. Simms Oliphant, Alfred Taylor Odell and T. C. Duncan Eaves, *The Letters of William Gilmore Simms* (Columbia: University of South Carolina Press, 1952), pp. lix–lxxxix.
9 A readily accessible sampling of his American criticism is his collected essays,

Views and Reviews in American Literature, History and Fiction (Cambridge: Belknap Press, 1966).

10   *The Yemossee* was reprinted Boston: Houghton Mifflin, 1961; *The Partisan*, Ridgewood: The Gregg Press, 1968; *Woodcraft*, Ridgewood: The Gregg Press, 1968.

11   Edd Winfield Parks, *William Gilmore Simms as Literary Critic* (Athens: University of Georgia Press, 1961), p. 110.

12   John Caldwell Guilds (ed.), *Long Years of Neglect: The Work and Reputation of William Gilmore Simms* (Fayatteville: University of Arkansas Press, 1988).

13   Mary Ann Wimsatt, *The Major Fiction of William Gilmore Simms: Cultural Traditions and Literary Form* (Baton Rouge: University of Louisiana Press, 1989).

14   James E. Kibler (ed.), *Selected Poems of William Gilmore Simms* (Athens and London: University of Georgia Press, 1990).

15   James E. Kibler, *Pseudonymous Publications of William Gilmore Simms* (Athens: University of Georgia Press, 1976), p. 2.

16   Not counting chronological treatments of his work that incorporate details of his life, for example, J. V. Ridgely, *William Gilmore Simms* (New York: Twaine Publishers, 1962). Note also that this chapter was written before the publication of John Caldwell Guilds, *Simms: A Literary Life*, Fayetteville: University of Arkansas Press, 1992.

17   Trent, *Simms*, p. 325.

18   John McCardell, "Trent's Simms: The making of a biography," in William J. Cooper, Michael F. Holt and John McCardell (eds.), *A Master's Due: Essays in Honor of David Herbert Donald* (Baton Rouge: Louisiana State University Press, 1985), pp. 179–203.

19   The stifling attitudes of the elite towards innovative artistry are widely documented; for example, the comments of the poet Paul Y. Hayne and Simms's close friend William Porcher Miles. See Jon L. Wakelyn, *The Politics of a Literary Man: William Gilmore Simms* (Westport: Greenwood Press, 1973), p. 195.

20   Vernon L. Parrington, *Main Currents in American Thought* (New York: Harcourt, Brace & World, 1927), pp. 125–36.

21   ibid., p. 136. Clement Eaton, *Freedom of Thought in the Old South* (New York: Peter Smith, 1951); Wilbur Cash, *The Mind of the South* (New York: Alfred A. Knopf, 1941). Perhaps the most balanced "traditional" account of the effects of Charleston's environment on its writers is Jay B. Hubbell, *The South in American Literature, 1607–1900* (Durham: Duke University Press, 1954), pp. 572–602.

22   McCardell, "Poetry and the practical"; Wakelyn, *The Politics of a Literary Man*.

23   Michael O'Brien, *Re-thinking Southern History* (Baltimore: The Johns Hopkins Press, 1988).

24   Kibler, *Selected Poems*, p. 197.

25   A character often compared with Falstaff; see, for example, Taylor, *Cavalier and Yankee*, pp. 292–3.

26   Herbert Marshall McLuhan, "Edgar Poe's tradition," *Sewanee Review*, 52 (1944), p. 28.

27   Jonathan Miller, *McLuhan* (London: Collins, 1971), p. 85.

28   Rollin D. Osterweis, *Romanticism and Nationalism in the Old South* (New Haven: Yale University Press, 1949); Raimondo Luraghi, *The Rise and Fall of the Plantation South* (New York: New Visions, 1978).

29   Michael O'Brien, "The lineaments of antebellum Southern Romanticism," in *Rethinking Southern History*, pp. 38–56.

30 On Simms rather than Poe as the representative antebellum Southern author see Mary Ann Wimsatt, "William Gilmore Simms," in Louis D. Rubin et al., *The History of Southern Literature* (Baton Rouge: Louisiana State University Press, 1985), pp. 108–117; and Hubbell, *The South in American Literature*, p. 572.

31 John P. Radford, "Testing the model of the preindustrial city: The case of Charleston, South Carolina," *Transactions of the Institute of British Geographers*, New Series, 4/3 (1979), pp. 392–410.

32 Steven M. Stowe, "City, country and the feminine voice," in O'Brien and Moltke-Hansen, *Intellectual Life*, pp. 297–324.

33 Richard C. Wade, *Slavery in the Cities: The South, 1820–1860* (New York: Oxford University Press, 1964), p. 61.

34 "Sonnet. Popular Misdirection," in Kibler, *Selected Poems*, p. 130. The spellings follow Kibler, who has restored Simms's original forms.

35 Quoted in Wakelyn, *The Politics of a Literary Man*, p. 123.

36 "Southern Ode," Kibler, *Selected Poems*, p. 186.

37 McCardell, "Poetry and the practical," pp. 186–210.

38 William Gilmore Simms, *Father Abbot or The Home Tourist* (Charleston: Miller & Brown, 1849), pp. 182–3.

39 Martin Greenberg, *The Hamlet Vocation of Coleridge and Wordsworth* (Iowa City: University of Iowa Press, 1986).

40 Taylor, *Cavalier and Yankee*, pp. 294–7.

41 Wimsatt, *The Major Fiction*, pp. 179–93.

42 ibid., pp. 188–192; also Anne M. Blythe, "William Gilmore Simms's 'The Cassique of Kiawah' and the principles of his art," in Caldwell Guilds, *Long Years of Neglect*, pp. 37–59.

43 Kibler, *Selected Poems*, pp. xi–xxii; also his chapter "Perceiver and perceived: External landscapes as mirror and metaphor in Simms's poetry," in Caldwell Guilds, *Long Years of Neglect*, pp. 106–25.

44 A similar viewpoint is occasionally expressed in his fiction, often in unlikely places, as for example in the introduction to a rather lurid domestic romance entitled *Confession* (rev. edn, New York: Belford, Clarke & Co., 1888), p. 8.

45 McCardell, "Poetry and the practical," pp. 186–210; quotation on p. 197.

46 Kibler, *Selected Poems*, pp. 130–1; on spellings see note 33.

47 Kibler, in Caldwell Guilds, *Long Years of Neglect*, pp. 112–13.

48 Simms, *Father Abbot*, p. 25.

49 The two sides of the debate are represented by Ridgely, *Simms*, pp. 97–8, and James B. Meriwether, "The theme of freedom in Simms's *Woodcraft*," in Caldwell Guilds, *Long Years of Neglect*, p. 29.

50 William Gilmore Simms, "The morals of slavery," in William Harper, J. H. Hammond, William Gilmore Simms and Thomas R. Dew, *The Pro-Slavery Argument as Maintained by the Most Distinguished Writers of the Southern States* (Charleston: Walker, Richards & Co., 1852), pp. 181–285.

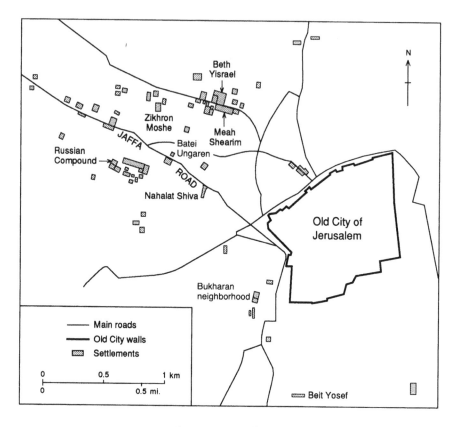

Map 11   Jerusalem

11

JERUSALEM IN S. Y. AGNON'S *YESTERDAY BEFORE YESTERDAY*

Yossi Katz

INTRODUCTION

The Jerusalem that Agnon describes in *Yesterday before Yesterday* is essentially the portion of the city located outside the Old City's walls. As opposed to the Old City of Jerusalem, which dates from ancient times, modern Jerusalem outside the walls originated from the nineteenth century, or more precisely from 1840 onwards. Nothing had been built in the area outside the city walls prior to that date, and the city was virtually imprisoned within its walls.

In the years 1840–55 the first harbingers of construction outside the walls appeared. Greek monks and Protestant missionaries began cultivating areas of land outside the walls, while a number of Christian institutions were erected outside the city walls at the same time. These activities were tied to the reforms instituted by the Ottomans upon their return to Jerusalem in 1840. The reforms encouraged foreign nationals to undertake a number of initiatives including land cultivation and the construction of various institutions. As a result of the Crimean War, these reforms were broadened during the 1850s and 1860s, which in turn facilitated a further stage in the development of the area outside the walls. Immigration to Palestine and to Jerusalem in particular accelerated during those years, as steamships began arriving in Palestine, the Suez Canal was inaugurated (in 1869) and a railway was laid connecting the port of Jaffa to Jerusalem. Jews occupied first place amongst the groups immigrating to Jerusalem. The number of Jews increased nine-fold from 5,000 in 1840 to 45,000 in 1910.[1] By comparison the general population of Jerusalem grew during that same period from 13,000 to 70,000 (a 540 per cent increase). This spurt of growth in Jerusalem's Jewish population stimulated the construction of many Jewish residential areas outside the city walls in order to cope with the paucity of housing opportunities inside the walls. The Jewish communal leadership took the initiative in starting the new neighborhoods. Prospective tenants committed

195

their entire, if paltry, life-savings to defray construction costs. Other neighborhoods were built by Jewish capitalists who donated the apartments to the tenants. Some neighborhoods were established by the leadership of the various Jewish ethnic communities and as such were restricted to members of that same community. This process gave rise to a number of Sephardic and Yemenite neighborhoods and a quarter for members of the Bukharan community. The Ashkenazic organizations (the Kollels) erected neighborhoods for Kollel members. As each Kollel was in charge of immigrants from a specific country of origin distinct neighborhoods arose. Examples of this pattern were the Batei Ungaren (Hungarian houses), for immigrants from Hungary, and the Batei Varsha (Warsaw houses), for immigrants from Poland. At the turn of the century a number of commercial housing companies arose who sought to supply the growing demand for housing.

If the growth of the Jewish population resulted in the establishment of new neighborhoods, the increase in the Christian population (from 3,350 in 1840 to 13,000 in 1910) found its spatial expression in the establishment of a number of Christian institutions. Christian growth was shaped by competition between the European powers for increased influence in Palestine and especially in Jerusalem. A number of duplicating institutions arose at the initiative of the competing European nations. For example, a German hospital, a French hospital, an Italian hospital and an English hospital were built during this period. The various European countries built post offices and banks as well as schools and other institutions. This burst of public construction was in addition to the feverish construction of churches during the same period.

The Moslem population, which during the years 1840–1910 increased by less than three-fold (from 4,650 to 12,000), also contributed to the development of Jerusalem outside the city walls. Most importantly the Muslims constructed a number of residential quarters. Muslim development differed from the Jewish pattern. Whereas the Jewish residential neighborhoods were intended for all strata of the Jewish population and were constructed in the entire region outside the walls, the Moslem neighborhoods were intended only for the affluent. The neighborhoods were located near the Old City walls and were thus adjacent to the existing Moslem centers within the city walls.

Jerusalem outside the city walls was built in a haphazard fashion without the benefit of any planning. A master-plan was devised for Jerusalem only during British Mandatory rule. Nonetheless, we can observe a pattern of segregation in the development of Jerusalem, since the Jewish neighborhoods, the Christian institutions and the Moslem neighborhoods were established in different areas. The main axis of Jerusalem's urban development was westwards, along the main road connecting the Old City with the port city of Jaffa and along the parallel arteries adjacent to this road. Additional clusters of construction arose north of the city on the Mount of

Olives and Christian institutions and religious edifices were erected nearby. The Jews also took advantage of the relatively moderate cost of real estate north of the city and built another group of Jewish neighborhoods. Construction was sparse to the south and east of the Old City, and Christians and Moslems were the primary builders.

On the eve of World War I, it was possible to perceive Jerusalem as a mosaic of communities and religions, and this was responsible for Jerusalem's special character. One could apparently find in Jerusalem the entire range of Christian and Jewish communities, and this diversity accounted for the city's unique architectural character. As opposed to the Old City, which could be characterized during that period as a backward pre-modern city, Jerusalem outside the city walls was a developing modern city. New Jerusalem gradually assumed a character of its own, which was independent of the Old City. This differentiation gathered further momentum in the years following World War I. The Jews were the major factor behind the growth and development of Jerusalem outside the city walls. They constituted the demographic majority of the city from the close of the 1860s onwards. The city's Jewish majority increased continuously and hence one can already by the nineteenth-century view Jerusalem outside the city walls as a Jewish city.

Much research has been devoted to Jerusalem, the eternal city, on topics dealing with its history, geography, society and the economy. Little research, however, has been awarded to Jerusalem as a background and location in literature. Scholars of literature have been almost exclusively responsible for the limited attention devoted to the subject in academic research. Historians and social scientists, whose interest generally focuses on objective realistic situations, do not deal with this subject, just as they generally tend to neglect the role of the city or the area of literature. This chapter is an initial attempt to examine the role of Jerusalem in literature from the perspective of historical geography, an approach which is apt for one literary text set in the early years of this century, Shmuel Yosef Agnon's *Yesterday before Yesterday*.[2]

Agnon, who was a Zionist Jew, and one of the greatest Jewish writers of the modern period, makes extensive use of the eternal city in his works. He was born in 1888 in Galicia, immigrated to Palestine in 1908 and lived there until 1913 when he left for Germany. In 1924 he immigrated for a second time to Palestine, and in 1966 he was awarded the Nobel Prize for Literature. He lived in Jerusalem for most of his life and died in 1970. Four sources, deriving from his own experience in Europe and Palestine, are of seminal importance to the works of Agnon; the first is Galicia, its Jewry and wellsprings of Jewish culture. Then there is modern Hebrew literature and Judaica. The "Second Aliyah" and its literature provides that stream of Jewish immigration which reached Palestine during the years 1904–14 and especially that elite component of idealistic, young and indigent pioneers, who sought employment in agricultural labor and in which can be found the origins of the labor movement. Finally, Agnon's work is influenced by the experience of German literature and culture and European literatures in general.[3]

A good deal of historical and historical geographic research has been undertaken on the period in Jerusalem's history covered by the novel.[4] Documentation on the subject is extensive and this helps me to implement my aim in this chapter: examining the degree of connection between the imaginative reality (including its urban, social and economic dimensions) of Jerusalem as depicted in the novel and the empirical reality that emerges from research. I hope also to examine how Agnon's subjective conception of Jerusalem emerges in the novel. This goal is predicated on the assumption that the author "is present" in his work, and that the manner in which the author occupies a specific space in his literary creation constitutes a statement of his ethical attitude towards the space that appears in his work.[5]

BACKGROUND TO THE PLOT AND ITS LITERARY CHARACTERISTICS

Yesterday before Yesterday (the English translation of *T'mol Shilshom*), a novel of about 600 pages, began taking shape in the 1930s and was published in full in 1945. The period covered by the plot is that of the Second Aliyah, and the geographical space in which the novel is set encompasses the cities of Jaffa and mainly Jerusalem. Agnon, who as a resident of Palestine personally witnessed the Second Aliyah, chronicles the life-style of the period's two major social strata – the "Old Yishuv" (traditional Jewish community) and the "New Yishuv" (modern Jewish community). The community dubbed "Old Yishuv" is the orthodox one which scrupulously and zealously observes the commandments of the Torah. This community had originated in Palestine towards the end of the eighteenth century and exemplified the values of the traditional ultra-Orthodox ("Haredi") society. Hence it occupied itself primarily with the study of the Torah, was concentrated mainly in Jerusalem and derived its principal sources of sustenance from outside support. The "New Yishuv" community originated in Palestine at the close of the nineteenth century with the waves of Zionist-inspired immigration. It personifies the values of the Jewish Enlightenment and the modern national movement and finds employment in productive occupations such as agriculture, trade and handicrafts. The "New Yishuv" is characterized by processes of secularization and abandonment of religion, although even the "New Yishuv" contained circles that strictly observed the religious commandments. It was geographically concentrated primarily in Jaffa, the agricultural villages, Haifa and only to a small extent in Jerusalem.[6]

Yesterday before Yesterday is a socio-psychological novel, steeped in unrealistic foundations and begging a mythical interpretation. Yizhak Kumer and the dog Balak occupy two distinctly separate areas at the beginning of the novel but the areas gradually merge into one in the course of the novel. The novel narrates Yizhak Kumer's immigration to Palestine. He immigrates due to Zionist ideological motives and an aspiration to work the soil, like his young pioneer compatriots of the Second Aliyah. Yizhak comes from small-town Eastern Europe, where the Jews maintained an Orthodox life-style. Again, like his fellow pioneers, Yizhak casts off the religious commandments immediately upon his arrival. He tries to join the life-style of the Second Aliyah but fails in his attempt at agricultural settlement. He proves no more adept at assimilating into the modern Jewish urban community of Palestine. Yizhak Kumer is still umbilically linked to his past and small-town roots. He finds himself inexorably returning to the values of religion and its commandments within the framework of Palestine's traditional Jewish community.

The story which chronicles the Second Aliyah must rely on extra-literary material to supply documentary authenticity. Agnon describes the social milieu of Jaffa and Jerusalem's neighborhoods and quarters as well as life in some of the agricultural villages. The author invokes a lengthy list of historical and fictional characters to portray Palestinian reality and provide a reliable backdrop to Yizhak's peregrinations over time and space.

The expanse epitomizes the psychological and social problems that assail Yizhak and create the bipolar contradiction that tears him apart. Jaffa represents the secular immigrant city while Jerusalem represents the religious city. Jaffa is an expanded version of Sonia with whom Yizhak falls in love and similarly the light-headed Sonia reflects the character of Jaffa. Shifra, Yizhak's future wife, lives in one of Jerusalem's most ultra-Orthodox neighborhoods. Yizhak's meandering between Jerusalem and Jaffa echoes his painful choice between Sonia and Shifra and his overall crisis of identity: is he to affiliate with the modern community or go back to the traditions maintained in the old community? His deliberations between the two poles cause Yizhak pangs of guilt. He has abandoned his parents' home for the sake of taking up agriculture. However, already on his voyage to Palestine, Yizhak chanced upon an alternate livelihood – house-painting. His betrayal of his father was thus compounded by his betrayal of Zionist values for whose sake he had originally betrayed his father. When Yizhak returns to religion in Jerusalem, his return must remain incomplete because it entails Yizhak's casting-off of his secular essence. This essence assumes a satanic yet satirical shape in the form of the dog Balak, who, as the personification of apostasy, gazes upon traditional Jerusalem.

Jaffa and Jerusalem, Sonia and Shifra, are two extremes and the twain will never meet. Yizhak Kumer may be regarded as a sacrificial victim of the eternal rift between exile and redemption, and between Judaism and Zionism. The novel concludes with a drought, and only after Yizhak Kumer's funeral does the land flourish again. A grave sin weighed oppressively on the land and this sin is expiated only after Balak has bitten Yizhak and caused his death. Yizhak's link to the dog leads to his demise. Since in a symbolic sense the link is genuine, Yizhak's death is also symbolically justified.[7]

YESTERDAY BEFORE YESTERDAY AS AN HISTORICAL SOURCE

Geographers reading *Yesterday before Yesterday* would find it difficult to free themselves from the impression that one of the goals of the novel was precisely to offer the reader a description of various aspects of the empirical reality of Jerusalem during the period between the last half of the nineteenth century until World War I, and it is quite clear that Agnon relied on historical sources.[8] It is not only by a comparison of the novel with historic reality that we can establish Agnon's intention to provide an accurate picture of Jerusalem in this period; Agnon himself alludes to this intention most explicitly when he describes in the novel a meeting of the Jerusalem intellectual set in the local "culture house." He has the participants say the following:

> When a great author will arise to write a novel about Jerusalem, a two part novel, the one part dealing with the ethereal upper Jerusalem and the second dealing with the corporeal lower Jerusalem ... the part dealing with the lower Jerusalem must mention built-up Jerusalem with all its functionaries and builders.[9]

It is difficult to liberate oneself from the impression that the major "urban geographic" topic with which the author sought to acquaint the reader was the development of Jerusalem outside the city walls beginning with the second half of the nineteenth century and the development of the Jewish part of the city outside the walls. There are many details about the squalid conditions experienced by the Jewish community within the walls – for example, living quarters

> in the outskirts of twilight bereft of the sun's illumination and lacking even a patch of garden or greenery. They who would do away with their eyesight, for want of light and distress their souls for lack of air to breathe.... Not every house had a window, and there were those houses who received some illumination through a chimney in the ceiling while other houses derived their light through a hole in the wall over the door-frame or via an opening to the air in the courtyard where the water cistern lay. Here the women washed their laundry while in the lower courtyard stood the privies which were cleaned out only once every few years.

(pp. 197, 198)

These conditions deteriorated still further as the immigration from abroad swelled, and necessitated an exodus outside the city's walls despite the precarious security conditions prevailing there.

> All the pathways outside the wall were perilous, for there was no settled area outside the Russian Compound and when they locked the gates of the city at night the life of anyone found outside the wall was forfeit.... all the areas that were teeming with people during the day turned into a wasteland and when they would lock the city gates from sunset to daybreak anyone remaining outside the wall placed his life in mortal danger at the hands of cut-throats.
>
> (pp. 195, 231–2)

Agnon gives details of the names and order of construction of the Jewish neighborhoods erected outside the walls (pp. 199, 273, 290, 301, 312, 336, 563, 569, 583). Emphasis is placed on the unique fact that Jerusalem outside the wall was in a sense an assortment of neighborhoods, each with its special attributes of size, organization and population, but also sharing common features such as the existence of a synagogue, a prevailing Orthodox character and the dynamics of their physical expansion. In Agnon's words:

> Jerusalem is recumbent like an eagle carrying her offspring on her wings. There are neighborhoods exclusively populated by Ashkenazim and exclusively populated by Sephardim and there are neighborhoods where both Ashkenazim and Sephardim reside. There are neighborhoods for Yemenites, Gruzians, Moroccans or Persians and there are those where a number of communities live together. But you will not find a neighborhood without a synagogue, and there are some neighborhoods where they have erected synagogues, Yeshivas for children and Yeshivas for adults [and the houses in the neighborhood] are in a state of growth; some add a room, some an upper story, some two rooms and some two stories, and there is no one to complain again that it is too cramped for me to reside in Jerusalem.
>
> (p. 200)

Certain neighborhoods which are important historically, symbolically and in terms of the plot are described in detail. Such is the case with the Meah Shearim neighborhood, the fifth to be established outside the wall. Agnon aptly describes it as a "city within a city"[10] which was built at a distance from both the main thoroughfare of Jerusalem and from the existing neighborhoods:

> Meah Shearim stood lonely and desolate at its beginning within Jerusalem's wilderness. Not a single house existed from Jaffa Gate to Meah Shearim, save for the seven houses of Nahalat Shiva [the third Jewish neighborhood built outside the walls] and the houses of the Russian Compound [the complex of institutions built by the Russians in 1860 near Jaffa Road]. Ten houses were built in Meah Shearim,

allotting each family one room and a foyer. Every night a candle was lit in these homes to ward off robbers and brigands and one member of the household would remain awake and study the Torah the whole night long, for the Torah afforded protection and safety. It did not take long for their companions to arrive and build houses of their own. Thus Meah Shearim filled up and even its environs were built up, until it appeared that it had been swallowed up by Jerusalem, but it kept to itself and did not blend into the adjoining neighborhoods and remains standing as a city within a city.

(p. 202)

The author is not required to provide an immediate explanation why he dubbed the neighborhood a "city within a city." Extensive parts of the novel display how Meah Shearim constituted the very heart of the "Old Yishuv," and was sealed off by the way it conducted its Orthodox, even zealous, communal life. The life-style of its inhabitants was sharply distinguished (and remains sharply distinguished even to this day) from the life-style of the other Jewish neighborhoods: thus it is "a city within a city."[11]

Other neighborhoods described at some length are Batei Ungaren and Zikhron Moshe. The Batei Ungaren neighborhood was built by Kollel Hungarin, the economic, cultural and political organ unifying Jewish immigrants from Hungary. (Other Jerusalemites similarly organized in Kollels in accordance with their place of origin.) Shifra, Yizhak Kumer's future wife, lives in this neighborhood. This provides Agnon with the opportunity to give his readers an idea of the uniqueness of a Kollel neighborhood which unlike others contains no permanent residents but experiences a turnover every few years in order to permit the maximum number of Kollel members to enjoy its comfortable conditions.

In the western part of Jerusalem which adjoins Meah Shearim and Beth Yisrael stand fifteen large houses with three hundred apartments for the Kollel members who lived there for three years and sometimes longer periods at the discretion of their benefactor and those in charge of administering the Kollel. All the houses are similar and each apartment has two rooms and a small corner for the women to do their cooking. A large cobblestone courtyard meanders between adjoining rows where one finds the water cistern. Just as the houses were similar to each other so all their residents were stately personages who observed the Torah and the religious commandments.

(p. 266)

Zikhron Moshe is symbolic and representative of the "New Yishuv" neighborhoods in Jerusalem. It was a modern neighborhood – the physical and cultural opposite of those comprising the "Old Yishuv." It was built at the beginning of the twentieth century when elements of the "New Yishuv" began proliferating in the city,[12] and it was precisely at this juncture that Agnon immigrated to Jerusalem. By that time the neighborhoods established in the 1860s and 1870s had "become obsolete" (p. 204). Zikhron Moshe, by comparison "excels all other neighborhoods. It was established according to a modern plan and incorporated sanitary laws. Every house stands by itself," in contrast to Meah Shearim where "every house clings to its neighbor" (p. 237). In Zikhron Moshe "an avenue courses through the middle and trees are planted alongside the road and are slated to provide shade during the summer months and roofing for the booths of the Sukkoth [Feast of Tabernacles] holidays" (p. 237), whereas in Meah Shearim

which had no ... gardens or orchards or anything else to provide a sense of commodiousness ... the streets are barely wide enough to afford passage to a loaded camel and there is no additional intervening space.... As befitting a quality neighborhood its population similarly bespoke quality ... they were pleasant company in religious matters and towards their fellow men and did not persecute each other for their opinions ... some are merchants and store-keepers, some are teachers and scribes, some are journalists and some secretaries to charitable institutions. And since their houses exceed their earnings in size, they tend to rent out a room or two. Students in the teachers seminary, modest Jewish youths, seekers of knowledge, Hebrew speaking, meek of spirit and ready to accept authority dwelled there.

(pp. 202–3)

Some details are provided about other neighborhoods, probably selected precisely because of their intrinsic uniqueness. The Nissim Beck neighborhood near the Damascus Gate was composed of a number of Sephardic communities: Gruzians, Syrian Jews from Haleb and Iraqis (p. 301). Beit Yosef, small, isolated and the solitary Jewish neighborhood in south Jerusalem, most of which was in the hands of European religious institutions, receives mention: "and there were fourteen houses and a synagogue and because of our multiple transgressions they sold these to non-Jews" (p. 563). The Bukharan neighborhood gained renown because of the wealth of its inhabitants and possibly because it was increasingly penetrated by a modern population (p. 274). It was therefore one of the few neighborhoods where one could find seated together "couple upon couple of youths and maidens ... who would relate to one another matters which even Adam did not confide to Eve in the Garden of Eden" (p. 279).

Agnon also draws the reader's attention to one of the interesting and unique phenomena connected to the establishment of Jewish neighborhoods outside the city walls: that people contributed funds to build houses in a neighborhood not for the purpose of living in them but in order to donate those houses to one of the Jewish public institutions. These houses furnished a source of income to these institutions once they were rented out. The name of the contributor and his work were immortalized on special plaques which were affixed to the face of the building. Thus "any person dedicating a house in Jerusalem affixes a memorial stone in order that he may be fittingly commemorated on its walls and he inscribes his name and his pious generosity for everlasting memory extending until the final generation" (pp. 272–3). To this very day such memorial plaques can be seen on houses in many of the original neighborhoods established outside the wall.

The central Jewish institutions erected outside the wall during the second half of the nineteenth century and the start of the twentieth century also receive attention in the novel. There are descriptions of the Central Synagogue "Yeshuat Yaaqov" in the Meah Shearim neighborhood (p. 560) and the mill-house built by Moses Montefiore in Jerusalem. This project, for grinding grain into flour and intended to be wind-driven, is of vast historical importance because it constituted an innovative attempt to improve the lot of Jerusalem's inhabitants outside the wall.[13] Agnon also mentions the school of the "Kol Yisrael Haverim" (the "Alliance" Association of France), the girls' school established by the Ahim Association of London, the Bezalel Art School, the Central Library and the Culture House (pp. 290, 302, 326–9). Although important Jewish institutions within the city walls are also mentioned by Agnon (p. 331), it is strikingly evident that Agnon lays emphasis upon those institutions outside the walls.

The author does not ignore the role played by non-Jewish elements in the building of Jerusalem outside the walls, and mentions all the Christian communities present in Jerusalem: Greek Orthodox, Greek Catholics, Gregorian Armenians, Catholic Armenians, Syrians, Maronites, Copts, Ethiopians, Franciscans, Presbyterians, Lutherans (p. 283). The survey of Christian activity is, however, limited and it seems that Agnon's intention is to shed light on the characteristics of the Christian exodus outside of the walls and the Christian role in building the new city in comparison to the Jewish role in these same processes:

> And what would the nations of the world proceed to do … they would take houses, courtyards and lands and would build houses and distribute to every member of their community, be he rich or poor, an apartment for nothing, and this was aside from the houses which were constructed on behalf of pilgrims which stand vacant all year round.…
> The Armenians purchased an area in the south-west corner in addition to fields and villages outside the [old] city … and they re-invested rental fees from the apartments and stores which they let out to Jews and purchased lands and built houses and stores. The Greeks purchased the north-west corner, aside from the fields and villages outside the [old] city, and they re-invested the rental fees on apartments and stores which they let to people who did not belong to their community, and purchased lands and built houses and stores until they surrounded

Jerusalem. The Russians purchased themselves that very same field and built houses for themselves, their priests, monks and their pilgrims who come every year from Russia in order to kiss the dust on the grave of their messiah.

(p. 198)

Agnon also observes the rivalry between the Great Powers in Palestine which found expression in consular activity: "It is the manner of consuls in Palestine that when they witness the actions of their counterparts they are quick to emulate them because they are most excitable about their honor and hasten to display their might to Jerusalem" (p. 219). Let us note that this competition also found expression in the Christian building projects in Jerusalem.[14] We can also discover a little about the Moslem exodus from the walled city in Agnon's descriptions (p. 191).

The reader also receives an impression of a number of important sites outside the wall, such as Jerusalem's main thoroughfare, Jaffa Road, which bisects Jerusalem from west to east. Regarding this road – the way it operated, its urban centrality and its colorfulness – Agnon writes:

On Jaffa Road there are banks, houses of commerce and people who have no spare moment for matters which do not offer pecuniary profit.... At that time of day there was no trace of a Jewish presence. There was not a peddler or merchant to be seen ... neither buyer nor seller ... not a bagel hawker nor a raisin vendor ... not a cookery chef nor a manufacturer of dainties ... not a drink tapster, or a sorcerer ... not a charity collector nor a nondescript collector, no one to utter incantations or unearth transgression, no kisser of mezuzot, no beggars, rabbinical fund raisers, or freelance solicitors of charity ... in short there was not a Jew to be seen on the streets. There were only Ishmaelites [Moslems] and Edomites [Christians].... These were smoking water-pipes over a cup of black coffee ... the ones harboring thoughts about Adam, the others about Eve ... all wearing striped clothing, they were holding chains in their hands, counting the rings, as out of one ear they listened to wondrous tales.

(pp. 301, 582)

Yesterday before Yesterday also deals with the city's economic and social geography, especially of its Jewish community. The economic base of the Jewish community was the "Halukah." The "Halukah" system arose from the fact that the traditional Jewish community in general thought its purpose in life to be the study of the Torah rather than engaging in any productive occupations (p. 43). The "Old Yishuv's" economic existence was made possible by virtue of the funds gathered by Jewish communities abroad, who viewed the Torah scholars in Jerusalem as their emissaries in the performance of righteous deeds and considered themselves morally bound to see to their needs, which were sent to Palestine and ear-marked for distribution amongst members of the traditional Jewish community. These "Halukah" funds were distributed by the heads of each and every Kollel:[15]

> They constituted the main source of sustenance for Jerusalem. At the time when the Temple existed, Jerusalem was sustained by the Temple. Now it is supported from the alms of charity, for Jerusalem, unlike all the other cities which host trade and industry, contains Torah and prayer.
>
> (p. 329)

Agnon divulges hints as to the manner in which the "Halukah" was distributed (p. 218). It was thanks to the "Halukah" money that the building of Jewish neighborhoods became possible (p. 232).

The "Halukah" monies sufficed with difficulty to support the traditional Jewish community and conditions deteriorated further with the intensification of Jewish immigration to Palestine. Additionally, some of the residents of Jerusalem, despite the fact that they belonged to the "Old Yishuv," supported productive labor on ideological grounds and sought an end to the total dependence on others. As a result, quite a few turned to agricultural work, trade and handicrafts (p. 522). However, the opportunities were very limited, given the economic condition of Palestine in general and Jerusalem in particular:

> For Jerusalem has no factories and no one to pay off arrears and presently there is nothing to do there.... Jerusalem is a city where people cannot earn a living on a day-to-day basis.... A person can't do everything he wants to ... this is doubly true in Jerusalem which was not blessed with handiwork.
>
> (pp. 224, 230)

Additionally, even as productive occupations were possible, they were still tied in one form or another to the "Halukah" money and the conclusion that emerged was

> that there is no form of occupation in Jerusalem that does not contain a trace of "Halukah." One must realize that save for the "Halukah" the enemies of Israel [Agnon means the very opposite, i.e., the Jewish people] would perish from hunger for there is no trade and industry in Jerusalem – and from what source could they earn a living?
>
> (p. 231)

Agnon also addresses the social consequences stemming from the reliance of the traditional Jewish community on the "Halukah" and the generally negative attitude in this community to occupations apart from Torah study. This was the antithesis of Jaffa whose entire Jewish community belonged to the "New Yishuv" and supported itself by its own labor. For example,

Agnon shows Yizhak Kumer, a house-painter by trade, going in to a local Jaffa restaurant:

> Here one does not look askance at a painter because here a profession confers respect. Here well-to-do persons who engage in handiwork dignify the labor, whereas in Jerusalem, a city where bans and excommunications imposed on a vocational school are made public, how can handiwork be praised in it? Therefore all the artisans in Jerusalem are held in low esteem by the people and their own self-esteem is low. They do not experience a moment of fulfilment and one cannot detect a rhapsody in their work.
>
> (p. 398)

Yet, in other respects, Yizhak

> was pleasant company to his friends and his friends in turn were pleasant company to him, as is the case with all the artisans in Jerusalem who offered good fellowship because their spirits are low and their opinion is downcast. So too they are regarded as lowly and downcast in the eyes of the trustees who distribute "Halukah" to them according to the number of souls in their family; they do not provide them with any additions from the special funds that arrive from time to time as they are want to add to Torah scholars and other worthies.
>
> (p. 218)

Criticism was leveled at the heads of the Kollels because too large a portion of the "Halukah" monies remained in their hands. Agnon takes a clear stand on this issue:

> As for those in charge against whom it is imputed that most of the money remains in their hands ... we have not witnessed them during their lifetime promenading in carriages through gardens and orchards and they do not bequeath capital and wealth upon their death. The elders of Jerusalem still remember Rabbi Yoshi Rivlin; for twenty-five years he was a scribe and a trustee. He would tire himself out by day dealing with the poor, and at night he would exhaust himself in the study of the Torah. During his entire lifetime he did not experience any comfort but he would steel himself to bear the burdens and sufferings with good heart and meek spirit, and all the money that arrived in Jerusalem he would distribute to the needy. ... He did not eat meat nor drink wine even on the Sabbath and festivals, but contented himself with black bread. And when he departed this world he did not leave enough money for his widow or orphans to suffice for a single meal, but his legacy consisted of eleven neighborhoods which he had added on to Jerusalem.
>
> (p. 231)

In general, in contradistinction to the leadership of the "New Yishuv," which did not spare any criticism regarding the life-style of members of the traditional Jewish community because they did not engage in productive labor and relied on the funds of the "Halukah," Agnon lavishes praise upon the traditional Jewish community:

> Who was responsible for the building of the new neighborhoods of Jerusalem? The residents of Jerusalem! The recipients of "Halukah," who found themselves mired admist a motley group of avaricious and blood-thirsty enemies. In order to settle Jerusalem they did not spare themselves or their families. We who are their contemporaries can see their deficiencies and not their virtues. However, were it not for those people who expanded the boundaries of Israel, Jerusalem would have been constricted between the walls and all these places would have remained desolate.

(p. 232)

These examples should suffice to demonstrate that *Yesterday before Yesterday* provides richly detailed descriptions of both the locales and the physical and human milieu. The spatial and social processes that the novel chronicles broadly resemble the historical processes that actually took place. This prompted Kurzweil, one of the foremost Agnon scholars, to contend that *Yesterday before Yesterday* constitutes "an invaluable historical source regarding the history of the Jewish community in Palestine."[16] This contention is unacceptable and the novel does not merit the status of an authentic source of information regarding Jerusalem's recent history. Agnon's descriptions leave the reader with the notion that the novel and historic reality are interchangeable; on occasion, Agnon's descriptions and authentic reality are in total accord and this only further strengthens the illusion. This false identity breaks down and the differences emerge[17] when Agnon's descriptions are subjected to a rigorous comparison with the empirical reality captured in the research of historians and historical geographers. The following points demonstrate the differences between "empirical geography" and the novel's geography.

Agnon, in describing one of the reservoirs located outside Jerusalem's walls, emphasizes that the reservoir ran dry during the summer and therefore "Baedeker [the tourist guide authored by Karl Baedeker in 1876] erred when he wrote that it was full of water, since he viewed it during the rainy season" (p. 285). However, a check of the guide-book reveals that Baedeker explicitly noted that the reservoir fills up with rain-water during the winter months but is empty during the summer and autumn.[18]

Agnon reports the intention of the founders of Meah Shearim to sow seeds and plant trees in the neighborhood. He adds that this plan was aborted because planting and sowing require noxious-smelling fertilizers, which the founders deemed incompatible with Jerusalem's unique holiness (pp. 202–3). These descriptions have no historical basis and the historical record shows precisely the opposite. Meah Shearim's by-laws (which served as a binding planning and legal document) made explicit provision for planting trees in the neighborhood and emphasized that the fears that Jerusalem's holiness would be impaired as a result were groundless.[19]

Agnon's data regarding the area of land purchased for building Meah Shearim are 20 per cent higher than the area that was actually purchased. In contrast, Agnon understates by 40 per cent the number of houses built in Meah Shearim. Similarly, Agnon supplies figures for housing in Beit Yosef that are 50 per cent below the number of houses actually built in the neighborhood.[20]

Agnon supplies a demographic sketch of Jerusalem's population: "Jerusalem is a city with a majority of old people arriving from all places in order to die here" (p. 248). This general impression of Jerusalem's traditional Jewish community is negated by historical research which shows that a substantial percentage of the "Old Yishuv's" new members were not aged.[21]

The novel surveys Jerusalem's Jewish community from the middle of the nineteenth century until the eve of World War I. Agnon treats this entire period, as well as the actions and policies of the various forces and personages who animated it, uniformly. However, historians divide the period into distinct sub-periods; accordingly, it is to be expected that the forces and actors functioned differently in each one of these sub-periods.

Finally, Agnon erroneously identifies two Jewish neighborhoods, although he was undoubtedly aware of their real names (pp. 312, 563).

On the basis of these discrepancies, one can convincingly argue that historical geography, which aims to reconstruct and analyze an urban setting, cannot utilize Agnon's descriptions as an *exact* source for research purposes. This is no way detracts from the novel's importance, since *Yesterday before Yesterday* still manages to convey a sense of Jerusalem's spirit and milieu during that period. Agnon, himself a resident of Jerusalem during that time, employs his literary gifts to bring the city to life and imparts an in-depth appraisal to the events. He can accomplish this *tour de force* only via a humanistic approach. Kurzweil perhaps had this aspect in mind when he made his claim for *Yesterday before Yesterday* as an invaluable historic source. It is necessary from the standpoint of historical geography also to understand the spirit and culture of a given historical period. Agnon's ability to instill life into the period being researched directs the historical geographer to profound insights. In this manner *Yesterday before Yesterday* renders an invaluable service towards our understanding of Jerusalem in the period between the mid-nineteenth century until World War I.

AGNON'S PERCEPTION OF JERUSALEM

Earlier, I stated the assumption that an author discloses his subjective value-orientation to a specific milieu by the way he chooses to portray it. The author is therefore "present" in the geographic milieu that he creates.[22] Agnon serves as a prime example of this tendency, and this is clearly the case where Jerusalem is concerned. Literary critics and researchers specializing in Agnon's work have already made this observation. Kurzweil, in attempting to sum up the centrality of Jerusalem as a theme in Agnon's stories, contends:

> Jerusalem in Agnon's stories is the entire purpose of his epic project, it is the reality of realities ... which is identical to completeness. This completeness holds within it most antithetical phenomena: but it manages to weld them all together – the unreal, the ethereal, the legendary are part and parcel of Jerusalem's reality.[23]

Agnon treats Jerusalem as an entirety allowing the contradictions and the unreal to coincide with the real. On the other hand, Agnon does not idealize Jerusalem. From a Jewish standpoint, argues Kurzweil, Jerusalem is the center of the world:

> The city of Jerusalem, the holy city, is the point of gravity in the Agnonian epic. It is both the goal and the spirit. It is the meeting-place between the divine presence and the Jewish people. It is also the point of collision and struggle between the forces of belief and apostasy. The nobility, eternal nature and holiness of Jerusalem combine to subdue the opposing and hostile forces and induces them towards integration – to a new unity. It is this completeness of Jerusalem that constitutes this new unity which raises contradiction to the level of a reality of realities, to a sublime reality.[24]

Weiss adds that Agnon felt that his stories about Jerusalem justified his entire literary efforts. Jerusalem, for Agnon, constituted the very essence of the Jewish nation: "The entire Jewish people are dependent upon the Land of Israel and the Land of Israel is dependent on Jerusalem which is the very heart of the Land of Israel."[25]

All these perceptions of Jerusalem in Agnon – Jerusalem as an entirety which encompasses contradiction, Jerusalem as a reality of realities and Jerusalem as the essence of the Jewish nation – are strikingly corroborated in *Yesterday before Yesterday*. Agnon reiterates the encomia and merits of Jerusalem and the advantages that she bestows upon her inhabitants throughout the entire novel. Thus for example: "Jerusalem the holy city" (p. 215); "Jerusalem which even in its state of devastation provides air which invigorates the soul" (p. 199); "Its men are hearty and hospitable people" (p. 207); "It is a good deed to hear the praises of Jerusalem" (p. 208); "Every four cubits of Jerusalem endow a person with knowledge and under-standing" (pp. 230–1); "Jerusalem is well versed in miracles. The eyes of the Lord rove through the city and He does not withdraw his protection from it even during its destruction" (p. 235); "Jerusalem, the city of G–d, most excellent of all other cities, the divine spirit never departs it" (p. 258); "They relate about Jerusalem that the divine spirit never departed from it and even the gentiles resident in the country do not contradict these facts even during Jerusalem's destruction. And the divine presence is always resident near the Western Wall, which though desolate, retains its holiness and remains impervious to all the actions and importuning of the nations" (p. 259); "The people of Jerusalem consider themselves superior since they have been privileged to live in Jerusalem" (p. 328).

However, alongside this impressive roster of advantages, we encounter the precise opposite when Agnon describes the day-to-day plight of Jerusalem and her Jewish inhabitants. Agnon informs us of the drought and hunger conditions prevalent in Jerusalem during this period (pp. 209–300, 312) and of the constant economic privation: "don't say I am satiated. I don't believe you. I have yet to see a person who has eaten his fill in Jerusalem" (p. 213). Filth, squalor and overcrowding plagued various places (especially in the old city); diseases were rampant; and the problems of poor and wretched people cast into the street were all-pervasive (pp. 312, 344–5, 557–8). Agnon unflinchingly describes Meah Shearim and the paths leading to the Western Wall:

> And in front of every store either a lame or a blind person sits down upon the ground, a person missing limbs or afflicted with sores and set before him is a charity box. Flies and mosquitoes flutter between their eyes and the sun broils their wounds and glistens upon their charity boxes.... On every step of the stone stairs leading to the Western Wall are strewn clusters of indigent lame and blind people, amputees of arm and ragged of leg as well as the assorted other maimed and impaired ... bits of inhumanity whose Creator has abandoned them in the midst of his work and did not complete their creation; and when he put them aside, he set his hand against them and multiplied their afflictions or he completed their creation and they were smitten by his attribute of severity.
>
> (pp. 204, 349)

Agnon manages to sum up in a nutshell his sense of Jerusalem as a contradiction between her unreal and ethereal aspects and her oppressive reality, while clinging to his concept that Jerusalem constituted the essence of the Jewish people:

> Is there any other city in the world as holy or as dear and as cherished as Jerusalem? Yet why do these misfortunes beset her so? And is it not

sufficient that she has been visited by these commonplace afflictions that we also have no water to drink?... Is there a nation as comely as Israel among the nations? Yet we are smitten and flogged.

(p. 603)

Agnon's Jerusalem is a place for prayer and Torah study, a bastion of orthodoxy but also of zealotry (pp. 217, 222–3, 234, 266, 294, 304, 329, 471). Agnon's esteem for "Halukah Orthodoxy" did not deter him from criticizing those negative aspects of the "Halukah" that promoted idleness, divisiveness and fractiousness (p. 33). Jerusalem, since it possessed a modern community, could by herself supply an antithesis to her traditional community. Agnon strikingly contrasts the Beit Ha'Midrash (house of study) – the seat of the Torah scholars and the symbol of orthodoxy – to the Beit Ha'Am (house of culture), the meeting-place for the modern community and the symbol of the young generation's revolt and abandonment of orthodoxy (pp. 533–4):

At Beit Ha'Am, Yizhak met various people, amongst them a number of Jerusalem youths who came there surreptitiously unbeknownst to their parents because everyone who came to Beit Ha'Am was considered by them to be a cult-member or an apostate. When they were small they studied at the Eitz Haim [Tree of Life] Yeshiva. As they grew older they stretched forth their hands to the tree of knowledge. They sought their path in life not only through the study of Torah and Halukah, because their eyes had already been opened and they saw how many corruptions befall the Torah students and recipients of Halukah.

(pp. 229–31, 326–7)

Agnon expresses his concept of Jerusalem – as a city that radiated completeness while encompassing contradictions – most engagingly by intermittently weaving comparisons between Jerusalem and Jaffa throughout the novel.[26] These comparisons, covering various aspects, show how differently Agnon perceives these two important Jewish cities. The chasm between Jerusalem and Jaffa runs deep and cannot apparently be bridged: "between Jaffa and Jerusalem lofty mountains extend upwards" (p. 221). Jaffa reflects the partial as opposed to the complete. It is totally unbeset by contradictions. It is not the holy city (p. 378). It symbolizes the "New Yishuv" with all the consequences that derive from it. In Jaffa one cannot find members of the "Old Yishuv" or "a poor person begging" (p. 204). One does not encounter "Halukah" but people who earn a living (p. 378). Jaffa likewise does not lack for drinking water (p. 488). Its inhabitants appear bronzed and sturdy rather than bent and wan (p. 328). Jaffa is even more egalitarian than Jerusalem ("for Jerusalem is quite unlike Jaffa. In Jaffa everyone sits together. In Jerusalem artisans sit apart and the intellectual professions apart," p. 256). Outwardly, Agnon casts Jaffa's penchant for the partial in a generally positive light, but he drops reminders that Jaffa's incompleteness is a veritable quicksand. Agnon, as mirrored by Yizhak Kumer, cannot find his place in Jaffa and he prefers Jerusalem. Agnon describes one of Yizhak's strolls through the sands of Jaffa:

> Yizhak would walk about the sands of Jaffa. Yizhak who was used to the terrain of Jerusalem would ask himself if he could ever make it on foot through the sand. He had hardly managed to extract one foot and the second foot was already submerged in the sand.... Upon his return to Jerusalem, despite the melancholy feeling which permeated the city on account of the traditional days of mourning [due to the destruction of the Temple] ... Yizhak suddenly felt the terra firma of Jerusalem where a man does not lose his footing, as in the sands of Jaffa.
>
> (pp. 377, 505)

Agnon writes at the start of the novel: "and if Jaffa was not sanctified with the holiness of the Land of Israel, it was nevertheless privileged to serve as a gateway to the Holy Land. For all those who ascend to Jerusalem, the Holy City, must first go up to Jaffa" (p. 98). He counterbalances this praise by reminding us elsewhere in the novel that Jaffa was also the point of embarkation for those leaving the land of Israel (p. 181). Sonia, the hero's first girlfriend, who lives in and symbolizes Jaffa, comes to Jerusalem for a brief period to study at Jerusalem's art-school. She fails to make a go of it and returns to Jaffa. On her return, she draws a compelling comparison between the two cities:

> The nights of Jerusalem are beautiful but the days are languid. The sun burns like a flame; the garbage exudes a stench and the city is suffused by sadness. The clods of hardened mud assault your legs and you skip over the rocks like those foul-smelling goats. At every turn you encounter either garbage and filth or a beard and side-curls, and when you approach one of them he flees from you as if he has seen a ghost. Whereas Jaffa ... is chock-full of gardens, vineyards and orchards, it has the sea and coffee-houses and young people and every day one sees new faces ... those arriving by ship from abroad and those coming in from the agricultural villages ... there are those whom you wish to see and those who desire to see you ... and there are those with whom you can promenade up lovers' hill and they don't talk to you ... about all sorts of creatures who died many centuries ago. They talk to you about people who are alive ... you can love them or you can hate them; they are intimate with you due to a propinquity of time and place. When the people of Jerusalem speak to you they don't call you by your name or address you directly but they speak in the third person ... the lady, her excellency, her worthiness. When you address an individual by his

name and speak to him in the second person, he gapes at you in bewilderment as if you have trespassed the bounds of good taste.

(pp. 155–7)

It would appear that when the comparison between the two cities is drawn on the obvious and superficial level rather than on a profound, fundamental and eternal level, it is Jaffa that enjoys the upper hand. Yizhak himself concedes the point when he visits Jaffa to wind up his affair with Sonia. Jaffa's "superficial" allure beckons him to remain:

Look what a person sees in Jerusalem during the summer months. Torrid dust covers the city's horizon ... even the bird in the sky, even the dog in the street are covered with dust ... a person takes but a single step and he sinks into pits and ditches of dust ... the roads are filled with holes and cavities, crevices and sharp objects. Save for the carcasses of cats and dogs, insects and reptiles, Jerusalem would seem to be a desert. A noxious odor emanates from the carcasses and all sorts of flies and insects swarm through the carcasses and scatter dust about so you won't notice them until they set upon you suddenly and sting you ... but here you find the sea of Jaffa which gladdens one's heart and the green orchards, a delight to the eyes, and the red pomegranates which distill charm and beauty and the palm trees swaying in the breeze amongst the orchards and vineyards which stretch on and on ... and Jaffa enjoys this advantage as well over Jerusalem. Every place you go you encounter friends. Go into an eatery and no one will turn up his nose because you are a house-painter, because here an occupation confers honor on the person performing it.

(pp. 396–7)

However, as soon as Yizhak completes his comparison, Agnon has him recall his friend and fellow house-painter from Jerusalem and the conversation that the two had at one of their first meetings. Yizhak's friend initiated him into Jerusalem's profundity, depth and hidden aspects and provided him with an inkling into Jerusalem's completeness. He congratulated Yizhak on leaving Jaffa and moving to Jerusalem:

For there is not a single moment in Jerusalem which does not partake of the world to come. But it is not everyone who is privileged to see this, since Jerusalem only reveals herself to her lovers. Come, Yizhak, and let us embrace each other because we have been privileged to live in Jerusalem. At first when I would compare Jerusalem to other cities, I found many faults with her, but when my eyes were finally opened I saw her as she truly was. What can I tell you, my friend? Can language do her even partial justice?

(pp. 214–15)

Agnon concurs with the assessment:

> a pact has been concluded with every city which leaves its imprint on
> its inhabitants and this is evermore the case with the City of G–d which
> excels all other cities and from whence the divine presence has never
> departed. And if the divine presence is cloaked and concealed, there
> remain those moments when even the humblest Jew who has been
> privileged to live in Jerusalem perceives it. Every person perceives it
> according to the capacity of his sensitivity, according to his merits and
> according to the light of mercy which illuminates his soul. He perceives
> it thanks to the ill fortunes that he has endured in the Land of Israel,
> which he accepted with a loving spirit and without demurral.
>
> (p. 258)

Agnon expresses Yizhak's deep empathy towards Jerusalem in comparison
with his casual attitude towards Jaffa as follows:

> In the course of walking with his friends in the outskirts of Jerusalem,
> Yizhak felt how beautiful the city was. How could Jaffa and even Jaffa's
> sea compare? There was only one Jerusalem. He would not consent to
> live in another city even if he was given the entire earthly void in
> return.
>
> (p. 331)

A comparison that Agnon draws between the intelligentsia of the "New
Yishuv" in both cities is similarly intended to show Jerusalem's sense of
rootedness to good advantage in comparison with the superficiality of Jaffa:

> what is the difference between the wise people of Jaffa and the wise
> people of Jerusalem? In literary matters the wise of Jaffa have the edge
> because they are familiar with the literature and are acquainted with the
> authors. In matters of science, the wise men of Jerusalem have the upper
> hand because they are fluent in German and can derive things from the
> original source, as opposed to the wise men of Jaffa who study from
> Russian texts which are translations of the original German. Not every
> translator is proficient in the language from which he translates.
>
> (p. 328)

CONCLUSIONS

Agnon has demonstrated in his writings that the task of describing Jerusalem
is an end in itself. Jerusalem is not merely a neutral setting for the novel; it
functions independently as one of Agnon's major and most colorful heroes.
In *Yesterday before Yesterday,* Jerusalem exerts a powerful influence upon
the unfolding plot. Jerusalem is the heart and mind of the novel. One cannot

subtract Jerusalem from *Yesterday before Yesterday* without dooming the entire novel to extinction.

I have attempted to show Agnon's deep interest in acquainting the reader with the socio-historical and geographic processes pertinent to the expansion of Jewish Jerusalem beyond the city walls during the latter half of the nineteenth century and until the outbreak of World War I. He considered this task as well to be an end in itself. Agnon's descriptions were based on historical sources of varying levels of reliability, but these descriptions cannot be utilized by historians and historical geographers as a scientific source. The descriptions can, however, be regarded as invaluable source-material for deriving a sense of Jerusalem's general atmosphere and acquainting oneself with the general processes that transpired within the city.

Agnon conceives Jerusalem as an entirety and he has therefore chosen to describe both the real and the unreal; the sublime and the ugly; the sacred and the secular; the old and the new, and so on. Agnon does not idealize the details of Jerusalem, which conveys that Jerusalem, in Kurzweil's words, is a "reality of realities." As such, Agnon's Jerusalem, enhanced by its depth and sense of rootedness, is an urban entity which enjoys preference over rival cities such as Jaffa. Nonetheless, since Agnon regards Jerusalem as the essence of the Jewish people, the Jerusalem of *Yesterday before Yesterday* in terms of its physical and human landscapes is first and foremost Jewish Jerusalem. Agnon does not give the Christian and Moslem areas the attention that they proportionally "deserve," given their share of Jerusalem's population and built-up areas.

NOTES

1   The statistical details about the size of populations noted above are from Y. Ben-Arieh, "The growth of the Jewish community of Jerusalem in the 19th century," in Y. Ben Porat, B. Z. Yehoshua and A. Keadar (eds.), *Chapters in the History of the Jewish Community in Jerusalem*, vol. A (Jerusalem: Yad Izhak Ben Zvi, 1973), p. 108.

2   S. Y. Agnon, *The Complete Tales of Shmuel Yosef Agnon*, vol. IX, *T'mol Shilshom* [Yesterday before Yesterday] (Schocken Publishing, 1952) (in Hebrew). Page references are from the original publication.

3   *The Hebrew Encyclopedia*, vol. XXVI (Jerusalem: Encyclopedia Publishing Company, 1974), pp. 728–734 (in Hebrew). Most of Agnon's works have been translated into English and German. *T'mol Shilshom* has yet to be translated into English; I have attempted the translations that appear in this chapter, which hopefully do some justice to the original.

4   See, for example, Y. Ben-Arieh, *Jerusalem in the Nineteenth Century, Emergence of the New City* (Jerusalem: Yad Izhak Ben Zvi, 1974; New York: St. Martin's Press, 1987). Aside from this volume, additional volumes and various articles have appeared which deal with Jerusalem during the nineteenth century and up to World War I.

5   Regarding the "presence" of the author in his works see W. C. Booth, *The Rhetoric of Fiction* (Chicago: University of Chicago Press, 1961); H. Fish,

Jerusalem and Albion (London: Routledge & Kegan Paul, 1964); S. Katz, "Yerushalayim KeMaba Amnuti B'Yezirat Hazaz veShahar" [Jerusalem as an Artistic Expression in the Works of Hazaz and Shahar], unpublished doctoral dissertation (Ramat-Gan: Bar-Ilan University, 1978), pp. 2–3 (in Hebrew).

6 Regarding the "New Yishuv" and the "Old Yishuv," see in detail Y. Kaniel, *Continuity and Change: Old Yishuv and New Yishuv during the First and Second Aliyah* (Jerusalem: Yad Izhak Ben Zui, 1981), pp. 21–34 (in Hebrew).

7 G. Shaked, *Hebrew Narrative Fiction 1880–1980 in the Land of Israel and the Diaspora* (Jerusalem: Keter, 1983), pp. 206–9 (in Hebrew).

8 See, for example, *Yesterday before Yesterday*, pp. 570–1, for the following historical sources: P. Grayevsky, *Ha'Harash v'Ha'Masger B'Yerushalyim* [The Craftsman and the Locksmith in Jerusalem] (Jerusalem: Memorial Volume for Artisans, Workshops and Early Industry in Jerusalem from the Inception of the Ashkenazic Community, 1930), pp. 6–7 (in Hebrew). Grayevshy's works are considered outstanding historical soures for the history of Jerusalem: *Ha-Magid* (January 13, 1859). Compare *Yesterday before Yesterday* p. 569, with P. Grayevsky, *Sefer Ha-Yishuv M'Hutz l'Homat Ha'Ir* [The Jewish Community outside the City Walls] (Jerusalem: 1939), p. 45 (in Hebrew); compare *Yesterday before Yesterday* p. 522. Compare also the letter sent by Mordecai Solomon to Moshe Montefiore in June 1839, which appeared in A. Yaari, *Letters from the Land of Israel* (Jerusalem: Massada, 1971), pp. 409–21 (in Hebrew). A comparison of the details that Agnon provides and the language employed in the text would attest to the extremely high likelihood that Agnon employed the historical sources. See also S. Hagar, "*Yesterday before Yesterday*: The unified structure takes shape," in G. Shaked and R. Weiser (eds.), *Shmuel Yosef Agnon: Mehkarim u'Teudot* [S. Y. Agnon: Studies and Documents] (Jerusalem: The Bialik Institute, 1978), p. 156, n. 5 (in Hebrew); A. Holtz and T. Holtz, "The old man of Jaffa-Moritz Hall," in *Tarbiz, A Quarterly for Jewish Studies*, 59/1–2 (1989/90), pp. 215–16.

9 *Yesterday before Yesterday*, p. 329.

10 ibid., p. 202. Meah Shearim was constructed in 1874 and was the fifth Jewish neighborhood to be built outside the wall.

11 On the place of Meah Shearim in the novel see also M. Tochner, *Pesher Agnon* [The Meaning of Agnon] (Jerusalem: Massada, 1978), pp. 67–71 (in Hebrew).

12 On the neighborhood see Ben-Arieh, *Jerusalem in the Nineteenth Century*, pp. 214–18.

13 ibid., pp. 74–5.

14 See for example N. Katzburg, "Features of the development of Jerusalem outside the City Walls," in H. Z. Hirschberg (ed.), *Yosef Y. Rivlin: Memorial Volume* (Ramat-Gan: Bar Ilan University Press, 1964), pp. 37–9 (in Hebrew).

15 On the "Halukah" see, for example, M. Eliav, *Eretz Israel and its Yishuv in the Nineteenth Century, 1777–1917* (Jerusalem: Keter 1978), pp. 110–29 (in Hebrew).

16 B. Kurzweil, *Massot Al Sippurei Shmuel Yosef Agnon* [Essays on the Tales of S. Y. Agnon] (Jerusalem and Tel Aviv: Shoken 1963), p. 103.

17 See n. 8.

18 K. Baedeker, *Jerusalem with its Surroundings* (1876; Jerusalem: Carta, 1973), p. 125; *Palestine und Syrien* (Leipzig: 6th edn, 1904), p. 61; *Palestine and Syria* (Leipzig: 4th edn, 1906), p. 61; *Palestine and Syria* (Leipzig: 5th edn, 1912), p. 68.

19 By-Laws for the Meah Shearim Neighborhood for the year 1889, *Jerusalem Municipal Archives*, p. 16.

20 Compare *Yesterday before Yesterday*, pp. 202–3, 563, to Y. Ben-Arieh, *A City*

Reflected in its Times: New Jerusalem – The Beginnings (Jerusalem: Yad Izhak Ben Zvi, 1979), pp. 155–7, 228 (in Hebrew).

21 A. Morgenstern, *Messianism and the Settlement of Eretz Israel* (Jerusalem: Yad Izhak Ben Zvi, 1985), pp. 92–3 (in Hebrew); H. Assouline, *A Census of the Jews of Eretz Israel (1839)* (Jerusalem: Dinur Center, 1987), pp. 9–20 (in Hebrew).

22 See n. 5.

23 Kurzweil, *Massot Al Sippurei*, pp. 301–10.

24 ibid.

25 H. Weiss, "Jerusalem's position in Agnon's works," in *Maariv* (September 9, 1977), p. 38.

26 Historical and geographic research also addresses the socio-cultural differences between Jerusalem and Tel-Aviv in the period that our chapter deals with. See Y. Kaniel "The conflict between Jerusalem and Jaffa over leadership of the Yishuv in Late Ottoman times," in J. Hacker (ed.), *Shalem*, vol. III (Jerusalem: Yad Izhak Ben Zvi, 1981), pp. 185–212 (in Hebrew); A. Kellerman, "Cultural uniqueness and conflict between Jerusalem and Tel Aviv: Does history repeat itself?," in S. Husson (ed.), *City and Region*, vols XIX–XX (Jerusalem: Ministry of the Interior, 1989), pp. 120–4 (in Hebrew).

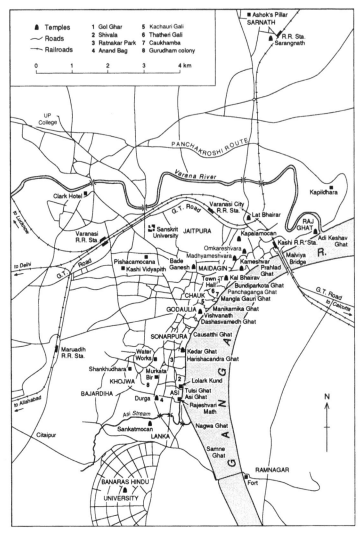

Legend:

▲ Temples
〜 Roads
⊢⊢ Railroads

1 Gol Ghar
2 Shivala
3 Ratnakar Park
4 Anand Bag
5 Kachauri Gali
6 Thatheri Gali
7 Caukhamba
8 Gurudham colony

0　1　2　3　4 km

Map 12a
The "real"
map of
Varanasi

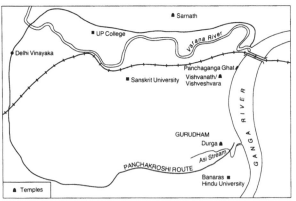

Map 12b　A
cognitive map of
Varanasi

12

MODERN VARANASI

Place and society in Shivprasad Singh's
Street Turns Yonder

Rana P. B. Singh

> I am Vishvanath, the Lord.
> Kashi is the light of liberation.
> The waves of the River of Heaven (Ganga)
> are the wine of immortality.
> What can these not provide?
>
> *(Kashi Khanda* 53.43)

INTRODUCTION

Literary geography is a relatively recent branch of cultural geography. Literature as a resource for geographic analysis has a strong base in India, but formerly only the "objective" use of creative writings was made to reconstruct geographic environments in historical context. It was only since the 1970s that the "subjective" use of creative writings has received attention in Indian geography to project the experiential feelings of place and people (see Dhussa 1992). If geography is to be defined as "the study of the earth as home of mankind," literature would prove to be a very rich resource (primary or secondary) to highlight landscape and environmental ethics (*earth*), built space and emotional bond (*home*) and culture, society and values (*mankind*). With its integrated triad of people, plot and place, literature provides a substantial base to reconstruct places, to decipher the images and to understand the socio-cultural order in terms of historicity (Herbert 1991: 194).

Reality is usually thought of as passing through the filters of the novelist and the reader (Herbert 1991: 195); nevertheless, the expression of reality depends on the imaginative grasp of the novelist, as to how the novelist integrates the tale, facts, narration and the final message. Many Indian novelists no longer live in the area where they were born or have written about, but Shivprasad Singh and his novel *Street Turns Yonder* (1991) are the product of the soil and deep experiences. The mythic frame of the city of

Varanasi is narrated through the medium of a historical context – a method very commonly used by Shivprasad Singh. When the novel was being written, the novelist frequently roamed the field with an aim to keep the tale and plot close to reality. Thus sometimes some plots became heavy and almost unnecessarily lengthy; however, the novelist's claim that he had never left Varanasi is valid.

VARANASI/BANARAS

Once Mark Twain stated "Benares is older than history, older than tradition, older even than legend and looks twice as old as all of them put together" (1898: 480). According to chemical and archaeological investigations, the city records continuous settlement since c. 800 BCE (Eidt 1977: 1332). Says Kane, "There is hardly any city in the world that can claim greater antiquity, greater continuity and greater popular veneration than Banaras. Banaras has been a holy city for at least thirty centuries" (1973, vol. IV, p. 618). Varanasi is not only a city, but a cultural and archetypal core in itself. The city presents an extraordinary location along the crescent-shaped and northerly bank which comprises 7 km. of the Ganga (Ganges) river. This physical deployment of the Ganga occurs only at Varanasi in its whole course.

There are 84 *ghats* (steps up to the river) made of stone along the Ganga. The unique river-front landscape is decorated with a series of palatial buildings. The sacred territory is delimited between the Asi stream in the south and the Varana river in the north; thus Varana and Asi together contain Varanasi. The ghats are linked with narrow lanes converging in the west into main and broad roads.

Mythology relates that in the ancient era Lord Vishnu (the protector), one of the deities of the Hindu trinity, was doing meditation here. Lord Shiva (destroyer and controller of time) came here with his consort, Parvati, to relax and finally settled here. Then the city became the city of Shiva, where he lives in 324 forms (12 zodiacs multiplied by 9 planets multiplied by 3 mythological realms).

No Shiva *linga* (an image) in Varanasi is called "Shiva"; rather all are represented with the suffix *ishvara* (old Sanskrit: "*i*", the center, and "*chara*," moving around, i.e., the center around which the cosmos moves) to the name. (This happens also with other gods: thus Vishveshvara.) This city is the axis mundi of the cosmos. Those who die in the sacred territory get relief from transmigration and finally get a seat in the realm of heaven of Lord Shiva. The first light of wisdom was reflected here; the city was known as "Kashi" from Kashya: the city of light. Over the course of time there developed varieties of the juice of life, always ready to serve the people, so it is known as "Banaras." It is never forsaken by Shiva ("Avimukta") and also provides the supreme joy in the natural settings of five forests ("Anandavan"). From time immemorial the funeral pyre was lit here ("Mahashmashan"). As a seat of learning, light, delight and *joie de vivre*, there also developed a unique culture of the carefree: a city of *masti*. These unique characteristics are preserved in about 526 fairs and festivals throughout the year, the existence of over 3,000 Hindu temples and about 1,400 Muslim shrines and sacred places. Vaishnavites (the devotees of Vishnu), Shaivites (the devotees of Shiva), Shaktas (the devotees of goddesses), Jains, Buddhists, Sikhs, Christians and several other minor religious groups have lived here together for centuries.

The city has developed on the archetypal plan. There are seven layers (sacred routes) of Ganesha images (the "elephant-headed god," the son of Shiva and directional guardian, popularly called "Vinayaka") and at each of the junctions in eight directions a Ganesha image protects the territory; thus his number reaches 56. Ganesha protects the sacred territory with the help of his 56 assistants (*Birs*, "brave heroes"). The twelve sun images provide the rays of hope for each of the months. A large mass of religious literature describing the glory and merit of the sacred sites and associated divinities is available (see Rana Singh 1993: 319–20), among them the *Kashi Khanda*, the fourth part of the *Skanda Purana* (c. twelfth century CE), which contains 11,624 verses describing the sacred sites and their myths related to the city. However, only limited literary works have been produced about the city and its people.

In its historical past Varanasi was famous as the most sacred place for Hindus, and so people from all corners of India came and settled here and brought with them their local religious and cultural traditions, which are preserved even today. The spatial patterning of the cultural groups and their associated shrines and temples with the passing of time mixed with the local soil. The varieties of gods and goddesses and their shrines and icons in many cases show the spatial transposition of the original sites lying in the different parts of India. Varanasi is identified as microcosmic India: it is said that "if you have seen Varanasi, you have seen the whole of India." Almost all the Shiva shrines scattered throughout India are spatially transposed in Varanasi.

Around 528 BCE the Buddha visited the city and gave his first sermon, "Turning the Wheel of Law." In the fifth century BCE the first grammarian, Panini, composed his *Ashthadhyayi* ("Book of Eight Chapters"). By the late fifth century CE the temple of Vishveshvara (or Vishvanath) (the patron deity of the city) was built and many Sanskrit schools and monasteries were already established. The two Chinese pilgrims, Fa-Hein (c. CE 405) and Hiuen-Tsang (c. CE 635), had visited the city and described its glory. By the late tenth century the city flourished as one of the great centers of learning and pilgrimage.

From the turn of the eleventh century until the sixteenth century the city had faced waves of demolition and plundering – it had been invaded at least eight times. Almost all the shrines and temples were destroyed but soon afterwards in more favorable times Hindus renovated and re-established the shrines and icons. The temple of Vishvanath was demolished four times between 1194 and 1669.

During the fifteenth to seventeenth centuries came a new wave of *Bhaktism* ("devotion") introduced by the great reformists and poets of the period like Kabir and Tulsi. In this period many religious treatises, epics and puranic mythologies were written which after the passage of time became the guidelines for the survival and continuance of Hinduism (see Rana Singh 1994).

By the late eighteenth century the East India Company gained control over the city, followed by British rule. Many ghats and temples were built or renovated during the eighteenth and nineteenth centuries, mostly under the patronage of Marathas from Western India. The early to mid-twentieth century was the period of the struggle for independence; as a shelter and seat of national leaders, the city had received its own special position in history. On August 15, 1947, India became an independent republic, and since then Varanasi has become the headquarters of the district and also of the commission of the same name.

Lying in the heart of the Ganga valley, Varanasi's urban agglomeration records over a million inhabitants at present. The city historically has been a seat of household industry, specializing in silk saris, toys, brass utensils, wall-hangings and similar handicrafts. The spinning and weaving industry is also famous in the city, involving the majority of Muslim workers.

THE NOVELIST AND HIS URBAN EXPERIENCE

Shivprasad Singh makes no claim to be a regional novelist; however, with contextual differentiation his novels contain the details of landscape and areas. He has published five novels, all in the Hindi language. Two of these are on the culture and landscape of Varanasi, *Street Turns Yonder* ("Gali Age Mudati Hai") and *The Blue Moon* ("Nila Canda"). His main intentions are to capture the temporal theme of student unrest in Varanasi and life as it was in medieval Varanasi, respectively; the description of landscape has been the essential subtext of both the novels. In this chapter, *Street Turns Yonder* is considered as a base onto which to project place and modern society.

While planning to write on Varanasi the novelist decided to prepare a set of three volumes dealing with the modern, medieval and ancient city. The city of Varanasi provided a backdrop and shaped the view of the Indian social world that he sought to convey in his three novels (the third one on ancient Varanasi is in progress). He worked on modern society in the 1970s using the students' reactions of the late 1960s against the present system of education and the use of English for teaching as his core themes. That is why there appears to be a discontinuity in describing the landscape and culture of the city. The novelist feels that since the city has been the strongest seat of educational learning, this theme should occupy the central position and that social degradation and students' reactions are the result of that time. He further stresses the need and demand of the society to know the modern culture of Varanasi. These are the main forces behind writing *Street Turns Yonder.*

The selection of Varanasi and the disturbed social scene for his canvas reflect the novelist's experiences from his early education to his employment in Varanasi. Shivprasad Singh was born on August 19, 1928, in a village about 50 km. north-east of Varanasi. He first visited the city in 1947 as a young student when he was admitted to the 11th grade in the UP College. He received his MA in Hindi literature in 1953 and his Ph.D. in 1957, both from the Banaras Hindu University. Soon afterwards he joined the department of Hindi there as a lecturer, and was promoted to professor in 1983. He finally retired in 1988. Currently he lives in the southern part of the city (Gurudham Colony) and still enjoys his childhood habit of wandering in the city lanes along the Ganga river-bank.

Before starting the novel, the novelist had devoted his energy to walking around Varanasi, meeting the people, watching the activities and rituals and also critically examining the basic writings on the city, ranging from mythologies and religious treatises to contemporary historical sources. With this involvement, the novelist cites such sources and experiences frequently, and so his narration becomes more relevant in understanding the culture.

It is commonly believed that *Street Turns Yonder* is mainly about the student movement. This, of course, is the setting; nevertheless, it is about something wider than that – the search for humanity in the holiest city of India, the uniqueness and distinct personalities of its dwellers, the horrors of terrorism which developed as a subculture and the philosophical and archetypal aspects of the sacred territory and the northerly flowing of the most sacred river, the Ganga (Ganges). The Ganga is taken as a simile for the life-world of Varanasi.

It is clear from the author's conversation that "the spiritual magnetism and contemporary culture" were the forces that inspired him. Predominantly, at first he felt that the feelings of the Hindus toward Varanasi as a beautiful, cosmic and sacred city are close to reality, and at different places in the novel this is fully described. The novelist feels that writing a novel about such a city needs a certain state of realization and experience, especially a struggle for survival among the accidents and sufferings in life. His own struggle to get a job in the Banaras Hindu University and the tragic death of his two children, including the death of his daughter in 1984 due to severe illness, were part of his repository experiences. He feels that in the dense darkness of his life, he had created a light through writing the novel to satisfy his own soul. Above all, "death" is the real truth of life; those who realize this fact can tell the truth explicitly. The novelist says that he had experienced personally the sorrow of the dying glories of this city but learnt to have peace in the darkness. Of course, the tale and narration are based on imagination, but the underlying realities are always preserved in *Street Turns Yonder*.

The novel is at most places strong on descriptive details and frequently refers to history, illustrated with mythologies. However, the novel has been criticized on this ground, one reviewer remarking that "this is an attempt to view myth from the vision of modernity, and modernity from the vision of myth"; in another sense this is the merit of the novel, since it projects the continuity of tradition.

In the field, both with the novelist and alone, the present author found most of the characters of the novel in and around the sites and spots described in the novel. Though they speak for themselves, the identity of the location at some places is, admittedly, inaccurate. Reising has rightly warned: "the text's actual meaning is not a position or idea articulated by an author through a text, but a struggle within the writer that the Lawrentian critic can discern by trusting the tale" (1986: 166).

Replying to his critics, the novelist says in the preface to *The Blue Moon*:

> Only those can understand the novel in full who have lived in Varanasi like the dwellers of the city. One can't perceive the eternity of this city by showering a few drops of the Ganga water on his body, or only watching the scene while sitting on the ghats. It is essential to have experience of the holy dip in the Ganga, lived experiences of the street-culture and participation in the festivities and sacred journeys as a *pilgrim* not as a *tourist*.
>
> (1988: ii)

Above all, every person has the right to say whatever he experienced, as an outsider or insider. The novelist is successful in narrating his experiences through his descriptive power and sheer virtuosity. In his own words the novelist says metaphorically:

> I attempted to project the varieties of images of this distinct city which are sometimes dark, muddy, swampy; however, there is a light and life too. Also, there is "the light of spirituality" shining at various places in the novel. "The turning of the street" is the "turning of life." Who can narrate it without deeper experiences of sufferings?
>
> (interview on October 7, 1992)

The cognitive map of Shivprasad Singh is developed from his own mental topography of the city. He was asked to draw the map in reference to the places mentioned in the novel (Map 12b). His perception of the edges in the four directions – the Ganga river in the east, the Asi stream in the south, the Varana in the north, and Panchakroshi route in the west – is clear and strong. In a physical sense, there is a contrast between reality and perception to a certain extent. However, directions, relative locations and relative scales are close to their actual counterparts. The novelist's cognitive map can be compared with a "real" map showing the important places mentioned in the novel (Map 12a).

In the novel itself the novelist has captured almost all the possible characteristics of the most common places and has also given more emphasis to the places in the central part and those along the Ganga river. Even the imaginary plots, characters and associated sites are (indirectly) closely related to reality. This shows the novelist's deeply rooted experiences and feelings.

CULTURAL ORDER: PLACE AND SOCIETY

The city of Varanasi, which has been continuously inhabited since c. 1000 BCE and attracted settlers from all over India, forms a multicultural center of a variety of images. Perseverance of the old traditions and acceptance of the new go together and still co-exist. Variety, distinction and assimilation are all woven to form the cultural personality of this city (known as "Banarasi," or "Banarasian" in English). The distinction and diversity related to place and society in Varanasi may be seen in *Street Turns Yonder*, under the following four broad contexts.

Holy and cultural center

Metaphorically the city of Varanasi is known as the abode of Shiva settled on his trident. These are symbolized with the three main sacred segments identified with the pilgrimage circuits and their patron deities whose shrines lie on the raised ground: from the north to the south they are Omkareshvara, Vishveshvara and Kedareshvara (see Rana Singh 1988: 5). In its physical milieu, Varanasi is famous for three unique properties: the northerly flow of the Ganga as base, the two cremation ghats as places of the last rites of life, and the golden temple of the overall patron deity, Lord Vishveshvara (Shivprasad Singh 1991: 32–3). Furthermore, three rivers delimit the edges of the city: the Varana in the north, the Asi in the south and the Ganga in the east (p. 38). The number three refers to the integrity of the three realms (heaven, the earth and the world below), the three qualities (good, moderate and bad), and the three eyes of Lord Shiva; altogether it shows "wholeness." Moreover, since the patron deity of the city is Shiva, the city necessarily assumes universal identity (see Rana Singh 1993).

According to the novelist, the Ganga river flows like an arc (p. 246), and the circle drawn from the radius between Panchaganga ghat and Delhi Vinayaka, the gate in the sacred territory, limits the most sacred territory (p. 113). His statement requires a small correction: "Panchaganga ghat" should be replaced by "the shrine of Madhyameshvar," which is the true center of the cosmic circle. In fact, this is the only sacred city demarcated by a route of five *krosha* (i.e. 11 miles/17.6 km.). The radius is like the thread of a garland and the Shiva lingas are like flowers (p. 343). In total there are 108 shrines and sacred sites on the route where pilgrimage has to be completed within the period of five days.

The pilgrimage to the outer limit of the sacred territory of the city as determined by the Panchakroshi route is commonly preferred during the intercalary month of the leap year (the *Purushottam masa*, p. 341). The architectural beauty of the Shiva lingas (phallic emblem) is metaphorically explained in the light of divine understanding. About the grand Shiva linga at Parma village the novelist says:

> Here is a human-sized Shiva linga. A great snake moves round and bounds the linga.... This shows that female energy (*yang*) is always keen to get male matter, Shiva (*yin*), in her arms, but Shiva accepts that touch and appeal as a divine ornament. The heat of infinity bounds him, but Shiva makes it cool and puts it on his forehead.
>
> (p. 344)

This description refers to the mystic and divine power of Shiva. It is this special nature of Shiva that attracted all the gods/godlings from various holy sites of India to come to Varanasi and establish the lingas.

After each two lingas on the Panchakroshi route there lies a weight-tower (*bojhateka*) to provide pilgrims with rest: pilgrims can relax and then proceed onward (p. 344). There are also theological explanations and the novelist comments:

> I thought earlier so, but now understood that these weight-towers are for those who were exiled from the city by the curse of the Kapaleshvara. Being not relieved from the sin of murdering Brahmins, and not allowed to enter the city, they stand as weight-towers along the circuit. This city is reserved for Brahmin sacrifices.
>
> (p. 344)

The high sense of sanctity and mysticism developed in the past at Varanasi attracted people from every corner of India. Devotees from Bengal, Gujarat and Bihar came and made their own distinct neighborhoods (p. 68). Bengalis and Gujaratis are famous for celebrating goddess-festivals on a grand scale with varieties of rituals (see pp. 71, 83). After Calcutta, the kaleidoscopic scene of the celebration of a goddess-festival (Durgapuja) is held only at Varanasi. With the integrity of faith, decorated lighting and dancing performances during that period (September–October) the cityscape is seen at its zenith. There are three regional representations of the festivities at that time: the goddess-festival of the Bengalis (the east), the Garaba of the Gujaratis (the west) and the Ramalila of the local Bhojpuri culture. This presents a microcosmic vision of India (p. 80).

Alienation

Varanasi has been famous from ancient times for strange and contrasting traditions. Sacramental glory and manipulative culture go side by side. Citing a myth of the second century BCE (*Tandukanali Jataka* I.5), the novelist relates that King Brahmadatt played a trick to purchase the best horses at the cheapest price. The king's secretary put a notorious horse among best horses when they were resting at night, and the next day some of them were found bitten. This enabled the king to purchase the bitten horses more cheaply (Shivprasad Singh 1991: 273).

Another myth (*Mahabharata* XIII.31.26–28) refers to a king of the third century BCE, Divodas, and explains how he got a boon from the Brahma, Lord of Creation, and removed all the gods, godlings and ghosts from the city. There was no fear of the gods and no terror of the ghosts among the inhabitants. The city had a very peaceful and pleasant life. Later on, to disturb the city life, gods and godlings came in other forms. Ultimately the Vinayaka came in the form of a Brahmin and advised the king to leave the city only for three weeks. During this period the whole city became disturbed, and finally the king left the city forever. The city was again recaptured by gods and ghosts. Shiva came and finally settled here (Shivprasad Singh 1991: 192–3). The peaceful life in the city had been replaced by upheaval. Since then upheaval prevails.

The characters of Bakkad Guru and Rajjuli are representative of disturbance-makers and rowdy persons (*Gundas*). They mostly work underground, while on the surface they always pose as people promoting good life in the city. The novelist has vividly narrated several incidents and tales revolving around them, both human and inhuman.

Most of the ancient places have mythic connotations. The oldest myth refers to a king, Harishachandra, who donated all his glories and properties for the cause of preserving truth and ultimately sold his wife and son to the funeral priest and himself became a servant working as a tax-collector in the cremation ground (*Markandeya Purana* VIII). At the climax of the test to see whether he would preserve truthfulness he did not allow even his wife to cremate his own dead son without paying the tax. The Lord of Heaven, Indra, finally appeared and declared the king's victory for truth (Shivprasad Singh 1991: 32–3). Ultimately the king, his family and all the followers got a special place in the heavens. The novelist expresses his disappointment with satire:

> Alas, O King! It had been better if you and your followers had not gone to heaven. Alas! You have not relieved yourself from the wish to leave this earth and go to heaven. If you and your people had stayed here on this earth it would not be controlled only by falsehood and flatterers.
>
> (p. 33)

The myth of the loss of the rivers Dhutapapa and Kirana (*Kashi Khanda* 59) refers to the curse of the beautiful divine girls who became rivers and eventually disappeared. The girls, Dhutapapa and Kirana, attempted to destroy the chastity of the Lord of Truth, Dharmaraja, with sex appeal, but finally the Lord saved himself (Shivprasad Singh 1991: 347). The site of this incident is known as "Panchaganga," which means "five rivers" (the two cited above and the three that mark the edges of the city: the Varana in the north, the Asi in the south and the Ganga in the east).

These mythological references are metaphors to explain the origin and maintenance of the cultural tradition: how manipulation was used to save money and maintain administration, how superstitions and terrors were removed (in spite of knowing that at the end again they may continue), how the truth was preserved but finally vanished and how sex had been shown to have less merit in the contest with chastity.

Even today the other face of the city is easily experienced. The most popular couplet in Varanasi says: "Widows, bulls, steps and ascetics; tolerate them; only then one could enjoy Varanasi/Kashi" (p. 141). The narrow lanes with human crowds and wandering cows, bulls and ascetics are the specialty of the city (p. 361). Following his common practice of contextual differentiation, the novelist exclaims:

> Brothers!
> Let's hear the story of the lifeless city.
> Here, on bricks and remains is written its sufferings and pains.
>
> (p. 80)

An analysis of the contents of newspapers also gives the impression that the city is now reserved for sinners: "It is a lifeless and unknown city famous for corrupt deeds and murder as reported in the daily newspapers. This has given me the impression to perceive Varanasi as the resort of *Gundas*" (p. 337).

The city has recorded the mythic history of manipulating tricks. Even in later days there are recorded several such examples which tell the stories of black markets and smugglers (p. 102). One of its root causes is related to the story of Harishachandra, who went to reside in heaven, leaving only liars and wantons in the city. Some of the writings of the medieval period refer to a similar condition (Rana Singh 1989: 213). Even as late as 1856, Dayanand, the founder of Arya Samaj and a great reformist, was declared defeated in the religious dialogue when he was tricked by being asked for an explanation of a false sacred metre from the *Upanishada*. "The art to make right from anything wrong has flourished well in this city. The city is well-known for the art of transferring anything into reality from the garbage of falsehood through tricky manipulation" (Shivprasad Singh 1991: 45).

Using the persona of the hero, Ramanand, the novelist expresses his own feelings about funeral homes where corpse-carriers come, sit, relax, wait, then proceed to the cremation ghats. These houses also have a different sense of value related to *joie de vivre* and delight:

> Formerly I never understood the function of corpse-carriers but when the city became an abode of *joie de vivre* for me, and I too became one among the many street wanderers, I slowly realized the act of natural happening lying therein.... Death is also a celebration here. If you could not believe it, watch the duties of corpse-carriers. After cremating the dead body and taking a purifying dip in the holy Ganga, they sit at the street-shops in Kachauri Gali and enjoy *puri-kachauri* [wheaten cake fired in deep butter]. While taking food they talk about various moments, scenes, ways and happenings during cremation, completely in a non-attachment mood.
>
> (p. 280)

In many places the novelist has commented on the notorious activities of the *pandas/ghatias* (priests at the ghats). Their self-praise (p. 255), their wine drinking, their association with low-grade people (p. 256), their plans to ensure getting a good place in heaven through sacred rituals (p. 174), and their involvement in promoting crimes at the ghats (p. 200) are some of their distinct characteristics. However, the majority of devotees solicit their help in getting the rituals performed at the ghats.

In the city incidents of terrorism can also be noted frequently. "Terrorism is a great social disaster. It is like a black cloud moving all around all the time" (p. 275). In fact, "It is like the germination of bacteria in the human body.... It is an infectious disease which spreads over a large mass" (p. 275). The novelist feels that living in this city requires a special training for passing daily life (p. 214). However, among the rowdy persons some are different; in fact, they always try to be humane. Such *Gundas* are open-minded and kind-hearted. "I am happy that still such kind of *Gundas* are living in Varanasi, of course in a pocket edition" (p. 351).

In spite of several such strange situations, the culture of *joie de vivre* and being carefree is maintained in the city. Remarking on the continuity of this tradition, the novelist cites the story of Tulsi (CE 1547–1623), a great Bhakta-poet: "Tulsi, the great poet of the people ... You know! Tulsi consumed poison throughout his life for the good of the society and gave nectar to the mass of the community" (p. 77). Tulsi was a symbol of the carefree; he says:

> Say me wanderer, ascetic, washerman,
> Or Sari-weaver, whatsoever may be!
> I would take food by alms, that's why there is no need
> To receive, or ask anybody, whosoever may be.
>
> (p. 57)

In CE 1620 when the shadow of Saturn fell on Pisces, the city faced an epidemic-like plague. Almost all the priests left the city, but Tulsi fought against it. He was serving the sick and became victim to the plague, and finally died in 1623 (p. 77). The novelist feels that at present similar astrological conditions prevail. Following Tulsi's idea the novelist says: "In this Dark Age neither asceticism nor wisdom pays. Both are false" (p. 78). This is comparable to Tulsi's saying:

> Now trickster flower, bear fruit and spread
> While holy ones at every instant perish.
>
> (*Kavitavali* VII.171.1)

Again, exclaims the novelist, "You are worried for your own! See the artisans, craftsmen, beggars, servants, thieves, artists and spies. They do not have even bread to sustain living! The 'demonic Ravana of poverty' has snatched our food and we become slaves of poverty" (p. 77).

Before Tulsi a social movement was initiated by another great reformist and poet, Kabir (CE 1398–1518). Kabir challenged the overseers of religion, the Brahmins (of Hinduism) and the mullahs (of Islam), and their fanatic and ritualistic religious patterns. He gave the message that peace comes from the soul, not from rituals (p. 292). However, "rarely one visits his shrine. And those who visit do business under his name. The chiefs of monasteries and elites both are enjoying only worldly pleasure" (p. 292).

Such contrasts in time and space give the city a strange environment. Since Varanasi is one of the most sacred places of pilgrimage, it attracts beggars also in search of livelihood based on charity, and they are an important part of the cityscape. An intelligent beggar says: "The Giver is only one, and the whole world is a beggar. I have seen how even *big* persons do anything, right or wrong, for money. Even we beggars cannot do that!" (p. 231).

The environments of monasteries and homes for widows are pitiful and filthy. Isolation, lack of means of subsistence, unauthorized occupation of the land and built areas and depression are the common factors there (pp. 23, 97). The weavers are also the victims of exploitation, demoralization and complicated ways of bargaining (p. 262). The novelist believes that good and bad have run side by side since the growth of human culture; they are counterparts (Rana Singh 1989: 218).

The city has long been famous for educational institutions and learning: however, at present the scene is polluted. There are four universities and an all-India-level institute. The single-minded teachers generally do not relate politics to teaching, but have to pay the price for this (Shivprasad Singh 1991: 30). In fact, in India the path of truth and worthiness is filled with thorny bushes and is always unstable. That is why people rarely try to follow the path of worthiness (p. 93). To adjust to the situation many teachers try to prove themselves ultra-modern by "taking tea, or coffee, breakfast and dinner with students. This, they believe, identifies them as so-called 'modern' intellectuals" (p. 165).

The main cause of student unrest is assumed to be the generation gap

(p. 270). "On every front there appears a tussle between old and young generations" (p. 163). The future for youth in India is not one of prosperity: "What will happen to the country where citizens are trying to seek solutions through magic, superstition and charisma!" (p. 177). Schemes to get fellow-ships (p. 93) and increased corruption in examinations (p. 39), together with strikes and late examinations, are part of a common routine in the universities in Varanasi and also in the rest of India (p. 140).

Twilight zones

The idea of twilight zones is related to the symbol of light and darkness, both in natural and cultural environments. The theological name of the city of Varanasi as "Kashi" is derived from *kashya*, referring to the reflection of light. The light-giving properties of sun, moon and stars also refer to transcendent values interpreted by artists to express their own insights. Light is explained as a "transcendent source or alternatively ... an immanent pervasiveness" (Dillistone 1986: 58). On the other hand, the city is also known as the "City of Death" symbolized by darkness. The novelist explains this idea in *Street Turns Yonder*:

> How grand is the city of Varanasi? Why did the Ganga turn in arc-form at this site only? And if she took the northerly flow in arc-form, how do its turnings get converted into this great city? The city looks like a chaste virgin girl putting the pot of Indian culture on her bent waist. Ah! How grand this umbrella-shaped, everlasting and immortal light in arc-form is!
>
> (p. 19)

This is based on puranic description:

> What is that divine light
> Reflecting over water in arc
> Which even during cosmic flood
> can be seen as it was?
>
> (*Kashi Rahasya* 2.13.38)

The myth refers to when Lord Shiva left Kashi and sent Surya (the sun) to make a disturbance in the city and also to search for a place for him. The sun came with this intention but fell deeply in love with the city and ultimately settled in the southern part, known as Lolarka Kunda (Shivprasad Singh 1991: 193). Later there were built twelve sun-shrines, each one representing a month.

The Ganga river symbolized Shiva's energy in the form of his second wife, and flows northerly in Varanasi. She is glorified in Hindu literature as "the supreme purifier" and the cradle of Indian culture and civilization; however, she has several other faces too. The novelist has vividly described light and dark sides of the Ganga in twenty-three "colors" or aspects, which are categorized into seven main groups.

In the first group are religious aspects: the morning scene along the Ganga is comparable to a young female-snake trying to bind the whole city (p. 126); in this way the Ganga river delimits the sacred territory (p. 341) and is well known for 84 ghats and a series of lofty buildings (p. 347).

In the second group are aspects of transportation: since ancient times the Ganga has also served as the channel of transport, means of subsistence, shelter and relief from the heat (p. 353). During floods patients are carried on boats from congested parts to the hospital (p. 308).

In the third group are aspects of water: the water level of the Ganga starts to increase after Ganga Dashahara (the tenth day waxing of Jyeshtha, May–June) and slowly its width expands (pp. 37, 57). When the water enters into city lanes, safety becomes a problem (p. 70) and there develops a special situational bond between flood and time (p. 78). During the flood it appears that the Ganga is trying to replace the city; it "enters and flows wherever it finds ways, lanes and streets" (p. 71). Everywhere muddy water is seen (p. 246).

In the fourth group are aspects of beauty: in different seasons the Ganga reflects rays of vivid beauty and inspires an ascetic sense and mood. During the rains the clouds kiss the river (p. 18) and in November the light of the rising sun reflects on the calm current of the Ganga and appeals to a divine feeling (p. 125). Similar conditions prevail in April (p. 232). In this period, together with the overall greenery and yellow mustard flower, the landscape is beautiful and fantastic (p. 267).

In the fifth group are aspects of ritual: since ancient times Varanasi has been the most sacred place of Hindu ritual and thus the Ganga river-front attracted people from all parts of India to settle along the ghats (p. 23). These dwellers daily took a holy dip in the Ganga and performed the morning rituals in their own ways (p. 295).

In the sixth group are spiritual aspects: the Ganga is also perceived as a life-symbol and a place of spiritual message. For boatmen, like a mother, the Ganga nourished them (p. 353); for others the silent dawn among the Ganga inspired self-revelations (p. 146). The northerly flow of the Ganga gives a spiritual message for "returning to the origin" (p. 357). The south is a realm of death and the north an abode of immortality. This is a message of the transcendental value of place.

In the seventh group are aspects of living: the Ganga also provides shelter which facilitates cheating and sins (p. 200) and provides beggars with sleep (p. 57). Sometimes people use boating during dawn to enjoy sex (p. 160) and sometimes the river-front at dawn looks like a mirage (p. 347). Due to light and darkness the same place changes into different environments at different times. Sarnath may be cited as an example (pp. 282, 285).

Describing the dawn and darkness as a condition for sexual pleasure, a wandering ascetic explains to the hero:

> Have you seen couples mating in the Sankata temple? What have you thought about these holy temples? ... All the corrupted deeds are performed there! Are these deeds right? You are living in the university. Tell me what happens during darkness in the temple there. Is that justified? Tell me the truth.
>
> (p. 198)

At dawn and in the darkness the hotel culture changes into a different world. The rich and westernized, tourists and businessmen come to the hotel in the darkness and enjoy the call-girls – this is another Banaras! (pp. 217, 224).

Gender

This woman has come like living soil: sow seed in her, ye men!
(*Atharva Veda* XIV.2.14)

In Hindu society a woman has four distinct roles in the life-cycle: as daughter and sister; as wife and daughter-in-law; as mother; and as mother-in-law. The position of mother is always given a special status with respect to her emotional bond with the family, children and husband. The mother, responsible for the education of her son, especially after the death of her husband, must take all precaution not to disclose the difficulties and troubles she faces. Sometimes feelings are exposed: "I have provided you with my blood in the place of milk to look after your progress" (Shivprasad Singh 1991: 39). The only wish of a Hindu mother is that her son may get a good and high occupation and receive a good sum of money (p. 40).

The divine sense of site and of the mother-goddess is vividly developed in Varanasi. The hero, Ramanand, expresses the transcendent value of a divine site: "Whenever my mind is stressful, my thoughts irrational and I feel mentally a vacuum, I pay a visit to Durga temple and certainly receive peace, hope and a new inspiration for leading life" (p. 41). Durga is the leading mother-goddess in Varanasi (p. 80). The dwellers of the city are fed by the Annapurna, "goddess of food," and overall life is watched by the patron deity, Vishvanath. The novelist laments that due to the loss of divine sense and human courtesy at present Vishvanath is blind and Annapurna is too old (p. 353). That is why the city is out of order, and beggars roam everywhere. On the occasion of the new-moon day of Karttika (October–November), the festival of Divali is celebrated with a view to welcoming the goddess of wealth, Lakshmi. For this purpose all the shops and houses are white-washed and decorated (p. 264). The icon of Mangla Gauri ("Auspicious White Goddess") always blesses her devotees for piety and well-being. And, without good will, how can one do good work? (p. 347).

In Hindu society inter-caste marriage is generally not accepted. Of course, there is also a group of educated people in the city who never accepted the old rules. The hero, Ramanand, says:

> I never like to see the relationship between woman and man through the lens of old-age rigid tradition. In no way do I like brutality in such a situation. Modernity may be accepted. Love may flourish.
>
> (p. 349)

However, most families and the rest of society do not accept such marriages:

> The prestige of the family has already fallen down. In whatever condition Arati had got herself married to a fellow of another caste, I cannot accept it even if I follow modernity. It has nothing to do with this. We lost our relationship to our village society. Uncle never even thinks to visit us.... How much hatred and reactions are in the eyes of countrymen? No limit at all!
>
> (p. 233)

Similarly the marriage of a rich girl to a poor groom is also an issue of social degradation. It is also difficult for a poor husband to meet the requirements of a rich wife (pp. 352, 356).

Woman is perceived as a life-giving substance, but more commonly as a means for relaxation and entertainment (p. 156). People frequently cite passages from the old religious texts which advise avoiding illegal and socially restricted sexual relations with other women in the place of one's wife (p. 222). However, there are many incidents of rape of poor girls (p. 183). A Harijan (cobbler) girl reacts:

> We are poor Harijans. That's why we have no right to value our prestige. The world measures prestige at the scale of caste hierarchy. The prestige of a Brahmin girl is considered higher than a Harijan girl. You are free to measure! I have nothing to say. Anyway, we have the same feeling of pain and suffering after rape. The same disappointment. The same desire to commit suicide. But who thinks in this way? Nobody! Nobody! How animal-like our lives are!
>
> (p. 186)

Challenging the dominance of Brahminism, a Harijan girl reacts in anger:

> Let these books of religious laws be burnt. All these books describe how to torture women.... How amazing are the regulations of the social world! How amazing the books of law are!
>
> (p. 150)

At another place a Harijan woman expresses the view that the sacred verses of the Brahmins are now dead; whatever blessings they give have no fruit at the end (p. 172). On the other hand, there exists an underground hotel culture where, with the company of call-girls, businessmen and rich people enjoy sex. The owner of a hotel explains: "These are the photographs of 'call-girls.' For each grade of girls pay the fixed rate of money and enjoy her in the way you like" (p. 226).

Most of the Harijans live in slums, a filthy and un-hygienic environment. Women in these slums face crucial conditions:

> Early in the morning women in the Harijan colonies go to toilets in the open spaces. Most of the males also do the same. Women can go to the toilets either around before four in the morning, or after around nine at night. Children use for their toilet garbage sites or garbage pits. Most of the women avoid taking excess food in order to avoid going to the toilet in the day hours.
>
> (p. 189)

The portrait of women in Shivprasad Singh's novel presents both sides of life, but gives more emphasis to the harsher aspects of the traditional picture.

POLYTHEISM

Most of the narration and plots of *Street Turns Yonder* are gleaned from the regional context; however, they show the reality of the pan-Indian scene of the Hindu social world – well known for its "multiple arrangements." In all spheres of society, Hinduism records paradoxical expressions and adjustments. Hindus are proud of their ancient glory but sorry for the ongoing cultural deterioration. These issues are expressed in several places in the novel. Since man is the most sublime creature on earth, "to know a man perfectly is essential. Only then could one intermix with others. But who knows the other perfectly?" (p. 313).

The quest and search for the other half-part (male and female energy) has continued since the ancient past, and hopefully will continue in future. Hindu mythology notes that Lord Shiva with his trident divided all life-organisms into male and female. Search from both sides is always in progress with a view to forming unity. This specific form of Shiva is known as "Ardhanarishvara" (half-male and half-female body). This divine form creates the mysteries in the ever-expanding universe (p. 196). However, illusion and material attachment have put man into darkness. Commenting on the Hindu faith, Mitra says that "absolute truth is a false notion. A little mixture of falsehood makes the truth a real appearance" (1985: 260).

The Hindu view of life also emphasizes acceptance of the guest: "Here everybody is a guest. Someone for a few days, others for a longer period, but everybody is a guest. There, the question of hostship does not arise" (Shivprasad Singh 1991: 93).

Hindu society is seen as deteriorating not only in terms of family affairs (p. 337) or social ties like inter-caste marriage (p. 299) but also in its politics.

The handling of politics in India becomes the means to get the "chair" (pp. 274–5). The political system in India follows the motto that "the Government is *off* the people, *bye* the people, and *far* the people" (paraphrased after the famous address given by Abraham Lincoln on November 19, 1863). An honest character shuns taking any part in politics (p. 275).

Adjusting to injustice and misdeeds and flattering to please powerful persons with a view to maintaining one's position and getting undue benefits are common in Indian society; of course, their roots go to India's ancient past (pp. 90, 318). Through the back doors there exist many job opportunities that do not exist through proper channels (p. 245).

The climax of depression may lead to suicide. At some stage, when all the ways to lead life are closed, a man tries to get relief through the act of suicide. Jamana, presently a wandering ascetic, was remembering his bad days in the past and expresses how he thinks afterwards:

> If life is felt to be useless, the realization between "to be" and "not to be" is lost; then the only way to eternal peace is by committing suicide! … With a great destiny we are born as human beings; however, if life becomes depressing and terrorized like that of beasts, then suicide is not a bad retreat.
>
> (p. 235)

The city of Varanasi is perceived to be a refuge for sinners – an extraordinary cultural notion in this city where "a thousand persons perform religious singing in groups (*kirtana*) honoring the murky lamp; there are few people who can tolerate the light of truth" (p. 243). Compare Chaudhuri (1979: 172), who cites a Sanskrit couplet based on Wilson's translation from the *Vishnu Purana*:

> For those who are ignorant of the revealed scriptures and the sacred traditions, who have abandoned purity and proper conduct, and for those who have nowhere else to go, for them Banaras (*Varanasi*) is the refuge.

The distress and disappointment make the novelist react with a sarcastic "utopian" vision. He opines that the mythological sage Vishvamitra will create a new form of man to annihilate human attributes that get in the way of man's social ambitions:

> I am Vishvamitra. I would make a man whose arms will be in the legs so that without bending down he may touch the feet of elders (so-called). One of the eyes will be in the back to watch the critics at the back. I will remove both the ears of man. No need to hear any person's prayer. No need even to do this as man never hears even with ears. I will put only a hole at the place of the nose – there is no need for facial beauty.
>
> (p. 303)

The novelist's new man is imagined with some of the basic "drawbacks" resolved through replacement of different parts of the human body; now man is freely adapted to his own degredation.

Despite these problems, a great country like India and its culture survive. When the hero Ramananda raises his voice, his experienced friend replies:

> Oh! Stupid! How could you be alive without combatting these smugglers, politicians, liars and rowdy persons living under white garments? Obviously you can survive, but life would be even worse than death.
>
> (p. 120)

Metaphorically, the novelist suggests, life is not only a circle or straight line, but rather a series of turnings. From one turning one goes out and reaches other turnings. Varanasi is famous for such a distinct network of street-turnings – symbolizing human life. The novel ends with this sense of perceiving the spatial structure of Varanasi and is associated with symbolism: "Be patient. Follow on the road, and at the turning go out: wherever you want to go, you can go – Maidagin or Chauk" (p. 361). Maidagin and Chauk are, respectively, the symbols of the "open plain" and the "road-crossing" and refer to the two life situations. Man is free to select one of the conditions for passing life. Human wills and wishes are the processes that enable us to control a path of passing life.

CONCLUSION

With closer examination of the characters, acts, places, and plots of *Street Turns Yonder* one can get an in-depth tour of the city: I personally feel that no regional urban geography has yet grasped the lived scenes of Varanasi so completely. Most of the writings on Varanasi are religious and emphasize its glories, ignoring the "city-life" as experienced by its dwellers. In this way the novel becomes a source of data where attitudes, values, perceptions, feelings and a sense of faith together present the overall picture of such a complex and historic city. Imaginations and realities meet so closely that they become two sides of the same coin.

Finally, to compare the modern scene with the medieval and the ancient, one has also to see two other novels by Shivprasad Singh, *The Blue Moon* (Nila Canda, 1988) and *Vaisvanar* (forthcoming). The novelist's march from modern to ancient is comparable to the northerly flow of the Ganga river: a search to the source!

REFERENCES

Atharva Veda (written *c.*1000 BCE; 1895), Bombay: Venkateshvara Press. In Sanskrit.
Chaudhuri, Nirad C. (1979), *Hinduism: A Religion to Live By*, Oxford: Oxford University Press.

Dillistone, F. W. (1986), *The Power of Symbols in Religion and Culture*, New York: Crossroad.

Dhussa, Ramesh C. (1992), "Trends of literary and humanistic geography in India" in *National Geographic Journal of India*, 38 (1–4), pp. 75–86.

Eidt, Robert C. (1977), "Detection and examination of anthroposols by phosphate analysis," in *Science*, 197 (September 30), pp. 1327–33.

Herbert, David (1991), "Place and society in Jane Austen's England," in *Geography*, 76/3, pp. 193–208.

Kane, P. V. (1973), *History of Dharmasastras*, 2nd edn., 5 vols., Poona: Bhandarkar Oriental Research Institute.

Kashi Khanda (the fourth part of the *Skanda Purana*) (written *c.* 12th century CE; 1961), Calcutta: Gurumandal Granthmalaya, no. XX, vol. IV.

Kashi Rahasya (written *c.* 16th century CE; 1957), Calcutta: Gurumandal Granthmalaya, no. XIV, vol. III.

Mahabharata (written *c.* 4th century BCE; 1933–59), critical edition by V. S. Sukthankar et al., 19 vols., Poona: Bhandarkar Oriental Research Institute.

Markandeya Purana (written *c.* 7th century CE; 1904), Trans. and ed. F. E. Pargiter, Delhi: Indological Book House. Reprinted 1969.

Mitra, Vimal (1985), *Literary Reminiscences*, Varanasi: Vishvavidyalaya Prakashan. In Hindi.

Reising, Russell J. (1986), *Unusable Past: Theory and the Study of American Literature*, London: Methuen.

Singh, Rana P. B. (1988), "The image of Varanasi: sacrality and perceptual world," in *National Geographical Journal of India*, 34(1): 1–32.

——— (1989), *Where Cultural Symbols Meet: Literary Images of Varanasi*, Varanasi: Tara Book Agency.

——— (1993) (ed.), *Varanasi: Cosmic Order, Sacred City and Hindu Traditions*, Varanasi: Tara Book Agency.

——— (1994, forthcoming), *Varanasi, the City of Holy Order: Sacred Cartography of Space, Time and Faithscape*, Varanasi: Tara Book Agency.

Singh, Shivprasad (1988), *The Blue Moon*, New Delhi: Vani Prakashan. In Hindi (Nila Canda).

——— (1991), *Street Turns Yonder*, rev. and expanded edn., New Delhi: Radhakrishna Prakashan. Orig. 1974, Delhi: National Publishing House. In Hindi (Gali Age Mudati Hai).

——— (forthcoming), *Vaisnavar*, New Delhi: Radhakrishna Prakashan. In Hindi.

Tandukanali Jataka (written 2nd century BCE; 1956), in *The Jatakas*, vol. I, no. 5, trans. B. A. Kausalyayan, Prayaga: Hindi Sahitya Sammelan.

Tulsi (written 16th century CE; 1985) *Kavitavali*, ed. H. P. Poddar, Gorakhpur: Gita Press.

Twain, Mark (1898), *Following the Equator: A Journey around the World*, Hartford: The American Publishing Company.

Vishnu Purana (*c.* 6th century CE; 1840), trans. H. H. Wilson, Calcutta: Punthi Pustak. Reprinted 1972.

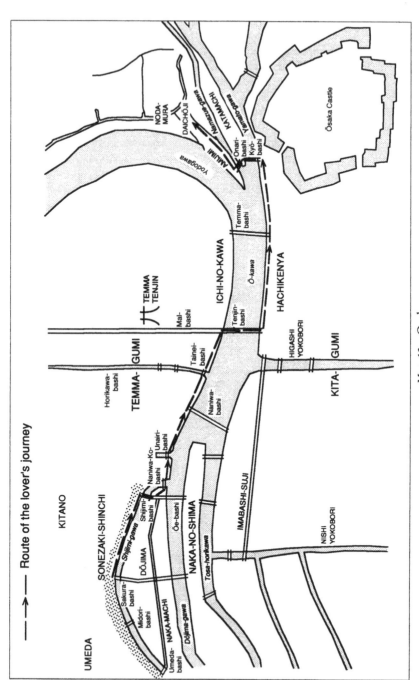

Map 13   Osaka

13

CHIKAMATSU'S OSAKA

Jacqueline Gibbons

INTRODUCTION

Japan's Tokugawa rulers established an exclusionary policy towards all foreigners that was to last until the 1850s. There was, however, one foreign physician who wrote extensively on Japanese culture and who penned the following in 1692 of Osaka:

> Osacca is one of the five Imperial Cities. It is both commodiously and agreeably situated in the Province Setzu, in a fruitful Plain and on the banks of a navigable river ... its length from West to East, that is, from the suburbs to the ... castle, is between three and four thousand common paces ...
>
> The river is narrow indeed, but deep and navigable. From its mouth, up as far as Osacca, and higher, there are seldom less than a thousand boats, going up and down, some with merchants, others with the Princes and Lords of the Empire, who live to the West of Osacca.... The banks are rais'd on both sides into ten or more steps, coarsely hewn of freestone, so that they look like one continued stairs, and one may land wherever one pleases. Stately bridges are laid over the river at every three or four hundred paces distance, more or less, all of which are built of ... the best cedar-wood of the country. They are rail'd on both sides, and some of the rails adorn'd at the top with brass buttons ...
>
> The houses are, according to the standing laws and custom of the country, not above two stories high, each story of one fathom and a half, or two fathoms. They are built of wood, lime and clay ...
>
> Osacca is extremely popular, and ... can raise an army of 80,000 men, only from among its Inhabitants. It is the best trading town in Japan, being extraordinarily well-situated for carrying on a commerce both by land and water.
>
> (Kaempfer 1910: 1–6)

The quotations from these journals of the day give us a powerful picture of a landscape and a city where prosperity and trade were dominant.

243

In the popular writing of the period one finds a particular playwright who stands out above others in his art of describing the culture of the city, and he is Chikamatsu Monzaemon (1653–1725). He was the playwright who can be said to have heralded in the new drama of the day, and for whom, in the eyes of the many Japanese scholars, the drama embodied a bold and new creative energy which reflected the contemporary way of life and the morality of the period.

The history of Japanese literature would not be the same if it were not for the drama of this educated and sensitive observer of life. Chikamatsu brought to Osaka and the Japanese stage the very flavor of city life, where the smells, senses, passions, fears, trials and tribulations of urban dwellers could be captured for the centuries that followed. Certainly his prestige and significance for posterity rivals Shakespeare, Dante or Chaucer in the western tradition, in the way he depicted human frailty and the human condition, and in his brilliant usage of prose and verse.[1]

His plays were immensely popular in Osaka, Kyoto and Edo (now Tokyo), during his time of writing, and he continues to be performed today in Japan and many other parts of the world. This attests to his appeal through the centuries and across cultures.

How is it that Chikamatsu managed (and still manages) to draw audiences (and the reader) into his world of the late seventeenth and early eighteenth century? First, his use of language united a style that draws from classical language and from colloquial everyday speech: this is especially characteristic of the dialect and lively speech used in Osaka. He also mixed classical songs, poems, Chinese couplets, Buddhist hymns, Confucian sayings, folk-songs, passages from nō plays and proverbial wisdom (Miyamori in Chikamatsu 1926: 48) in remarkable combination and sophistication, utilizing rhetoric and a profound knowledge of the Japanese tradition in literature in complex and brilliant ways.

Second, he drew from the genres of the "historical" and "domestic" plays, but it is in the latter that his genius has withstood the test of time, for it was this that he developed to a great height. Unlike the historic plays, which some have argued lacked "realism" (Shively in Chikamatsu 1953: 18) the domestic play drew from real life in the day-to-day examination and exploration of love, work and obligations (giri) as these relate and interconnect with one another. "If the history play may be said to appeal to the eyes and ears, the domestic play appeals to the ears and the heart" (Shively in Chikamatsu 1953: 18). The domestic play has been described as "natural" and "realistic," with enhancement of emotionality depicted through the evocative use of the music. It explores people's social problems, the issues of the day and, importantly, the dilemmas of contemporary life (Shively in Chikamatsu 1953: 19). The lives of the Osaka merchants are delineated with comedy and tragedy.

Chikamatsu wrote twenty-four domestic plays. Almost all deal with a love-affair, and over half address love-suicides. His plays draw on popular tales of the day and described life of the time and actual events in the lives of Osakans in ways that brought crowds to the theater. More often than not, the love-affairs that he portrayed – particularly in *The Love-Suicide of Amijima* (1720) and *The Love-Suicides at Sonezaki* (1703), each set in the city of Osaka – were real-life situations or events that had taken place just before the writing and performance of the plays. These immensely popular plays were, it seems, newsworthy as well as touching. Indeed in 1722 plays on this theme were banned for a short while because it was said that they were causing too many love-deaths in real life (Shively in Chikamatsu 1953: 13).

Third, and more to our purposes here, Chikamatsu lived for a significant part of his life in Osaka, where some of his most important and popular plays were performed. Despite the fact that the city has changed immeasurably, some of the districts, the bridges and the "routes" that he describes remain. Thus the Temma and Tenjin bridges still exist and contour the contemporary city-scape.

Chikamatsu often depicts Osaka's pleasure quarter, where men could escape from family and work, to be entertained by other women, spend their money lavishly if they wished and feel "freed" from the normal constraints that daily life laid upon them. The city had a well-developed pleasure quarter and it was close by here that the theater stood where Chikamatsu's plays were performed.

It is clear that in delineating the cultural and physical geography of Osaka, Chikamatsu drew audiences who understood his plots and their settings.

THEATRICAL AND SOCIAL BACKGROUNDS

Before we look at the daily life of Osaka portrayed through a dramatic text, it is necessary to explore a brief history of its theatrical genres. The history of Japanese drama goes back to the earliest of the religious dances, the *Kagura* of the Shinto temples. The legends record that the Sun Goddess decided to hide in a cave because of her disapproval of her brother's behavior (Miyamori in Chikamatsu 1926: 10). This darkness throughout the world worried the other gods. During this gloomy period the goddess Uzumē did a comic and witty dance outside the cave entrance to cajole the Sun Goddess out. At last, the sun emerged and ever since then the sacred dance has been performed through the centuries.

As it developed, this dance incorporated pantomime and masks: it has had singing as accompaniment, and also instrumental music, especially in the past, particularly the flute and drum. Actors would imitate the different deities and demonstrate their whims, their characters and their deeds.

From this early dramatic form four significant classes of drama emerge (Miyamori in Chikamatsu 1926: 11). These comprise the *yōkyoku* or *nō* play; second, the play that represented a comic interlude – the *kyogen*; third, the *kabuki*; and a fourth, the *jōjuri* or puppet play. Though Chikamatsu drew from the broad spectrum of the dramatic tradition of Japan, he refined the art of *kabuki* and *jōjuri* to new heights, and it is to these last two genres that we will look, in his renderings of a drama of contemporary urban life.

By looking at the development of drama in Osaka and Japan in the decades that surround the Genroku period (let us say 1680–1740), we observe a general consolidation of political control. The period was also marked by peace and unprecedented prosperity throughout the cities. This new interlude was a remarkable contrast to the warfare of previous times, and it gave time and space for the development of all the arts. These new times were also marked by an economy that was expanding and developing at a great rate: thus goods circulated with increasing vigor, the money markets multiplied and credit houses and the famous Rice Exchange of Osaka represented the increased development of the economy and made for a rise of wealth that touched most classes of life.

The major development of affluence occurred for the merchant class who acted as middle-men in much of the business doings of the city. As professionals, they took on government projects, moved commodities from one city or region to another and acted as money-lenders and speculators in the burgeoning of an affluent economy. Of all cities in Japan at this time, Osaka represented the greatest development of a bourgeoisie. The townsmen in Osaka had great autonomy in the conduct of their business and organization of their lives. However, it should be remembered that the official class system in Japanese culture at the time did not take economic status into consideration and, in theory, the merchants were the fourth in a class hierarchy that comprised, above them, from the top, samurai, farmers and artisans. Like a new class of *nouveau riche*, the merchants sought to command respect with the money they earned and with the wealth that was accumulating among the townspeople.[2]

Social codes

These political and economic conditions play an important part in the dramas that unfold in Chikamatsu's plays. He depicts the class confrontations in terms of life-style, obligations or *gīri*, human love and frailty. He illustrates the strict code of *gīri* which incorporates role obligation, expectations about social relations, moral principle and law and the embodiment of tradition in the life of an individual as he or she is woven into this cultural matrix. Chikamatsu suggests on many occasions that these obligations oppress the individual; they divide people's loyalties and interfere with human passion. At the same time as admiring the samurai code for its following of Confucian principles and its stalwart adherence to the centuries-old feudal code, he argues through his characters and plots that these comprise principles that derive from a warring past when life was lived through fealty to one's lord and the might of the sword. Chikamatsu points to the dilemmas of contemporary society where peace reigns, yet where the family and the obligations of loyalty threaten to crush the passions of individual men and women as they fall deeply in love or are deeply in debt in a world where there may be only one way to go: namely, a "love-suicide."[3]

We might argue that the social code had been transformed from one of a military format to a civil or civilian format. Previously, a military adherence to social rules meant no major personal dilemma because discipline and loyalty questions were so clear-cut and "institutionalized." With national peace and the development of a new culture for the townspeople, there was a questioning of the old order and there was now cultural permission to indulge in one's feelings. Chikamatsu deftly explores the realm of human feeling (ninjō). He demonstrates how gīri could break a man, and where Buddhism forgave human frailty (Shively in Chikamatsu 1953: 28). He showed the compassion that the Buddha would exercise as love-suicide was contemplated in the plays, and he showed how this contrasted with the harsher principles of Confucianism through the centuries.[4]

Chikamatsu's plays were especially interesting to the local townspeople because during the current long peace there were many unemployed samurai (rōnin) in Osaka, a number of whom altered their professions to become writers, painters, teachers or doctors, while others were poor, or unemployed, or married their children into the merchant class to alleviate their upper-class poverty. People came to Osaka from nearby Sakai, and the growth of Osaka was linked to its being the base and center for famous families of the time. We know that the writer had served with noble families and scholars have generally accepted that he was himself of samurai birth (Shively in Chikamatsu 1953: 13).

It was during the Genroku period that the pleasure quarters in particular cities became complex and varied. The "Shimmachi" quarter of Osaka was one of the most sophisticated, and certainly it vied with the "Shimabara" in Kyoto and the "Yoshiwara" area in Edo, now Tokyo. This special district was exempt from the restrictions of other parts of the city and the rest of peoples' lives. The pleasure quarter was where money was king and class origin could be put aside. When samurai would enter, they would usually disguise themselves, because otherwise their social standing was compromised – a point that Chikamatsu avidly and often humorously illustrates in the plays. In the city "proper" there were laws to limit ostentatious living such as the size of one's house or the way one could dress; in the pleasure quarter, ostentation knew no bounds, and money could be spent with a beautiful woman who could ruin one's family and one's business, if the heart ruled the head (this happens in several of the love-suicide plays).

The pleasure quarter was a "floating world," as the artists described it, where pleasure, leisure and indulgence were connected. The women of those quarters had many different gradations of status, and those with high class were often talented in song, dance, poetry recital, calligraphy, flower arrangement, the great art of the tea ceremony and, of course, sophisticated companionship. Indeed, unlike the western tradition of prostitution, these relationships often comprised institutionalized "affairs" that lasted over long periods of time. Problems, in the east as in the west, emerged when jealousy and love entered into the relationship, and when the couple might find that they could not live without each other.

Chikamatsu shows women of Osaka's pleasure quarter as morally upright, virtuous, respectable and upstanding. Extra-marital relationships were institutionally tolerated by the wives of the time, and marriage was defined in traditionally patriarchal terms (thus the notion of *giri* in its many contexts could also mean family economic obligation and fealty to male lineage). But not only could a love-affair in the pleasure quarter interfere with the wife and children; it could also affect the whole extended family in various ways, and, as in *The Love-Suicide at Amijima*, the male lover's parents threaten to break their son's marriage, since they are affected by the chaos that is being wrought in the economic and emotional life of their family.

Kabuki and *jōjuri*

Chikamatsu's dramatic art was especially developed in *kabuki* and *jōjuri* theater. The *kabuki* tradition was said to have derived its origins around 1603 in a Shinto shrine dance performed by a woman, Okuni, who came to Kyoto to perform. She drew from Buddhist and Shinto religious dance along with folk-dance tradition, when she danced the "Okuni Odori." Her rendition has been referred to as a form of prayer dance. This was received with much acclaim and later, with the collaboration of a male friend who helped choreograph her work, there began the early form of what was to be called *kabuki* (Yoshida 1977: 133).

"Okuni kabuki" came to be performed by troupes of professional women. In these performances, women acted as women and as men to such a degree and in such ways that the Tokugawa Shogunate saw this as a moral disruption and the dance troupes were repressed, even "stamped out" (Yoshida 1977: 133). The *kabuki* theater was born to stay, however, and it re-emerged with all-male casts in the form of "Wakashu kabuki" where men acted male and female parts. This time around, however, it was seen as a moral disruption, encouraging male homosexuality, and so the plays re-emerged, newly censored, with the homo-erotic components fitted more carefully with the current community standards.

Having left its overtly suggestive and "disruptive" origins behind, *kabuki* emerged around the 1680s as a major theater-form in the Osaka, Kyoto and Edo areas, and several great actors became associated with the plays, including the renowned Sakata Tojuro (1647–1709) who performed in Osaka and in other cities in a number of Chikamatsu plays. He was especially well known for his dramatic style of *wagato* – the techniques of courting women (Yoshida 1977: 134).

Kabuki also drew on the classic *nō* dramas and from *kyōgen* farces of previous centuries (Shively in Chikamatsu 1953: 8). It often incorporated the poetic "journey" derived from *nō* plays and symbolically integrated in many of Chikamatsu's love-suicide dramas. *Kabuki* stressed liveliness on stage and extensive use of dialogue in the vernacular, which broke from the traditions of the past; these techniques Chikamatsu harnessed to much acclaim. He wrote some thirty *kabuki* plays.

The dramatic genre that had developed from and entwined with *kabuki* traditions was the *jōjuri* or puppet play.[5] Chikamatsu wrote between 100 and 140 *jōjuri* (Shively in Chikamatsu 1953: 12): these were typified by a particular metric style in recitation and involved elaborate puppetry techniques where characters had the freedom to do superhuman leaps and tumbles, twists and turns, and appearances and disappearances because of their special animation. The history and tradition of puppetry had been well established in several parts of Asia, and the development of *jōjuri* by Chikamatsu took on great sophistication technically, poetically and aesthetically, in its association with the theater of the day. In fact, Osaka was famous for its *jōjuri* theater, called the Takemot-za, and from 1685 onwards (Shively in Chikamatsu 1953: 11) many of Chikamatsu plays were first performed here; these included *The Love-Suicides at Sonezaki* and *The Love-Suicide at Amijima*.

Love-suicides were first dramatized in *kabuki* around 1674 and the first major plays occurred from the end of that century onwards (Shively 1953: 56). The theme was so popular and close to the hearts of so many that theaters would vie with one another to present a play based on the most recent such "mutual death." People also followed the "scandal sheets" of the day with hunger for the latest "news" on such events. The principle of vows of undying love was taken so seriously in real life that some would shave their heads, cut off fingers, remove toe-nails or otherwise inflict wounds to their own bodies to demonstrate their faith in that special union.

Chikamatsu, in emphasizing religious salvation via the Buddhist idea of rebirth on the calyx of the lotus (Shively in Chikamatsu 1953: 26), artistically moves the illicit love-relations and questions of morality to another plane. Dying for one's love could be connected with death taken for one's feudal lord, as with the samurai past: thus Chikamatsu wove past with present and other-worldliness with this world.

THE PLAYS

We now take two of the love-suicide plays which take place in Osaka. They illustrate but a microcosm of Chikamatsu's dramatic style. The *Love-Suicides at Sonezaki* is about a courtesan, Ohatsu, and her lover, Tokubei. The plot revolves around money that was borrowed and not returned, and blame falls on the innocent hero. In addition, the hero is expected to marry a woman of his family's choice, which is certainly not where his heart is set. Here the contemporary world of business, commerce, family life and obligations are contrasted with the love-affair between Ohatsu and Tokubei.

The mood of Osaka is shared by the narrator as he describes the pleasure quarter at the edge of the Shijimi River:

> The breezes of love are all-pervasive
> By the Shijimi River, where love-drowned guests
> Like empty shells,[6] bereft of their senses,
> Wander the dark ways of love.
> Lit each night by burning lanterns,
> Fireflies that glow in the four seasons,
> Stars that shine on rainy nights.
> By Plum bridge,[7] blossoms show even in the summer.
> Rustics on a visit, city connoisseurs,
> All journey the varied roads of love,
> Where adepts wander and novices play;
> What a lively place this new Quarter is![8]

(Chikamatsu 1961: 47)

Chikamatsu paints a scene of the city where night brings forth all manner of people, yet where there is mystery and magic. There is also the ephemeral, because the *Dōjima* is the new "playground" for the men of Osaka.

Chikamatsu describes the lovers' journey from Dōjima and the Umeda bridge to the Sonezaki Shrine and shares the following note with his audience:

> Farewell to this world, and to the night farewell.
> We who walked the road to death, to what should we be likened?
> To the frost by the road that leads to the graveyard,
> Vanishing with each step that we take ahead
> How sad is this dream of a dream!

(p. 51)

As the lovers prepare for their final few moments,

> Their strings of tears unite like entwining branches, or the pine and palm that grow from a single trunk,[9] a symbol of eternal love. Here the dew of their unhappy lives will at last settle.
> *Tokubai.* Let this be the spot.

(pp. 54–5)

The act of committing suicide is carefully and painstakingly described:

His eyes grow dim, and his last painful breath is drawn away at its appointed hour. No one is there to tell the tale, but the wind that blows through Sonezaki Wood transmits it, and high and low alike gather to pray for these lovers who beyond a doubt will in the future attain Buddhahood. They have become models of true love.

(pp. 54–5, 56)

We learn from this play about the physical shape of Osaka at the turn of the eighteenth century, and we see how the pleasure quarter of town, adjacent to the river, married sentiment and sadness.

Some seventeen years later, Chikamatsu wrote *The Love-Suicide of Amijima*. Many scholars argue that this work represents an apogee in his writing and of the genre of domestic play. It also places a real event that took place in Osaka onto the stage.

There is an anecdote associated with this play which lends color to the writing of it, although the tale is not proven. It recounts that Chikamatsu was to be found at a restaurant close to Osaka when a messenger rushed to him to tell of a love-suicide that had occurred the previous night at Daichōji on Amijima. The tale of the incident was so compelling to the writer that he made great haste and immediately returned to Osaka, taking up his brush and beginning with the phrase "hashiri-gaki" which means "run and write" or "running writing," which is also a pun on the idea of the "journey" or "run," besides referring to the flowing style of calligraphy.

When performed in 1720 at Osaka's Takemoto-za, the play met with immediate success and was followed the next year with a performance in Edo on the *kabuki* stage. It has experienced a number of revisions in *jōjuri* and *kabuki* form since then, and to the present day.

The plot is centered around the not-so-uncommon social problem of a husband torn between a lover and his family. The play appealed to the audience because it raised the problems that occur when social classes are not supposed to mix, yet do. The hero is a paper-dealer called Jihei who is losing his fortune as a result of spending money on his beloved Koharu. She resides in Osaka's pleasure quarter. The plot examines the merchant affluence of Osaka, and the *gīri* of the hero, since this colors his dilemmas of wife, family complications and financial disaster.

Chikamatsu opens the play with yet another evocative scene on Osaka's Shijimi River:

The prostitutes of the Shijimi River [Quarter][10] have deep affections. This Shijimi River is indeed an ocean of love which cannot be ladled dry. [As men walk along] singing songs of love full of recollections, their hearts stop them, as if the characters on the lantern at the entrance [of their favorite house] were a barrier. Sauntering in good spirits about the Quarter, they chant improvised *jōruri*, imitate the intonation of actors, and sing the popular songs of the riverside pavilions.[11] Some visitors are enticed in by the music of the samisen coming from second-floor rooms. Other guests are trapped unwittingly, like the one who, wishing to avoid festival days[12] so that he will not be obliged to spend more than he can afford, hides his face and assumes a concealing guise.

The narrator continues to describe the scene by enumerating the "plum" and the "cherry" bridges (Umeda-bashi and Sakura-bashi) (Chikamatsu 1953: 101).

The play continues and the hero is introduced by a rival, Tahei, who chants the following ditty:

> The paper-dealer Jihei
> Is infatuated with Koharu;
> And his property has become
> As thin as rice-paper
> And thinner than toilet paper.
> Jihei is worthless as wastepaper
> Which is not good enough
> Even to blow the nose with
> Namaida! Namaida! Namaida!
>
> (Chikamatsu 1926: 228)

Paper was one of the major products of Osaka. In this translation, we observe the play and playfulness with words that have held Chikamatsu's reputation in high esteem through the centuries. There is wit, sarcasm and challenge writ large by this rival and competitor. We note how a man is scorned and chastised for running his business and family into economic ruin because his love makes him take leave of his "senses." This was particularly mortifying in Osaka, which prided itself on its upright merchant dealings.

The story in the play unfolds and we watch the couple make their irrevocable decision to die with one another. At the beginning of the third act, when they meet in the dead of night to embark on their lonely journey along the Osaka river-banks, Chikamatsu echoes the words of the beginning of the drama, playing with these to embroider the unfolding scenario:

> The shoals of love and affection are here at the Shijimi River [Quarter]. Its flowing waters make no sound, nor does the passing of people at three in the morning. In the sky, the moon of the fifteenth night is clear, while the lights have burned low in the gate lanterns, ... the wooden clappers of the night patrolman [beat] drowsily in time with his staggering walk. "Be careful [of fire], be careful [of fire]," and his voice also makes it [seem] late.
>
> (Chikamatsu 1953: 87)

We, the reader/audience, even without the play before our eyes, can feel the mood of Osaka at night, the furtiveness of the couple's exit, the suggestive "stagger" of a watchman – a stagger to death of the lovers? He cries out a safety warning at night – an irony in view of the mutually agreed-upon personal violence that will take place with the lovers shortly. And the fifteenth night is the full moon, which in itself connotes something unusual. The author continues

> Even the teakettle of a *chaya* rests for a two-hour period at night between two and four. The only thing moving is the flame of the short lamp grown thin in the late night. The river wind [blows] coldly, and the frost grows thick.
>
> (Chikamatsu 1953: 87)

There is here an eerie silence in the pre-dawn of the city, where even in the pleasure quarter, archetypically the place of night-life, there is a pause between night and day. The couple start their journey along the river, crossing Osaka's familiar bridges and landmarks. Jihei remarks:

> If I should walk toward the north, I could have another glimpse of my home, but I do not look that way. I suppress in my breast [thoughts of] my children's futures and my wife's pitiable lot. Toward the south stretches the bridge[13] (which we cross) with its innumerable piles [the other shore has] (innumerable) houses, so why should it be named "Eight Houses"?[14] We must hurry along the road before the boat from Fushimi *lands* there.[15] [I wonder] who are (sleeping) together on it. To us, who are abandoning this world, it is frightening to hear of the demon.[16] The two rivers, the Yodo and the Yamato, flow into one stream, the Ōkawa.[17] The water and fish go along together [insep-arably], [just as] the two, Koharu and I, go together, [and cross] the River Styx by the blade of one sword. I should like to receive [this river water] as an offering of water [from the living].[18]

Jihei shows his concern about the wife he leaves with all his problems, to say nothing of the children who are about to lose their father. He talks of the fish and water being as one, like the couple, who, we could argue, are fish out of water if they do not make this journey to death. There is symbolism in the flowing rivers as they pass the bridges, an inevitable flow that is a pathway to ocean and a symbolic eternity for both the rivers and the couple. The reference to the Styx in the translation is really the river with three fords that is crossed after death in Buddhist tradition. Osaka's river-system, her bridges, the embankment walks, her river-traffic and the major landmarks of the time are graphically shared with the audience. Body and soul of the couple are shared by the spirit of the river, her pulses and rhythms.

The narration continues, noting "their tears are falling [fast] that they can hardly see each other's faces, and they seem to flood the bridges over the Horikawa" (Chikamatsu 1953: 92): this is the canal that leads north from the Ōkawa. It is probable that they could see several of the other five bridges that cross this canal – thus the powerful allusion to a swell of tears like a flood-water passing under the bridges (Chikamatsu 1953: 128).

CONCLUSION

Chikamatsu, as poet and playwright, demonstrates textual artistry. The two plays we have described illuminate the cityscape of Osaka as they spell out love and intrigue. The plays demonstrate the real emotions that Osaka's townspeople felt between the idea of *gīri* or obligation, and the exuberance of real emotion and passion.

These moods and values have been shared by audiences for three centuries, and the contrasts between past and present values shared then, as now, continue to fascinate cultural observers of today. Chikamatsu situated his plays in the city that he knew and loved. Osaka was both text and "meta-text" as he depicted the nature of human frailty.

REFERENCES

Chikamatsu Monzaemon (1926), *Masterpieces of Chikamatsu*, transl. Asataro Miyamori, London: Kegan, Paul, Trench, Trubner & Co. Ltd.

—— (1953; orig. 1720), *The Love Suicide at Amijima*, ed. Donald Shively, Cambridge, Mass.: Harvard University Press.

—— (1961), *Major Plays of Chikamatsu*, ed. Donald Keene, New York: Columbia University Press.

Kaempfer, Engelbert (1910; orig 1692), *The History of Japan: 1690–92*, vol. III, trans. J. G. Scheuchzer, Glasgow: James MacLehose & Sons.

Sansom, George (1963), *A History of Japan*, California: Stanford University Press.

Yoshida, Chiaki (1977), *Kabuki: The Resplendent Japanese Theater*, Tokyo: The Japan Times Ltd.

NOTES

1   We must acknowledge here the well-known playwright Ihara Saikaku (1641–93) who was born and raised in Osaka. Saikaku's writings precede Chikamatsu's œuvre. His literary presence would have been well known to Chikamatsu.

2   One of the ironies of Osaka's Dōjima district was that many of the upper-class courtesans were the daughters of poor or unemployed samurai (Morris 1963: 9), which gives us some indication of the importance of money that superseded traditional class structure inside the pleasure quarter.

3   The love-suicide (*shinjū*) was legally called by Tokugawa officials *aitaishi* or *aitaijini* (i.e., "mutual death"). The contemporary term *jōshi* appears in usage after the Meiji restoration.

4   Chikamatsu draws from the Buddhist term *inga*, "fate" or "karma," whereby what we have done determines where we will go in the next life; thus through love-suicide, the couple will receive salvation through death and reach the Western Paradise (Shively in Chikamatsu 1953: 29).

5   At the end of the seventeenth century the typical puppet was 2 ft. high, and held and manipulated from behind (Shively in Chikamatsu 1953: 54).

6   Shijimi means "corbicula," a small shell-fish, thus linking the river's name with shells.

7   Umeda bridge, literally "plumfield."

8   The Dōjima new Quarter of Osaka was opened around 1700.

9   Such a tree actually existed, as it appears from contemporary accounts of the Sonezaki shrine.

10 Square brackets indicate the translator's added words to the text; parentheses give an alternative translation using the complexity of puns or pivot-words (*kakekotoba* or *iikakekotoba*) which, in their classical form, are related to *haiku* (Shively in Chikamatsu 1953: 45, 62).

11 These were sheds or commercial warehouses in Osaka which were located along the river bank. The popular songs usually refer to the pleasure quarter (Shively in Chikamatsu 1953: 100).

12 Festival days meant that the rates went up and special tips were expected (Shively in Chikamatsu 1953: 100).

13 This is the Tenjin bridge, one of the longest in Osaka, with many pylons.

14 It seems that in olden times there had been eight houses located on this stretch of the river.

15 There was passenger-boat traffic along the *Yodo-gawa*, and the night boat would land at Hachikenya early in the morning.

16 There is a pun here because *temma* means "demon" and "Temma Tenjin" is the deity after which the bridge is named.

17 Ōkawa meant "Great River" and was the popular name for the short stretch of the Yodo river. (Today, the Yodo is re-directed around the north of Osaka, and the branch that follows the course mentioned here is the Ajigawa.)

18 Water is given as thanks to Buddhist and Shinto deities; thus people will pray for the couple with this water and connect life with death.

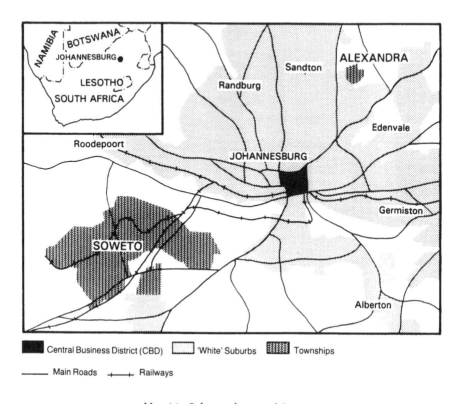

Map 14   Johannesburg and Soweto

14

GAZING ON APARTHEID

Post-colonial travel narratives
of the golden city

Jonathan Crush

Johannesburg, the inhuman city, with no river and no mountain
(Robbins 1987: 89)

INTRODUCTION

In 1881 the South African industrialist Sammy Marks made a business trip on horseback between Pretoria and Kimberley. On the way he and his companion, Dumont, passed through a sparsely populated farming region of little apparent consequence. A local farmer offered Marks a farm called Driefontein for a trifling £800. Marks was keen to buy but Dumont convinced him that he owned enough of the Transvaal already. He was to regret profoundly this error of judgment. Driefontein, it soon turned out, straddled the Witwatersrand main gold-reef upon which the modern city of Johannesburg was built (Mendelsohn 1991). Marks, a shrewd and calculating man with considerable business acumen, gazed on the parched and waterless farm that would become Johannesburg and was persuaded he saw nothing of interest. For the rest of his life, whenever he visited the city, he was haunted by his failure to see. The landscape of Johannesburg became a painful rebuke, a moral commentary on the failure of his own desires.

In a sense, Marks was the first in a long line of travelers to gaze on Johannesburg. He was also the first to reconstitute the city, to rewrite its landscape, as a moral text. Johannesburg's landscape has always borne the weight of such moral commentary. Travelers to South Africa visit Johannesburg, and many write about it. As they do, they bring narrative order, significance and meaning to the city's landscape. Johannesburg is a city of signs, of moral coherence, of essences. Its landscape encodes stories about its origins, its inhabitants and the broader society in which it is set. The traveler's task is to discover and write these truths. Rick Johnstone (1978: 102) calls Johannesburg "the play within the play." In the mind of many recent travelers, to explore and comprehend Johannesburg is to understand

South Africa as a whole. For much of its history, however, travelers wrote of Johannesburg as a place of Europe, not Africa.

Greater Johannesburg (the Witwatersrand) is a comparatively modern city, dating from the first gold discoveries in the central Transvaal in 1886. Gold mining accelerated a profound process of industrial transformation which had begun earlier at the Kimberley diamond fields (Worger 1987). Regional and international flows of labor, capital and commodities were redirected towards the new mining complex of the Rand. By 1910 Johannesburg was the site of fifty working mines and the home of over 200,000 people (Crush 1987: 3). For many years, most of the city's black population were temporary sojourners, migrant workers who came from and returned to rural homes scattered throughout the subcontinent (Jeeves 1985; Crush, Jeeves and Yudelman 1991). The city was always intimately connected to its rural hinterland, reaching deep into the countryside for the labor that sustained its mines and industries (Beinart 1982; Murray 1986; Moodie forthcoming).

The British and the Boers fought a war over Johannesburg's gold between 1899 and 1902, some of it in the city streets themselves (Cammack 1990). Thereafter, as the city expanded, its slumyards, suburbs and compounds experienced recurrent strife between the urban underclasses and the white rulers. Social historians and historical geographers have begun to document the rich social and cultural life of the city's subaltern spaces. They have also identified the cleavages of power and privilege that divided the city population (Bozzoli 1979, 1991; Marks and Rathbone 1982; van Onselen 1982; Bonner, Hofmeyr, James and Lodge 1989; Crush and Ambler 1992). In the 1930s and 1940s the city grew rapidly as people poured in from an impoverished countryside to find work. By the 1950s the mining companies had worked out most of Johannesburg's gold deposits while the locus of the industry moved east and west. The ubiquitous mine-dumps that still scar the urban landscape serve as a continuous reminder of the city's origins.

The Witwatersrand, by now South Africa's major industrial complex, continued to expand. After 1948 the Nationalist government began systematically rewriting the geography of Johannesburg, driving black and white apart into "obscenely ostentatious villas" and "grindingly spartan slums" (Parnell and Pirie 1991: 145) and forcing the inner-city population out to vast peripheral townships such as Soweto. Since 1986, and the abolition of controls on black mobility, the city has experienced a new rush from the countryside. Some 20 per cent of South Africa's population of 34 million currently live in the vast sprawling agglomeration of the Witwatersrand.

THE TROPES OF TRAVEL

Over the last two decades travel writing has again become a prominent feature of the popular literary landscape of Europe and North America: "the signs of an invigorated contemporary interest in travel are everywhere to be seen" (Kowalewski 1992: 6) Travel writing about South Africa has experienced a revival in Europe (Hope 1988; Crawford 1989; Mallaby 1992), in North America (see below) and within South Africa itself (Robbins 1986, 1987). Indeed, argues Robin Hallett (1988: 15), "South Africa cries out for explorers." The volume of recent South African travel writing – and the fact that most of the authors are journalists – suggests, however, that we may not be dealing as much with a general cultural renaissance of a lost art-form as a specific manifestation of a new cultural interchange between Euro-America and South Africa.

Rob Nixon (1991a) argues that since the Soweto uprising there has been a profound "cultural crossover" between South Africa and the United States. A decisive manifestation of the crossover is the spate of big-budget anti-apartheid films which appeared in the late 1980s. A second sign is the wide readership and international renown enjoyed by a small coterie of white South African novelists, particularly Gordimer, Brink, Breytenbach and Coetzee (Ward 1989). Third, there has been a veritable deluge of books aimed at the mass market attempting to decode South African apartheid for North American audiences. It is in this context that travel writing on South Africa has recently experienced its notable renaissance.

At first glance, it may seem unremarkable that temporary sojourners in South Africa should simply recount their experiences and adventures in the order in which they occurred. In fact, there is no requirement that one's experiences in a foreign place should necessarily be represented in this way. Many popular accounts of contemporary South Africa published overseas rely on devices other than travel and exploration for their narrative continuity. A number of recent British books on South Africa, for example, attempt to combine historical exposition with personal anecdote (Attwell 1986; Lapping 1986; Manning 1987). South African writers such as de Villiers (1987), Heard (1990) and Sparks (1990) adopt a similar approach to explaining the psychology of white racism in South Africa. Johannesburg, too, has been mapped in a variety of textual modes: as historical narrative (van Onselen 1982; Mandy 1984; Lang 1986), as visual representation (Kallaway and Pearson 1986), as autobiography (Mathabane 1986; Mattera 1987; Gready 1990) and as fictional setting (Ward 1989).

The appeal of the travelogue as a way of writing South Africa is thus considerably more than a matter of convenience. In part, as Dennis Porter (1991: 20) points out, it is

a form of cultural cartography that is impelled by an anxiety to map the globe, centre it on a certain point, produce explanatory narratives, and assign fixed identities to regions and the races that inhabit them. Such representations are always concerned with the question of place and placing.

Travel writing is also a powerful vehicle for a liberal North American (or

European) to exoticize and comment on South Africa's landscapes and peoples – to write the South African city as a moral text. Such travel narratives paint South Africa as a place of strangeness and difference. Simultaneously, the travelogue allows the author in the guise of exploration to unveil the essence of the society traveled through. The travelogue is a seductive medium in which to address an audience receptive to understanding as a journey of progressive unveiling.

There are three varieties of contemporary American travelogue on South Africa. The first is autobiographical or semi-autobiographical in character, a retelling of the journey out of apartheid by the exiled South African writer (Mathabane 1986; Mattera 1987; Gordon 1989; Diash 1990). Second, there is the pseudo-ethnographic account. Here the American traveler takes up residence in a South African community and narrates the community as a simulation of the country as a whole (Crapanzano 1985; Stengel 1990). The final form adopts a more familiar trope – the journey of exploration (North 1985; Lelyveld 1986; Meyer 1986; Aronson 1990; Cowell 1992; Mallaby 1992; Streeten 1992). In this chapter, I focus on the latter, more classical, style of travel writing. The chapter considers three recent travelogues of South Africa which also journey through the South African city of Johannesburg (Finnegan 1989; Hochschild 1990; Malan 1990).

William Finnegan's *Dateline Soweto* was first serialized in *The New Yorker*. The consistent appeal of the travel-writing genre for Finnegan, a staff writer for *The New Yorker*, is obvious. Earlier, he published an account of his adventures as a supply teacher in a Cape Town school (Finnegan 1987). More recently he has written a dramatic account of his journey through the blighted landscapes of war-torn Mozambique (Finnegan 1992). The second book is Adam Hochschild's *The Mirror at Midnight: A South African Journey*, which recounts the author's experiences in South Africa during the 1980s. Hochschild returns periodically to South Africa from his home in San Francisco where he was co-founder of *Mother Jones* magazine. *Mirror at Midnight* recounts one of these journeys in 1988. The final book under consideration is Rian Malan's *My Traitor's Heart*. Malan is also a journalist based in North America (and earlier published a condensed travelogue under the same title in *Cosmopolitan* magazine) but grew up in South Africa. He therefore portrays his journey as a discovery of lost roots, of a suppressed conscience.

Colonial-discourse analysis, following Edward Said's seminal work *Orientalism* (1978), has devoted considerable effort to understanding how travelers construct other peoples and cultures for home consumption. The enduring appeal of travel writing as a mode of representation has been remarked upon in a host of recent studies (Leed 1991; Mills 1991; Webb 1991; Clifford 1992; Crocker 1992; Driver 1992; Greenfield 1992; Hutchinson 1992; Kowalewski 1992; Pratt 1992; Van den Abbeele 1992; Spurr 1993). To represent the unfamiliar in the language and tropes of exploration is to lend authority to the account, to represent as fact what is often fiction, to rewrite exploration as discovery of the exotic and to appeal to deeply racist and/or Eurocentric notions of other peoples and places. The appeal of the travelogue for writers and audiences lies in its revelatory character. In the contact zone, the strange becomes familiar, danger is tamed, difference is reconstituted as inferiority, the unknown becomes knowable. Travel writing reveals, as J. M. Coetzee has pointed out, the great intellectual schemas by which Europe has thought about others (Coetzee 1988; see also Comaroff and Comaroff 1991; Haarhoff 1991; Noyes 1992).

Yet there is no one colonial discourse and no single genre of travel writing. As Lisa Lowe (1991: 8) observes, "discourses operate in conflict; they overlap and collude; they do not produce fixed or unified objects." Mary Louise Pratt (1992) has argued, for example, that the conventions of representation in European travel writing both unify and fragment the genre. Her study ranges across several centuries of writing, showing both what joins and what separates. She shows the resilience of the "dehumanizing western habit of representing other parts of the world as having no history" and of the monarch-of-all-I-survey scene. But there is also difference: between those who narrate the conquest of others by Europe and those who narrate an "anti-conquest" – seeking, that is, to secure their own innocence in the same moment as they assert European hegemony.

In his analysis of colonial travel writing, Rob Nixon (1992: 53) contrasts the "decisive racism" of the Victorian travel writers with the "indecisive bohemian wanderlust" of the Georgian writers of the colonial 1930s. He goes on to argue that travel literature is versatile because it admits both ethnographic and autobiographical possibilities (p. 67). In similar vein, Dennis Porter's psychoanalytical account of travel writing (1991) unveils the strain of self-disclosure in many accounts and the constant desire to transgress the conventions of the genre itself. In a more narrowly conceived study, Sara Mills (1991) has pointed to the gendered differences between male and female travel accounts in the late nineteenth century.

Nixon (1992: 51) points out that the Victorians expected far more from their travel writers than whimsical and bizarre tales of the unexpected; travelers had to be "intensely opinionated" about the places they visited. Travel writing turned into a vehicle for substantial political and moral commentary, carrying what Nixon calls the "freight of political commentary" and Porter (1991) "the mission of political witness." Nixon (1992) and Pratt (1992) both suggest that much post-colonial travel writing has to be a vehicle for political commentary and witness. Post-colonial travel writing therefore unintentionally mirrors the multiple conventions of colonial discourse – an observation of profound importance for this chapter. Nixon, for example, argues that V. S. Naipaul has more in common with late-nineteenth-century travelers than his immediate predecessors, the Georgians (such as Waugh, Lawrence and Green). In an atmosphere of faltering imperial resolution, the Georgians were "self-absorbed, inward, capricious, and drained of missionary purpose." The earlier sense of Victorian political purpose is recaptured in Naipaul, a displaced author relentless in his pursuit and representation of Third World decrepitude.

As 1980s liberals writing primarily for an anti-apartheid, liberal, North American audience, Hochschild, Finnegan and Malan write from a post-colonial vantage-point. The fundamental challenge for the post-colonial writer is to escape the seductive precedents of colonial travel writing. Pratt (1992: 219) argues that many post-colonial metropolitan writers "exemplify a discourse of negation, domination, devaluation, and fear that remains in the late twentieth century a powerful ideological constituent of the west's consciousness of the people and places it strives to hold in subjugation." I argue that the three texts under consideration are ultimately unable to withstand the "imperial eye" and transcend the enduring legacy of the colonial travelogue.

The "freight of political commentary" also weighs heavy on the new genre of travel writing about South Africa's urban landscapes. Nixon (1991a) argues that the abiding dilemma in the representation of South Africa to Americans is how to communicate the experiences and values of a radical liberation struggle to a predominantly liberal American audience. This question "touches on fundamental issues of cinematic [and literary] convention, ideological compatibility, cross-cultural comprehension, and political effectiveness" (Nixon 1991a: 500). The quest for compatibility, comprehension and effectiveness has prompted many authors to turn to travel writing. Only one American travelogue makes any serious effort to understand the radical character of the domestic anti-apartheid struggle on its own terms (North 1985). The rest write and interpret within a liberal North American paradigm of race relations, racial conflict and white morality (although see Peckham 1990; Ó Tuathail 1992 for alternative American constructions).

CONSTRUCTING THE CITY

European travel narratives on South Africa date back to the arrival of the country's first white settlers in the mid-seventeenth century (Coetzee 1988; Pratt 1992). Pre-modern travelers' accounts of the South African interior constructed moral geographies that both emptied the landscape of its human inhabitants and denied them any legitimate claim to it (Comaroff and Comaroff 1991). The "depressive emptiness" of the landscape or the gross stereotyping of its indigenous inhabitants effaced the pre-existing reality of social structures and cultural practices – a reality to which the interlopers had no access (Coetzee 1988; Comaroff and Comaroff 1991). By the late nineteenth century, Africans were admitted to the landscape but only as temporary inhabitants, inept cultivators or uncivilized children wanting the manners and eruditeness of the evolutionarily more mature European.

Representations of African peoples and practices delivered deeply racist messages about European superiority (see also Miller 1985; 1990; Brantlinger 1988; Spurr 1993).

Beginning in the 1870s South Africa's industrial revolution produced a profound rupture in European representations of the country. As the forces of industrialization and urbanization reshaped the urban landscape, a new language had to be found within which to write the South African city and its inhabitants. What transpired was a language suffused with the dynamic imagery of progress and advance – a vocabulary that praised the forces of modernity, while simultaneously suppressing their darker side. Johannesburg, at the geographical center of the industrial revolution, epitomized modernity and progress.

The narration of Johannesburg as a journey of exploration became a recurrent feature of the city's representation by outsiders. From the city's earliest days in the 1890s through to the present, Johannesburg has generated a voluminous body of travel writing. Some of this writing focusses on Johannesburg per se; more often the city is a central element in a broader South African travelogue. In the late nineteenth and early twentieth centuries, the city landscape was viewed as a testament to projected European power, a thoroughly unAfrican creation with little connection to the vast rural hinterland that surrounded it. Johannesburg's landscape was read primarily as a moral text – a testament to European ingenuity, energy, technology and civilization.

Few European travelers were not seduced by early Johannesburg. In their writings, the estheticized city landscape is a canvas of industry, motion and masculinity – an island of imperial progress in a sea of wilderness and waste. Andrew Lees (1985: 16) points out the ambivalent attitudes of most Victorian commentators towards their own cities (see also Williams 1973; Himmelfarb 1984; Sharpe and Wallock 1987; Caws 1991). No such ambivalence compromised the representation of Johannesburg as vehicle and symbol of modernity. To Fuller (n.d.: 176), Johannesburg was nothing less than "Little England beyond sea and plain." Yet it lacked "the unspeakable misery of the East and West slum and the degradation of the workhouse." This is not to say that slums, misery and degradation were, in fact, missing (van Onselen 1982; Robinson 1992). Rather, their existence is deliberately suppressed by most travelers. Here, then, is a second silencing of the indigenous inhabitants of South Africa's landscapes. The imperial eye sees and accentuates only the motifs of progress and technological advance in Europe's outposts of empire. Conversely, the European urban poor are consistently depicted in the language and imagery of "darkest Africa" (Comaroff and Comaroff 1992: 285–8). To admit, however, that progress also victimizes and destroys abroad would be to indict, not justify, modernity and European colonial expansion. Depictions of "unspeakable misery" have little place in the literary canon of the imperial city.

Travel writing about Johannesburg continues in this high imperial vein until well into the 1930s. Thereafter, the writing of the city by outsiders is increasingly inscribed within a broader post-War discourse – a late colonial world in which many old certainties are crumbling. The city is still a medium for the articulation of political and moral exhortation but the content of the lesson has shifted. Johannesburg is no longer a testimony to progress but an indictment of modernity's baser instincts. The playwright, Martin Flavin, passed through Johannesburg *en route* to the Congo in the late 1940s:

> The core of it is an ugly, throbbing, congested business section, sprawling planlessly, and glutted with traffic, depressingly unequal to its present needs; fringed with residential sections in process of, or threatened with, absorption ... the whole ringed round for part of its circumference, at a respectful distance antipodal to the suburban area, with wretched Native slums – the most appalling slums I've ever seen. Finally, there are the mine dumps, rearing from the landscape – ocher-colored hills growing into modest mountains, pyramids of powdered stone geometrically designed, not devoid of beauty of a cold, forbidding sort, certainly impressive against the deep blue sky, which seems deeper and bluer in Africa than elsewhere: the monuments of Joburg, night soil of successful enterprise, excrement expelled in sixty years of eager, unremitting gluttony.

> (Flavin 1949: 2)

Flavin's Johannesburg remains an outpost, an alien imposition disconnected from its surroundings, a thoroughly "unAfrican" place: it is "an urban island in the midst of nowhere, a pin point in the vast, sad wilderness of Africa." But no longer is the city a testimonial to modernity, a source of wonder and a site of beauty. It is a place of dystopian bleakness, sprawling, planless, wretched, glutinous, excremental. Flavin travels through the city as supercilious critic, finding no pleasure and little to praise in its people or landscapes.

The black inhabitants of the city are similarly alienated – foreigners in a strange landscape, passive, accepting, "docile and resigned, alien and sad" (p. 15). Flavin admits nothing of the rich social life of the black slumyards and townships he passes by, the vibrant oppositional cultures of the city or the intense political struggles of the period. His purpose is to represent Africans as passive victims of this deeply alienating, European city. Here – embedded in a still pervasive colonial discourse – is a third silencing of the indigenous inhabitants of the city. To travel to Johannesburg is therefore to confront "a civilization that seemed mechanistic and soulless and to have come adrift from the causes of progress and power" (Nixon 1992: 53). Flavin's Johannesburg provides the stage for the elaboration of a thoroughly metropolitan set of concerns and commentaries – a Georgian disillusionment with the social debris of industrialization and urbanization. He nevertheless still writes with deep imperial arrogance. Africa has largely become a set of personal inconveniences and discomforts for the traveler.

With apartheid comes a new style of travel writing, a new post-colonial morality. The city is still read for its moral lessons but no longer are these about European modernity or empire. Johannesburg's landscape is no longer an outpost of imperial expansion but an increasingly African place. The city reconnects with its hinterland. Its residents are no longer trapped within its spatial bounds. They come into the city and leave. Johannesburg is read for the lessons and stories it tells about South Africa itself. Its black inhabitants and their struggles begin to be increasingly foregrounded. The city's whites are parodied and stereotyped, residents of "Sunny South Africa" – deeply racist, hedonistic, blinkered and completely dissociated from the suffering they cause. The city's landscape is scripted by a new form of political witness – one that dwells on the visible divisions between white and black, privilege and poverty, light and darkness.

ENFRAMING JOHANNESBURG

"I wanted to make a journey both in geography and time" explains Adam Hochschild (1990: 15) at the beginning of *The Mirror at Midnight*. The central motif of the book is the Great Trek – the movement of Boer farmers into the South African interior in the 1830s and their subsequent wars with and dispossession of the indigenous inhabitants. Hochschild retraces the overland journey from the Cape, juxtaposing epic incidents from the Great Trek saga with tales of his own discoveries. The revelation is progressive. After passing through the "dream landscape" of Natal, Hochschild articulates his first great moral discovery. He now understands that even as there are two rival versions of the Great Trek so "are there two contending versions" of South Africa's present. One of these versions is correct; the traveler's task is to discover which. From there the text assumes a tone of increasing moral assurance as the author moves towards Johannesburg.

The Great Trek provides no obvious narrative entry-point into Johannesburg; it had, after all, preceded the discovery of gold by 50 years. Another kind of journey is necessary, as is a guide. Hochschild is suddenly riding through the northern Transvaal with Francis Wilson, "an economist who is South Africa's leading expert on rural poverty." Wilson appears unannounced and is erased just as quickly from the narrative. His rhetorical presence is needed to direct the traveler to the start of a new journey. The sure-footed guide brings the intrepid traveler to the scene of discovery, to what Pratt (1992) labels a "classic promontory description":

As the sun goes down we stand looking out on a scene of great beauty. The village stretches down the steep hillside to a valley, with the mountains opposite us in deep shadow; there are round huts with mud

265

walls and thatched roofs; three brown goats wander by. Friendly children in bright clothes smile at us; the sound of drumming and singing comes from the valley floor. I smell corn meal cooking – a woman is stirring a big iron pot of it that bubbles over a wood fire. Other women walk up the hill with jars of water on their heads.

(Hochschild 1990: 104)

This is, of course, a highly stylized rendering, a framing designed as prelude to the moment of discovery: "It takes several minutes of looking at this picture-postcard scene of Africa before you notice that only one thing is missing: men." Francis Wilson, a long-standing white critic of South Africa's migrant labor system, is the perfect guide.

Timothy Mitchell (1990) has argued that the technique of enframing is fundamental to the way in which nineteenth-century Europeans gazed on other peoples and places. Enframing remains a powerful narrative trope in contemporary travel writing. From the moment of the enframed discovery of the missing men, Hochschild is caught by the power of his own gaze. His journey into and travels within the city of Johannesburg comprise a shifting series of pictorial representations. Hochschild himself rarely directly engages with the city gazed upon. His journey into Johannesburg, for example, supposedly traces the route of the missing men, the migrant workers. He gazes on the crowded buses and trains that carry the migrants from the Northern Transvaal, but he does not travel with them. For many migrants, the first taste of Johannesburg means "the mine shafts and corridors that reach far beneath it." This is their Great Trek (Hochschild 1990: 106). Hochschild travels through the encapsulated world of the South African gold-mine, taking verbal snapshots from the privileged vantage-point of an official Chamber of Mines tour.

Hochschild later describes a pilgrimage into the black township of Soweto. He is there to interview Dr. Nthato Motlana. In many South African travelogues and documentaries Motlana has become the icon of angry and articulate black middle-class sentiment. After the statutory interview with Motlana, Hochschild gazes on a Soweto framed by the back-window of a diplomatic Mercedes Benz in which he has hitched a ride:

The bluish haze of smoke that spreads over Soweto from tens of thousands of coal fires makes it hard to see more than a mile or two. The tan brick houses we pass have privies in their backyards; each four-room matchbox home holds an average of seventeen to twenty people and sometimes as many as thirty, sleeping on couches and floors. More than 2 million people live in Soweto now; once a family has a foothold here, friends and relatives from the countryside come flooding in to stay, hoping for work. The houses are in even rows on streets built wide enough for armoured cars to make U-turns. Children thus use these streets as soccer fields, daringly stepping just a few inches out of the way, like nonchalant bullfighters as the diplomats' limousine hurtles by.

(p. 130)

There is obviously more here than can possibly be known simply by gazing. Hochschild inserts the statistics, the description of in-migration, the absent presence of armored vehicles to give the landscape meaning as a moral text. This is much more than a promontory description of a new discovery. It is a landscape that must be made to convey a moral commentary on apartheid.

The relieved diplomats are deposited at the luxurious Carlton Hotel in downtown Johannesburg and hurry off into the hotel's spacious lobby. Where Hochschild goes we are not told; we never are. A city gazed at cannot simultaneously be lived in. To live in the privileged spaces which by dint of the color of his own skin he is entitled to (and may well have inhabited in fact) would be to compromise his moral authority as an objective enframer of the divided society he is passing through. The muted juxtaposition of the squalor of Soweto and the luxury of the Carleton Hotel typifies Hochschild's version of the South African city. *The Mirror at Midnight* consists of a series of such juxtapositions of people, places and landscapes. Each landscape has its own bounded space. Each space has its opposite close by.

Hochschild's Johannesburg is thus a city of divisions, of juxtapositions. Sometimes these are brief images caught by the lens of a single sentence: "A freeway leads to Johannesburg's international airport; ten minutes away from it homeless Africans are living in abandoned mine tunnels" (p. 127). Sometimes he captures the contrast in a single gaze: Johannesburg's white apartment dwellers have black servants living in cramped quarters on the roof: "the windows are always very small, and at transom height, seven or eight feet above floor level. Thus the servants are sure not to have what whites would surely get if *they* were on that floor – a view" (p. 128). Apartheid even denies South African blacks the power to gaze, perhaps the greatest deprivation of all.

Within the opulent northern suburbs, malls jostle walls. Whites spend their "grossed-up earnings" in "an expanse of elegant shopping malls, sleek mirror-glass corporate headquarters with courtyard fountains, and restaurants." Here too are the "spacious supermarkets where whites shop, pausing at the butcher's counter to buy cheaper cuts for their servants"; the "comfortable homes on quiet winding streets lined with plane trees or the brilliant blue flowers of jacarandas"; and the walls:

> Everywhere in the white suburbs are eight-foot high fences or cement walls. You see all these fortifications and think they must be protecting a military post, then come closer and find it's someone's backyard. Yet except for the ones with razor wire, the fences are easy enough to climb over. A friend and I do so with no trouble when trying to get to a political meeting whose main entrance is being watched by the police.
>
> (p. 134)

A political meeting? In the luxurious white suburbs? Hochschild enters his own picture to subvert the absolute division of the city landscape. He never inhabits the white landscape; he enters from outside only to undermine it.

In a similar vein, Hochschild depicts a series of brief forays onto the other side of the divide. Again, these side-expeditions have a moral authority-giving, because subversive, intent. The writer visits black activists, and attends political funerals and meetings in the townships. His visit to Alexandra township, with which the book opens, is typical:

> From a distance, Johannesburg's township looks as if wisps of fog had collected above it, despite the sun-scorched day. Coming closer, I see that it is not fog but dust, for most streets here, unlike those in the white suburbs that surround Alexandra, are unpaved.... As we drive into Alexandra, we pass a ramshackle bus station: cracked and battered open-air concrete platforms with destination signs whose very names
> – Rosebank, Ferndale, Parktown – speak of the leisured white suburbs of living pools, tree-shaded streets, and well-sprinkled lawns....
> Through the open windows of the car, I can smell sewage ditches at the side of the road.... Small, tin-roofed homes are cramped together. Desperate for housing, some families have even moved into a row of abandoned buses. Their wheels stripped of tires, have sunk into the soil. Scattered about are reminders of recent street fighting: smashed windows, a few burnt-out cars.
>
> (pp. 3–4)

Written, like most of Hochschild's adventures, in the present tense, this passage embodies all the central elements of the author's gaze. The scene is viewed, once again, from a passing car. The picture carries considerably more background information and interpretative weight than a promontory description could ever do. White and black spaces are repeatedly juxtaposed.

What is Hochschild doing? He has traveled into Alexandra in the company of white activists to attend a black political funeral in a highly charged atmosphere. He is again transgressing the boundaries which his gaze simultaneously creates. When crossing and subverting such spatial boundaries, the nature of the reception from the other side becomes crucial. Is this action approved of, and therefore legitimate, or is it rejected and therefore invalidated? In what becomes a virtual parody of the colonial "happy native," every time that the author crosses the line his reception is generous and warm. "As our white caravan jounces slowly over the rutted road toward the funeral we are cheered. Older people clap from the roadside. Children smile and wave from doorways. Young men give the clenched-fist salute" (p. 4). Later (yet earlier in real time) the reception is the same in Port Elizabeth's New Brighton Township:

How, how, from that desolation, from these dirt streets with a single faucet, have these two hundred people emerged clean and in their Sunday finest, the choir in bright robes? And how can their hands so strongly clasp mine, the colour of the world which has shunted them into this slum, shouted at them, tear-gassed them, failed to send trucks to pick up their garbage? How can they welcome us with such warmth?

(p. 49)

And later, "as we slip out the door in mid-song, two hundred hands shoot into the air to wave goodbye." For most contemporary travelers, the black townships of South Africa are places of uncertainty, of trepidation, of fear. For Hochschild they are places in which to confirm the moral certainty of his own gaze.

Hochschild reads a particular parcel of Johannesburg's urban space as a new kind of moral text. One entire hilltop shoulder of Johannesburg – the Hillbrow district – is a center of vitality, a place of the unpredictable interweaving of cultures, a place with an open gay community, the place where apartheid is crumbling: "This part of Johannesburg gives you the feeling engendered by great cities everywhere; a whole that is greater than the sum of its parts." Johannesburg not only embodies the racial and spatial divisions of the past and present; it is a moral signpost to the future. In Hillbrow's cultural life, "you can glimpse the embryo of a new society already alive within the stubborn shell of the old." "I like this city," writes Hochschild with unexpected enthusiasm, "I could imagine living here" (p. 218).

For much of the book, Hochschild writes the city as a moral commentary on apartheid. But here, the unexpected inversion – Johannesburg as a place of cultural interweaving, a great world city, a place where the author himself could live – prefigures a more fundamental textual inversion. Breyten Breytenbach, the poet and author, has written that "looking into South Africa is like looking into the mirror at midnight ... A horrible face, but one's own." Hochschild borrows and extends the intensely personal metaphor:

> The reflected image that South Africa holds up to the United States is not an exact duplicate. Parts of the image are greatly distorted and magnified. But it still reflects something back, not only about our relation to the Third World, but about how we are developing into two separate societies at home. Is the social and economic distance from Soweto to Johannesburg's leafy white suburbs that much greater than the distance from Harlem to New Canaan, or Watts to Beverly Hills?
>
> (pp. 243–4)

Here, neatly encapsulated, are the two defining elements of Hochschild's gaze. Johannesburg is again enframed but this time simply as reflected image.

Second, Johannesburg's landscape is ultimately another moral text – a place to learn about America.

The image invoked by this kind of post-colonial travel writing is therefore the reverse of that peddled in the self-confident European travelogue written at the height of empire. While superficially exhibiting the traits of the older genre – the intrepid adventurer, the purposeful journey, discovery as revelation, what you see is what there is – the gaze is no longer imperial. Pratt (1992: 217) has argued that the impulse of post-colonial metropolitan writers such as Moravia or Theroux is "to condemn what they see, trivialize it, and dissociate themselves utterly from it." Here though is a different and deeper impulse. Johannesburg is again severed from its surrounds; it is New York and Los Angeles writ large. Its geography has meaning and purpose because it teaches Americans uncomfortable truths about their own society.

MAPPING THE UNMAPPABLE

William Finnegan's *Dateline Soweto* (1989) is a better example of the impulse that Pratt identifies in Moravia and Theroux. Like Hochschild, Finnegan's book intentionally subverts many elements of the travel writing genre, but pushes still further. For Finnegan, the South African city is unmappable and unknowable to the outsider. In part, this is because of the strange geography of the apartheid city; in part, it is because the outsider is inadequate, by dint of color and culture, to the task. One journeys only to discover the undiscoverable. Traveling unveils the traveler's own inadequacies. Finnegan's book is therefore something of an anti-travelogue, parading the personal weaknesses of the traveler, discovering nothing for himself. The most telling sign is the structure of the text itself. Despite the suggestion of the subtitle – *Travels with Black South African Reporters* – only a very small part of the book is devoted to travel description.

Dateline Soweto is the only text under consideration set exclusively in Johannesburg. Yet it too is scripted in context. The journey begins outside the city – in the bantustan of KwaNdebele to the north-east (a place already made familiar through Joseph Lelyveld's *Move Your Shadow*, 1986). For Lelyveld, KwaNdebele was a place to discover. For Finnegan, it is the opposite. He is lost – indeed "it was an especially bad time to be lost in KwaNdebele." He had wanted to bring a map; the black reporter, Qwelane, insisted that they leave it behind: "we wandered all over the blasted, frightening landscape of KwaNdebele." The basic comfort of the modern white explorer – cartographical power/knowledge – is irrelevant on this journey. One must rely instead, totally on black companions/guides:

> I had been lost a lot over the preceding six weeks. I had been spending time with some of Johannesburg's black newspaper reporters, trying to understand something of what their lives were like, and the one thing

I had learned was that they spent a large part of their professional lives asking people for directions. Nearly always, their beats were the black townships and the bantustans – places not known for street signs or house numbers. *Maps tended to be useless as well as dangerous* [my emphasis]. Black reporters and their drivers either developed remarkable powers of intuitive navigation or else they missed their deadlines. Just asking people for directions required special skills. Around Johannesburg, it had to be done in any of six or seven languages, and, in the extremely tense atmosphere of the past couple of years, it had to be done very sensitively.

(Finnegan 1989: 4)

Here most of the basic themes of *Dateline Soweto* first surface: to travel is to be continually lost, to rely totally on another (elsewhere we learn of Qwelane's "six languages" and "intimate knowledge of every street in Soweto"), to discover the complete unmappability of the landscape, the total inadequacy of the traveler to the task and the lurking sense of unease and menace. Being white in a black township is dangerous: "the complexities of black politics and black society – not to mention the intricacies of township geography – are really comprehensible only to those who live among them." Not only is Finnegan perpetually lost, he is also a liability to his black guides. To have a white in the car is dangerous: "my ability to observe black reporters going about their work was limited. My company slowed them down. Worse, it endangered them" (p. 153). The bulk of the book is therefore not about traveling at all – it is about sitting in the newsroom at *The Star*'s Johannesburg office, trying to win the confidence of the guides; about the lives and stories of the black journalists as told to Finnegan; about what the racial cleavages of South African society mean for "ordinary" black journalists. Johannesburg is only episodically gazed upon and there is no chronological thread that might suggest a progressive revelation (the book begins where the journey itself ends). Johannesburg provides no point of reference for America: "there are really no models for what is occurring in South Africa" (p. 228).

Dateline Soweto, like the *Mirror at Midnight*, scripts a divided city landscape. "On my first morning back" writes Finnegan,

I went walking through the Johannesburg suburb where I had arranged to stay. The winter light was intensely pure and fresh, the earth richly red, the highveld sky heartbreakingly blue ... the ineffable natural pleasures of southern Africa. Then I rounded a corner and saw, spray-painted in big black letters on a brick wall opposite a police station: VIVA UMKHONTO WE SIZWE!

(p. 39)

Finnegan depicts the white landscape much as in Hochschild with adjectival modifiers suffused with precision, purity and cleanliness. The black city is also a place for dense landscape description. Soweto, for example:

> has the same higgledy-piggledy, fear-edged, endlessly incongruous quality that the situation as a whole in South Africa has. It is the largest city in southern Africa, but it is also not a city at all. Amenities are few and furtive: practically all the bars are illegal, located in private homes; there are three cinemas to serve the estimated two million residents.... There are vast "hostels" where migrant workers live in gulag conditions, strictly segregated from more permanent residents. There are crowded, filthy shantytowns stalked by disease, hunger and evil-looking dogs. Only a stone's throw away (*the* measure of distance in the townships these days) from these hells-on-earth are modern ranch-style houses with sliding glass doors, indoor plumbing, and sports cars in the driveways – the houseproud, upmarket neighbourhoods, known locally by mordant nicknames like Beverly Hills and Prestige Park, where the Tutus and the Mandelas live. Schools are numerous but pokey and rundown and, while I was around, they were usually vacant.
>
> (pp. 48–9)

Here too Soweto is gazed upon. Like Hochschild's city, this Johannesburg, this Soweto, is the play within the play, the landscape that "reflects the separate realities that black and white South Africans inhabit" (p. 35). But unlike Hochschild's homogenized poverty and overcrowding, Finnegan's gaze fixes on the inequalities and contradictions within Soweto. The two best-known black South Africans to Americans, Mandela and Tutu, are complicit in this outrage. This is a new way of reading the landscape – attempting to subvert the absolute division between white and black, to disaggregate the black population itself. The landscape is still read for its moral content, but a new kind of commentary is emerging – one that begins to unbalance the liberal American sense of South Africa as a place of race and racial conflict alone.

Any appearance of tranquility and order must, however, be undermined with a rhetoric of deceptive appearances, of hidden dangers, of imminent chaos and mayhem:

> The atmosphere in the Soweto streets could be deceptively peaceful, even somnolent, with donkey carts clopping down the roads, monotonous rows of matchbox houses, small children pushing steel chair frames around in the dirt, women hanging laundry on backyard lines.... And then suddenly the air would electrify around a group of youths gathering stones to greet a delivery van or a Hippo or a "mellow yellow" – the absurd name that blacks have given to the yellow police vans that, completely encased in shells of heavy steel mesh, patrol the

townships. Like so many other South African townships, Soweto is a bedroom community and a war zone. At night, the violence escalates wildly, with rape, robbery, and nonpolitical murder keeping pace easily with the more era-specific mayhem of firebombings, assassinations, police shootings, and raids on the homes of activists.

(p. 49)

Soweto is a war zone with two wars. One is between black youth and white police; the other is an uncivil war between the residents themselves. Both are wild, chaotic, out of control – doubly threatening to the white outsider who ventures inside.

One might ask how a place that is so dangerous to whites can be gazed upon at all. Elsewhere in the book we learn that the writer usually only went to Soweto when the night hid his white skin. Yet this is also the time when the violence "escalates wildly." This is the fundamental tension throughout the book – how a traveler who by his own confession is unprepared, incompetent and terrified can gaze on the black city at all or have the authority to write about what he cannot, logically, see. Finnegan's way out is to suggest that by braving the threat, by confronting the fear, by finally emerging unscathed from the dangers, the triumphant traveler acquires the power to write. The trope is as old as the earliest colonial travel account.

Blacks, we learn, are in an "apocalyptic frame of mind." In the Katlehong township the most disconcerting thing about the comrades (or young activists) is "their eyes, which were absolutely expressionless":

I wondered how many of them had been tortured. I wondered how many of them had participated in a necklacing. I wondered how long it had been since they were last in school. . . . And I wondered how well these kids understood the democratic ideas for which they were ostensibly fighting.

These "brutalized youths" lived in a world, a city of "guns, bombs, terror, pain, death and funerals" (p. 65). Finnegan himself is thoroughly disconcerted as the gazed-at gaze back: "My curiosity about their politics, my tentative questions, plainly failed to register among such primal considerations. Only my race, my strangeness, seemed to interest them." The enormous irony of the final sentence is completely lost on Finnegan. Unable to deal with their withering post-colonial gaze, he retreats into a thoroughly colonial reconstruction of the encounter.

Johannesburg is, for Finnegan, a place of implacable otherness. It may look "much like" Akron, Ohio, but that is all:

With its streets devoid of Africans – who call the place Egoli, the city of gold, which at least evokes the great warren of tunnels beneath one's feet, and the millions of miners whose dark, dangerous, mind-wrenchingly meaningless work created the wealth that built and sustains the city – its dullness is actually frightening.

(p. 191)

The American civil-rights struggle bears some obvious similarities "but that comparison soon breaks down." South Africa differs in basic ways too from other African countries. "There really are no models for what is occurring in South Africa, no precedents" (p. 228). Finnegan's is a post-colonial vision of a city that, in Pratt's terms (1992: 218), embodies "ugliness, grotesquery and decay." The city symbolizes nothing except its own madness, its own barbarism, its own primal instincts. It is a landscape the American outsider can and must distance himself from utterly. Finnegan's book ends with a convoluted essay on change and reaction in South Africa. It is a last-ditch attempt to map the unmappable and to bring some objective explanatory order to a society that is consistently depicted as out of control. Even here Finnegan is unable to suppress the feelings of terror that constantly intrude into the narrative.

THE VORTEX OF FEAR

Pratt (1992: 225) argues that a key ideological matrix of post-colonial travel writing in the 1980s is terror. This matrix is virtually absent from Hochschild, lurks menacingly in Finnegan and reaches fruition in Rian Malan's *My Traitor's Heart* (1990). Porter (1991) argues that the most interesting writers of non-fictional travel books are those who manage to combine outer and inner exploration. Hochschild and Finnegan fail almost completely in this regard; Hochschild's book is an exercise in self-effacement except insofar as the author intrudes into the text to establish the authority of his own gaze. Finnegan's self-debasement is also a rhetorical inversion for effect. The book reveals far less about the author even than his earlier work *Crossing the Line* (1987). Malan's journey is explicitly scripted as a narrative exploration of the self.

My Traitor's Heart tells of an unorthodox journey into the country and the soul. Malan, unlike the other two authors, is returning to the country in which he grew up and from which he went into self-exile when confronted with the prospect of conscription. The first half of the book is an exploration of his own childhood memories, growing up as an Afrikaner in Johannesburg. The second half attempts to describe what he found out about South Africa and himself on his return from exile. The two parts of the book stand in an uneasy relation to one another – the first, retrospective and nostalgic, explores what for Malan is the central paradox of his life: he loves blacks and hates Afrikaners; yet as an Afrikaner he also fears blacks. The second part of the book is a series of chapter-length reconstructions of events that Malan investigated – the result is a macabre and unrelenting exploration of "the way South Africans kill each other."

Several of Malan's vignettes are set in Johannesburg. The book therefore involves, in part, a textual juxtaposition of a remembered Johannesburg and a traveled Johannesburg. Malan's city (in both its past and its present incarnations) differs in two major ways from that of Hochschild and Finnegan. The primary difference lies in the author's place and role in the landscape. In the earlier Johannesburg, Malan places himself at the center of the account. In contemporary Johannesburg he hovers on the margins of the narrative, intruding only with moral asides about the significance of his discoveries to the process of self-disclosure. Nonetheless, in contrast to Hochschild and Finnegan, Malan's Johannesburg is a city lived in rather than gazed upon.

Malan represents his youth and teenage years in Johannesburg as a study in idealistic and foolish rebellion against the conventions of the Afrikaner family, the master race, the white suburb and bourgeois culture. The young rebel is motivated by his love for blacks and his reading of the messages and symbolism of the cultural crossover, the "hand-me-downs from the great white [American] culture":

> Johannesburg lay in Africa, but that was more or less incidental. Johannesburg had skyscrapers, smart department stores, cinemas and theaters. It was part of a larger world. There was no TV in my boyhood, because the Brotherhood feared it would cause what Mao termed "spiritual pollution." We had radio, though, and all the characters in the boys' serials were British or American. Randy Stone was the night-beat reporter in an American city. Ricky Roper, the Sunrise Toffee Junior Detective, lived in an ambiguous someplace where everyone had BBC accents. Chuck drove a cab in Brooklyn. Mark Saxon came from outer space, but even his world seemed more immediate than Afrikanerdom to me. I mean, I didn't realize *Reader's Digest* was a foreign magazine until I was at least ten. I have no recollection at all of the Sharpeville massacre, in which sixty-nine black people were shot dead while protesting against the pass laws.... I remember the day John F. Kennedy died, though.
>
> (Malan 1990: 28–9)

Here, then, is a further extension of the colonial trope of Johannesburg as a city out of Africa. Its location is "more or less incidental"; its landscape depicted a "larger world." And, Malan suggests, the white middle class – the new Afrikaners – took their cultural bearings not from the Africa around them but from British and American media culture: "I had a more or less generically Western childhood, unfolding in generic white suburbs where almost everyone subscribed to *Life* and *Reader's Digest*, and to the generic Western verities they upheld" (p. 47).

Malan constantly implies that the city (and rural) landscape always conceals much more than it reveals. Appearances mean nothing; behind the tranquility of the scene lie darker truths. His text contains a series of arresting narrative violations of the visible landscape. The earliest examples fix the author himself as the violating body. One night Malan and friends spray-paint a "monumental graffito, six feet tall, forty yards long," bracketed by twin sets of hammer and sickle, on a "lily-white" suburban wall: "Say it Out Loud, I'm Black and I'm Proud." When he proudly exhibits the resulting newspaper photograph to the family servant, she tells him to "get lost." Later the teenage Malan has furtive sex with a black woman in the domestic quarters of a neighboring white Johannesburg mansion. He leaves home and lives, not in the generic white suburbs, but in a "dank, verminous" flat where "green slime oozed out of the ancient water faucets." Malan recalls these incidents poisoned, in his own words, by cynicism about his own actions.

Leaving school, Malan becomes a crime reporter for a local liberal newspaper. Here the juvenile violations of the apartheid city are replaced by something altogether more menacing. The job puts him in contact with people most white liberals never met, and took him to places they never saw, including Soweto:

> In my memory it is always winter in Soweto; the setting sun is an orb of cold orange in a gray sky, and gray smoke is drifting across a gray desolate landscape. The houses are gray; their asbestos roofs are gray; the wastelands are strewn with gray ash and rubble. The roads are clogged with morose men streaming home from the railroad station in cast-off gray overcoats.... I was always relieved to leave Soweto, to see the lights of white Johannesburg looming up ahead – always. Once I was back in the world of the light, the fear would dissipate, and I'd be ashamed of myself, to think that my own psyche was riddled with irrational racist phantasms. And yet, whenever I returned to Soweto the fear fell in behind me.
>
> (p. 64)

Here again is the divided city. Superficially, Malan's Soweto is not dissimilar from Hochschild's or Finnegan's. The contrast of endless gray with the comforting light of white Johannesburg echoes the juxtapositions of *The Mirror at Midnight* and *Dateline Soweto*. Malan's Soweto is, however, also the place for personal confession. This is not a Hochschildian landscape – gazed upon, but not engaged with. Nor is it the landscape braved by Finnegan, traveled through incognito, and emerging unscathed and triumphant. This is a landscape of fear, a landscape that prompts "irrational racist phantasms."

Johannesburg's divided landscape is unmasked; the city is "chin-deep in gore." Incidents of brutality and mindless violence pile on top of one another:

In my imagination, Soweto came to resemble Europe in the Dark Ages, a place where humble people barricaded their doors at darkness and trembled through the night while werewolves howled outside. It was not an entirely fanciful vision. Soweto was a charnel house. Its murder rate was four or five times higher than New York's. Its trains were infested with *tsotsis*, young gangsters who immobilized their victims with a sharpened bicycle spoke in the thigh, and then made off with their cash or packages. On Friday nights, pay nights, wolf packs of gangsters lay in wait for incoming trains, and picked off the breadwinners on their way home from the station.

(p. 67)

Fanciful imaginings? Irrational racist phantasms? Malan wades through the carnage, to convince that Soweto *is* a charnel house, a place of relentless danger not only to white visitors but to ordinary black inhabitants as well.

In the second part of *My Traitor's Heart*, Malan makes himself more anonymous, though the voice of the seasoned investigative reporter is ever-present. The author's earlier childish and meaningless violations of the white landscape, his subversion of the orderly divided landscape of the apartheid city, become the backdrop for a far more serious and deadly set of violations – what Malan calls "tales of ordinary murder."

Observe two scenes. The first is a typical white South African social event – the weekend braaivleis (or barbecue). "You must understand" says Malan, "that a braaivleis is no mere barbecue. It is a profound cultural ritual.... It evoked all that was finest about the sweet white life in the land of apartheid." In most post-colonial South African travel narratives, the braaivleis has an iconographic and highly stylized place – symbolizing at once the generosity of whites to other whites and to white outsiders, and their indifference and obliviousness to the black world on their doorstep. The braaivleis invariably takes place on sun-splashed and verdant lawns close to the crystal blue swimming-pools in which children laughingly cavort. Malan's braaivleis is identical, with the added twist that at least one family has just returned from church.

The idyllic scene is interrupted by the arrival of a distraught black woman, the servant of the church-going family, seeking refuge from an angry boyfriend. "And now you must steel yourself" warns Malan "for what is to come is hideous." In lurid and painstaking detail, Malan describes the verbal assault, physical battery and eventual murder of the boyfriend while the white women scream and the white children jump encouragingly up and down. "I was unhinged by the terrible image that lay at the heart of the story – that quintessentially South African tableau of braaivleis, rugby, sunny skies, and torture. It was all so fucking, heartbreakingly traditional."

The second scene is a gold-mine south-west of Johannesburg "on the western arm of the gold reef that made white South Africa rich." It is the tale of a man who died on "the boundary between rival kingdoms of consciousness" (p. 192). The tale begins with a description of the coercive geography of the mine landscape:

> It was like a city down there, he said, or a giant underground factory, so noisy that your brain seemed to rattle inside your skull. Trains thundered, machines clattered, drills yammered, and the shock waves of muffled explosions came rumbling out of the mouths of dark, claustrophobic tunnels. It was a frightening place to work, he said, what with a mile of rock between you and the sun and cave-ins an ever-present danger.... Back up in the sunlight, black miners lived in bleak all-male compounds, far from their homes and families. Themba's compound looked rather like a prison to me – a model prison, to be sure, with sports fields and recreation facilities, but still: not the nicest place to live. It consisted of nine single-story buildings, strewn in a crescent across several acres of lawn. In the center stood a giant mess hall, and off to one side was an open-air amphitheatre, where movies were sometimes shown. The whole was surrounded by tall fences, and accessible only through an elaborate gate flanked by houses teeming with security men.
>
> (pp. 194, 196)

For Malan, this scene is simply the backdrop for "a parable about industrial relations." There follows a brief description of the history of the National Union of Mineworkers (founded in 1982) and its growth at the mine in question. The narrative builds to a bruising climax, describing industrial action and violent confrontation between mineworkers and management over working conditions at the mine. Conciliation fails; the mine slides into anarchy.

Wildcat strikes follow. The work-force splits into two factions, each claiming union support. The strikers attack scabs and management evicts them from the hostels. Mine security moves in and forcibly crushes the strike. Malan's description, drawn primarily from a mining-house report, depicts the events with a full measure of modern industrial-relations jargon. It is a peculiarly South African confrontation – violent and coercive on all sides – but the language is resonant of industrial strife in any advanced capitalist country. Again, the effect is illusory. Under the surface of the recognizable, lurks something more sinister. "Early one Sunday morning" observes Malan "an alert security guard noticed something odd happening in a marsh just off mine property" (p. 204).

The appearance of a modern industrial action, however violent and confrontational, begins to dissolve, replaced by a far more menacing, and unknowable, threat. The workers at the marsh had earlier consulted a renowned "witch doctor" who had provided them with shoe-polish tins of a "very powerful battle medicine." With this smeared on their bodies, they would be immune to the white man's bullets. Returning from the marsh, they have tiny cuts on their bodies (p. 207). Confronted by mine security with guns and a water cannon, the workers "go mad" and seem "utterly fearless" in the face of an onslaught of rubber bullets. Malan places the reader on the top of a mine dump gazing down at two ordinary white policemen (one likes Fleetwood Mac and Van Halen, the other looks like Buddy Holly) attempting to disperse a growing crowd of miners:

> As the tear-gas grenades fell hissing in around them, Hlangana leapt to his feet yelling *"Bulala amaBunu!"* – Kill the Boers! "The medicine is strong!" he shouted. And so a great cry arose from the assembled mine workers, and they rose and charged as one. They came out of a cloud of tear gas in the oxhorn formation of nineteenth-century African warfare, the center attacking while the horns rushed to outflank the enemy and close in on him from behind. Sergeant Pretorious unslung his shortgun and shot one of the charging miners in the chest, blew his head off the next at point-blank range. Constable Koekemoer was standing behind the open door of the truck, blasting away with his service pistol. The guns were going off in their very faces, but the miners seemed utterly unafraid. They kept coming.... Pretorious stagger[ed] as someone caught him a blow on the head from behind, and then he was overwhelmed and disappeared under a tide of black men. Koekemoer fired until his gun was empty, and then ducked back inside the Chevy's cab.... And then the mob grabbed his legs and hauled him out of the cab, and he, too, was stomped, hacked, and kicked to pieces. Literally. Once the cops were dead, the ferocious mob tore their truck apart with their bare hands and overturned it on Koekemoer's corpse.
>
> (p. 213)

Malan's descent into the language and imagery of nineteenth-century colonial discourse is striking – the witchcraft, the Zulu fighting-machine, the invincible tide of black men, the rampaging mob, the violence and the blood. The illusion is complete with a clinical description of the subsequent vengeance exacted by "an endless convoy of police vans and armored personnel carriers" (p. 15).

The narrative of *My Traitor's Heart* snakes into and out of Johannesburg. Many of the other incidents and stories recounted in the book carry similar messages about the contrast between surface, and comprehensible, appearance and the unknowable forces of darkness and rage just beneath. Malan, like so many other travel writers of Africa, is seduced by the most enduring colonial travel trope of all. Rob Nixon argues that Conrad's *Heart of Darkness* has exerted a powerful centripetal pull over western representations of Africa "unequaled in this century by the sway of any other text over the portrayal of any single continent" (Nixon 1992: 90). The heart-of-darkness trope has, in the process, acquired a rhetorical force only distantly connected to the geographical context and cultural form of Conrad's initial usage (Nixon 1991b). For Malan, Africa is at once utterly foreign, yet very familiar. His job as a crime reporter enabled him to ask the right questions, "the questions that cut to the very darkest heart of the matter" (Malan 1990: 81):

> That passage in Conrad's *Heart of Darkness* where Kurtz rears up on his deathbed and croaks, "The horror! The horror!" – you remember it, don't you? Critics have always wondered what Kurtz was looking at as he spoke those words, but I know, I know, I know. I was out there for a while in the winter of 1986.

Malan's whole journey into Africa, like that of many contemporary writers (Gruesser 1992), is scripted by this neo-Conradian vision. Africans "continue to be seen widely as the 'tribes' at the remote source of the river." Whites in Africa irrationally descend to unspeakable acts of barbarism. Johannesburg lies in Africa, true, but that is not incidental. Yet the city's landscape is a modern façade, barely concealing the primitive, dark African heart within (see Robbins 1987 and Ward 1989 for similar discoveries). Around the heart of darkness is a vortex of fear and terror.

CONCLUSION

This chapter analyzes a particularly pervasive and enduring form of representation of South Africa and its cities – the travelogue. In its origins and enframing gaze, travel writing is a fundamentally imperial creation. Yet it retains its attraction in a post-colonial era among a variety of post-colonial writers. The iconographic post-colonial travel writer is undoubtedly V. S. Naipaul. For Dennis Porter (1991: 306) "that European-male gaze which from the eighteenth century on could claim virtually the whole globe and its people for its object finds in Naipaul ... an answering gaze." Nixon's (1992) more skeptical and persuasive conclusion is that Naipaul is unable to escape the seductions of the imperialist discursive traditions within which he writes.

The recent upsurge in western media interest in South Africa has spawned a whole industry of visual and literary representation. In this context, the writing of South Africa as a journey of discovery has experienced a remarkable renaissance. In this chapter I have attempted to draw out some of the tensions in ostensibly post-colonial American journalistic writing about the South African city. The travelogue seems to have more appeal to contemporary American commentators than it does their European counterparts. Yet these authors are unable to escape from the imperial discursive context that originally spawned the modern travel writer. I have tried to suggest that a post-colonial vision of the South African city is compromised at virtually every turn by what Pratt (1992) calls "the imperial eye."

A common feature of all contemporary travel writing of the South African city, and Johannesburg in particular, is the way in which the landscape is scripted by a broader moral discourse about race, oppression and inequality. These are all justly post-colonial concerns. In *The Mirror at Midnight* Johannesburg's intimately divided landscape becomes the vehicle for an answering commentary on the inequalities in American society. In *Dateline Soweto* Johannesburg's confusing and frightening landscape can only be rendered tame and comprehensible through total reliance on those with local knowledge. In *My Traitor's Heart* Johannesburg's blighted landscapes represent the divisions, the combination of desire and transgression, in the heart of the oppressor.

Yet all three Johannesburgs are unable to subvert the imperial discursive tradition in which the South African city has always been written. Hochschild enframes the city within a series of imperial promontory descriptions of the city's divided landscape. Finnegan's gaze cleverly inverts the trope of discovery, portraying a city with which he has no affinity and in which he always claims to feel lost. Yet he too cannot escape the need to gaze and invest the landscape with moral coherence. Malan's is potentially the most subversive vision of the three. Here the visible landscape is drained of any visible sense of moral coherence and meaning. Yet in another sense Malan's gaze is the most colonial of all. Beneath the surface lurks the horror of a particularly African barbarism and primordialism. The city – for all its appearance of modernity – cannot contain or disguise the primitive impulse that empowers and destroys its residents, black and white.

ACKNOWLEDGMENTS

I would like to thank the Social Sciences and Humanities Research Council of Canada for its research support.

REFERENCES

Aronson, R. (1990), *"Stay Out of Politics": A Philosopher Views South Africa*, Chicago: University of Chicago Press.

Attwell, M. (1986), *South Africa: Background to the Crisis*, London: Sidgwick & Jackson.

Beinart, W. (1982), *The Political Economy of Pondoland, 1860 to 1930*, Cambridge: Cambridge University Press.

Bonner, P., Hofmeyr, I., James, D., and Lodge, T. (eds.) (1989), *Holding Their Ground: Class, Locality and Culture in 19th and 20th Century South Africa*, Johannesburg: Ravan Press.

Bozzoli, B. (ed.) (1979), *Labour, Townships and Protest: Studies in the Social History of the Witwatersrand*, Johannesburg: Ravan Press.

—— (1991), *Women of Phokeng: Consciousness, Life Strategy, and Migrancy in South Africa, 1900–1983*, New York: Heinemann.

Brantlinger, P. (1988), *Rule of Darkness: British Literature and Imperialism*, Ithaca: Cornell University Press.

Cammack, D. (1990), *The Rand at War, 1899–1902: The Witwatersrand and the Anglo-Boer War*, Berkeley: University of California Press.

Caws, M. (1991), *City Images: Perspectives from Literature, Philosophy and Film*, New York: Gordon & Breach.

Clifford, J. (1992), "Traveling cultures," in L. Grossberg (ed.), *Cultural Studies*, London: Routledge, pp. 96–116.

Coetzee, J. M. (1988), *White Writing: On the Culture of Letters in South Africa*, New Haven: Yale University Press.

Comaroff, J., and Comaroff, J. (1991), *Of Revelation and Revolution: Christianity, Colonialism and Consciousness in South Africa*, Chicago: University of Chicago Press.

—— (1992), *Ethnography and the Historical Imagination*, Boulder: Westview Press.

Cowell, A. (1992), *Killing the Wizards*, New York: Simon & Shuster.

Crapanzano, V. (1985), *Waiting: The Whites of South Africa*, London: Granada.

Crawford, R. (1989), *Journey into Apartheid*, London: Epworth Press.

Crocker, M. (1992), *Loneliness and Time: British Travel Writing in the Twentieth Century*, London: Secker & Warburg.

Crush, J. (1987), *The Struggle for Swazi Labour, 1890–1920*, Montreal & Kingston: McGill-Queen's Press.

Crush, J., Jeeves, A., and Yudelman, D. (1991), *South Africa's Labor Empire: A History of Black Migrancy to the Gold Mines*, Boulder: Westview Press.

Crush, J., and Ambler, C. (eds.) (1992), *Liquor and Labor in Southern Africa*, Athens: Ohio University Press.

Diash, P. (1990), *The Sharpeville Six*, Toronto: McLelland & Stewart.

Driver, F. (1992), "Geography's empire: Histories of geographical knowledge," *Society and Space*, 10: 23–40.

Finnegan, W. (1987), *Crossing the Line: A Year in the Land of Apartheid*, New York: Harper & Row.

—— (1989), *Dateline Soweto: Travels with Black South African Reporters*, New York: Harper & Row.

—— (1992), *A Complicated War: The Harrowing of Mozambique*, Berkeley: University of California Press.

Flavin, M. (1949), *Black and White: From the Cape to the Congo*, New York: Harper.

Fuller, R. (n.d.), *South Africa at Home*, London: George Newnes.

Gordon, S. (1989), *The Middle of Somewhere: A Story of South Africa*, New York: Orchard Books.

Gready, P. (1990), "The Sophiatown writers of the fifties: The unreal reality of their world," *Journal of Southern African Studies*, 16: 139–64.

Greenfield, B. (1992), *Narrating Discovery*, New York: Columbia University Press.

Gruesser, J. (1992), *White on Black: Contemporary Literature about Africa*, Urbana: University of Illinois Press.

Haarhoff, D. (1991), *The Wild South-West*, Johannesburg: University of Witwatersrand Press.

Hallett, R. (1988), "Explorers," *Southern African Review of Books*, 1/3: 15–16.

Heard, A. (1990), *The Cape of Storms*, Fayeteville: University of Arkansas Press.

Himmelfarb, G. (1984), *The Idea of Poverty: England in the Early Industrial Age*, London: Faber & Faber.

Hochschild, A. (1990), *The Mirror at Midnight: A South African Journey*, New York: Penguin Books.

Hope, C. (1988), *White Boy Running: A Book About South Africa*, London: Martin Secker & Warburg.

Hutchinson, S. (1992), *Cervantine Journeys*, Madison: University of Wisconsin Press.

Jeeves, A. (1985), *Migrant Labour in South Africa's Mining Economy: The Struggle for the Gold Mines' Labour Supply*, Montreal & Kingston: McGill-Queen's Press.

Johnstone, F. (1978), "The labour history of the Witwatersrand in the context of South African studies," *Social Dynamics*, 4/2: 101–8.

Kallaway, P., and Pearson, P. (1986), *Johannesburg: Images and Continuities*, Johannesburg: Ravan Press.

Kowalewski, M. (1992), *Temperamental Journeys: Essays on the Modern Literature of Travel*, Athens: University of Georgia Press.

Lang, J. (1986), *Bullion Johannesburg*, Johannesburg: Jonathan Ball.

Lapping, B. (1986), *Apartheid: A History*, London: Grafton Books.

Leed, E. (1991), *The Mind of the Traveler*, New York: Basic Books.

Lees, A. (1985), *Cities Perceived: Urban Society in European and American Thought, 1820–1940*, New York: Columbia University Press.

Lelyveld, J. (1986), *Move Your Shadow: South Africa, Black and White*, New York: Penguin.

Lowe, L. (1991), *Critical Terrains*, Ithaca: Cornell University Press.

Macmillan, A. (1934), *The Golden City: Johannesburg*, London: Collingridge.

Malan, R. (1990), *My Traitor's Heart: A South African Exile Returns to Face His Country, His Tribe, and His Conscience*, New York: Atlantic Monthly Press.

Mallaby, S. (1992), *After Apartheid: The Future of South Africa*, New York: Times Books.

Mandy, N. (1984), *Johannesburg: A City Divided*, New York: St. Martin's Press.

Manning, R. (1987), *"They Cannot Kill Us All": An Eyewitness Account of South Africa Today*, Boston: Houghton Mifflin.

Marks, S., and Rathbone, R. (eds.) (1982), *Industrialisation and Social Change in South Africa: African Class Formation, Culture and Consciousness 1870–1930*, London: Longman.

Mathabane, M. (1986), *Kaffir Boy: Growing Out of Apartheid*, New York: Macmillan.

Mattera, D. (1987), *Sophiatown: Coming of Age in South Africa*, Boston: Beacon Press.

Mendelsohn, R. (1991), *Sammy Marks: The Uncrowned King of the Transvaal*, Cape Town & Athens: David Philip & Ohio University Press.

Meyer, C. (1986), *Voices of South Africa*, New York: Gulliver Books.

Miller, C. (1985), *Blank Darkness: Africanist Discourse in French*, Chicago: University of Chicago Press.

—— (1990), *Theories of Africans: Francophone Literature and Anthropology in Africa*, Chicago: University of Chicago Press.

Mills, S. (1991), *Discourses of Difference: An Analysis of Women's Travel Writing and Colonialism*, London: Routledge.

Mitchell, T. (1990), *Colonising Egypt*, Cambridge: Cambridge University Press.

Moodie, T. D. (forthcoming), *Going for Gold: Men, Mines and Migrancy in South Africa*, Berkeley: University of California Press.

Murray, C. (1986), *Families Divided: The Impact of Migrant Labour in Lesotho*, Cambridge: Cambridge University Press.

Nixon, R. (1991a), "Cry white season: Apartheid, liberalism and the American screen," *South Atlantic Quarterly*, 90: 499–529.

—— (1991b), "Preparations for travel: The Naipaul brothers' Conradian atavism,"

Research in African Literatures, 22: 177–90.

Nixon, R. (1992), *London Calling: V. S. Naipaul, Postcolonial Mandarin*, New York: Oxford University Press.

North, J. (1985), *Freedom Rising*, New York: Macmillan.

Noyes, J. (1992), *Colonial Space: Spatiality in the Discourse of German South West Africa 1884–1915*, Philadelphia: Harwood.

van Onselen, C. (1982), *New Babylon: New Nineveh: Studies in the Social and Economic History of the Witwatersrand*, London: Longman.

Ó Tuathail, G. (1992), "Foreign policy and the hyperreal: The Reagan administration and the scripting of 'South Africa'," in T. Barnes and J. Duncan (eds.), *Writing Worlds*, London: Routledge, 155–76.

Parnell, S., and Pirie, G. (1991), "Johannesburg," in A. Lemon (ed.), *Homes Apart: South Africa's Segregated Cities*, Bloomington: Indiana University Press, 129–45.

Peckham, L. (1990), "Ons stel nie belang nie/We are not interested in: Speaking apartheid," in R. Ferguson, M. Gever, T. Minh-ha and C. West (eds.), *Out There: Marginalization and Contemporary Culture*, Cambridge: MIT Press, 367–76.

Porter, D. (1991), *Haunted Journeys: Desire and Transgression in European Travel Writing*, Princeton: Princeton University Press.

Pratt, M. L. (1992), *Imperial Eyes: Travel Writing and Transculturation*, London: Routledge.

Robbins, D. (1986), *The 29th Parallel: A South African Journey*, Pietermaritzburg: Shuter & Shooter.

—— (1987), *Wasteland*, Johannesburg: Lowry Publishers.

Robinson, J. (1992), "Power, space and the city: Historical reflections on apartheid and post-apartheid urban orders," in D. Smith (ed.), *The Apartheid City and Beyond*, London: Routledge, 292–302.

Said, E. (1978), *Orientalism: Western Conceptions of the Orient*, Harmondsworth: Penguin.

Sharpe, W., and Wallock, L. (eds.) (1987), *Visions of the Modern City: Essays in History, Art and Literature*, Baltimore: Johns Hopkins University Press.

Sparks, A. (1990), *The Mind of South Africa: The Rise and Fall of Apartheid*, London: Heinemann.

Spurr, D. (1993), *The Rhetoric of Empire: Colonial Discourse in Journalism, Travel Writing, and Imperial Administration*, Durham: Duke University Press.

Stengel, R. (1990), *January Sun: One Day, Three Lives in a South African Town*, New York: Simon & Schuster.

Streeten, P. (1992), *Paul Streeten in South Africa: Reflections on a Journey*, Innesdale: Development Society of Southern Africa.

Van den Abbeele, G. (1992), *Travel as Metaphor*, Minneapolis: University of Minnesota Press.

de Villiers, M. (1987), *White Tribe Dreaming: Apartheid's Bitter Roots*, Toronto: Macmillan Canada.

Ward, D. (1989), *Chronicles of Darkness*, London: Routledge.

Webb, T. (ed.) (1991), *Voyaging through Strange Seas*, London: Ashfield Press.

Williams, R. (1973), *The Country and the City*, London: Chatto & Windus.

Worger, W. (1987), *City of Diamonds: Mine Workers and Monopoly Capitalism in Kimberley, 1867–1895*, New Haven: Yale University Press.

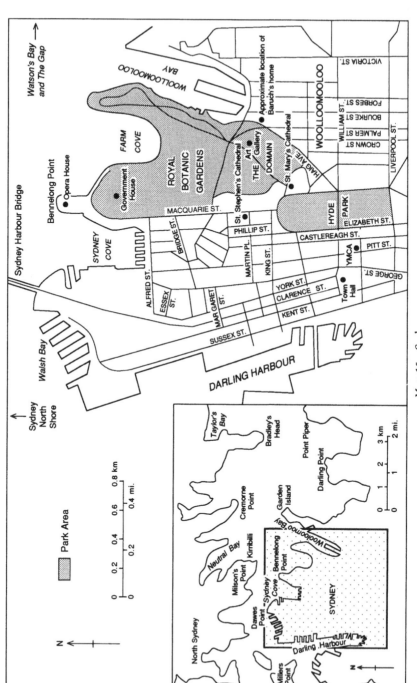

Map 15 Sydney

15

TINSEL TOWN

Sydney as seen through the eyes of Christina Stead

Deborah Carter Park

Why are we here? Nothing floats down here, this far in the south, but it is worn out with wind, tempest and weather; all is flotsam and jetsam. They leave their rags and tatters here ... why can't we run naked in our own country, on our own land, and work out our own destiny? Eating these regurgitated ideas from the old country makes us sick and die of sickness.... This land was last discovered; why? A ghost land, a continent of mystery: the very pole disconcerted the magnetic needle so that ships went astray.... Its heart is made of salt: it suddenly oozes from its burning pores, gold which will destroy men in greed, but not water to give them drink. Jealous land! Ravishers overbold! Bitter dilemma! And lost region! Our land should never have been won.

(Stead 1935: 358)

INTRODUCTION

Australia has a relatively short literary history but one worthy of close examination in its treatment of local subjects, themes, landscapes and sense of geographical marginalization. With a history of approximately 200 years, there is a great deal of diversity within the literary tradition of this country. How Australian writers have responded to the land in general, and Sydney in particular, is especially varied.

Before Australia's emergence on the international literary map, the majority of Australian literature was rooted in a specific historical, geographical and social context. For most of the nineteenth century, the motif of the "lost person" (Bennett 1991: 12) entrenched a strong sense of disorientation and loss of purpose in artistic and literary representations. Similarly, the suspended fictional text displayed a preoccupation with the past, where a sense of "exile" rather than a sense of "place," was the overriding universal theme (Bennett 1991: 12). The Australian novel tended to pivot on themes of suffering, violence, isolation, loneliness and alienation, especially in its depiction of the oppressive system in place to handle a transplanted criminal population. Kiernan (1971: 181) pushes this thematic connection of death and total alienation further: it was a situation where the individual was resolved to finding "some idyllic compromise between the conflicting forces of a society dehumanizing in its demands and a nature that offered no refuge for the solitary."

This tendency was exacerbated by Australia's unique geographical and topographical situation, where remoteness from the rest of the world engendered a distinct sense of marginalization. The Australian settler was seen to be "severed from the past more drastically than those who people the United States and Canada owing to the greatness of the isolating distance between the colonies and the mother country" (Moore 1971: 267). In turn, the clash between European civilization and an antipodean geography out of which Australian culture evolved proved geography the more powerful (Eaden and Mares 1986). But all is not relegated to themes of exile and the "lost person" motif, for in the inter-war years and more particularly, since the 1960s, the mythologizing of Australia as an empty country, as a land without owners, has lost currency.

This chapter will look at the work of Christina Stead (1902–83), focussing specifically on *Seven Poor Men of Sydney* (1935), her only novel exclusively based in her native Australian city. The chapter's first task is to justify the selection of Stead and to provide an overview on the growth and development of Sydney reflected in *Seven Poor Men of Sydney* and then to shift focus to Stead herself. The chapter then moves to an analysis of this fictional text in which twilight zones, alienation and stereotypical gender relations and ideology figure prominently. From this, insights are gleaned into the character of Sydney.

The reader may be forced at this point to ask why Christina Stead and why *Seven Poor Men of Sydney*? Why not look at the works by Patrick White or David Ireland or David Malouf? It is worth conceding that there are indeed other Australian writers who could have been selected for review, but essentially the principal motivation (besides personal preference) for selecting Stead stems from three concrete criteria. First, the individual had to be an Australian writer who had produced literary works illuminating Australia and Australians in their urban environment; the writer had to present the urban experience and depict that place comprehensively – as Stead did. Second, the writer had to be truly reflective of the period in which he or she wrote – as Stead was. Third, the writer had to be representative of the place writers occupy within this literary domain – and Stead truly was.

Stead is able to delineate in *Seven Poor Men of Sydney* the urban experience and the inherent sense of alienation, oppression and ghettoization of a young working population. The reader is afforded a view of the ongoing class struggle between a group of young people at a turning-point in their lives in the growing city of Sydney in the 1920s. The novel provides a composite of what it "meant to be young in a particular medium, that is, in Australia at a particular stage in its development" (Green 1968: 153) and of the inherent spatial implications of *laissez-faire* economics.

DEVELOPMENT OF SYDNEY

The growth and development of Sydney is an important contextual framework underlying *Seven Poor Men of Sydney*. Nestled on one of the world's finest natural harbors near the south-eastern tip of Australia, Sydney began life as a convict settlement in 1788 (Kramer 1981; Hughes 1986), with a population of 1,024 inhabitants (Spearritt 1986). It emerged as a "tortured town" (Kelly 1986), one of the largest penitentiaries on earth for the "outcasts of Mother England" (Hughes 1986: 18). This convict settlement was the "beachhead for British imperial expansion on the Australian continent" (Wells 1986: 65) and came into existence due to penal laws and official public policy in distant England.

By the 1850s Sydney had grown fifty-fold and served as the engine for the colonial economy, functioning as a port city and trading entrepôt "where the merchant adventurer, the jack of all trades and master of some, made the money go round" (Kelly 1986: 43). Paralleling this growth, commodification of pastoral production granted Sydney a central economic role: export commodities and local food supplies flowed through Sydney. With commodification came food-processing functions such as "brewing, milling and baking ... along with tanning, clothmaking and soap production. Merchants with their interests in warehousing, retailing, port handling, auctioneering and short-term financing met the marketing requirements and supplied provisions for the inland pastoral producers" (Wells 1986: 66).

During the last half of the nineteenth century, and more particularly after the depression of 1890, new trends emerged which form the backdrop for Stead's novel: "Australian cities and urban life generally had become vitally important in determining the nature of economic, social and political life in Australia" (Woodward 1975: 115). With a "reformist labour movement championing the claims of state and private workers" and a "continental market that transcended the colonial economies," Sydney rose as an industrial city (Wells 1986: 69), unlike other industrializing cities in North America and Europe. Similarly, Asa Briggs (1963: 73) opined that Australia, at the end of the nineteenth century, was one the most highly urbanized of what he coined the "nearly developing countries." Between 1881 and 1901 the population of Sydney climbed from 225,000 to 482,000, and between 1911 and 1933 Sydney's population almost doubled from 667,000 to 1,235,000 (Spearritt 1986). More fundamentally, the economy moved into an intensive phase of commodity manufacturing and urban development, and as a corollary the urban phenomenon began to be expressed by specific interests concentrated in the city (Wells 1986).

Up until World War II, Australia was afflicted with stagnant economic conditions, despite periodic cycles of boom and bust. In the 1920s sustained industrialization and the establishment of the large modern factory forced employment into the urban industrial suburbs, more specifically, the "inner suburbs of Sydney, followed by the southern and western areas of the city" (Wells 1986: 71). During this time, Sydney grew in a rather haphazard fashion, governed by a "combination of difficult topography and cavalier economics" essentially free of "legislative or civic restriction" (Kelly 1986: 49).

In the context of Stead's novel, it should be noted that in its geography, Sydney embodies a complex array of water and land relationships. The proximity to open ocean, harbors, bays, estuaries and meandering coastlines (Buckley 1986) creates the environment of a seaside city. Blessed with one of the finest and most flourishing harbors in the world, Sydney's proximity to water provides a whole ecology and texture of its own. It has been suggested that the harbor is the life-blood of the city and provides Sydney with a distinct sense of place, unique scenery and landscape which is refracted through *Seven Poor Men of Sydney*. Reflecting on her earlier years in Sydney, Stead commented that

> all the ocean liners and other ships, mercantile ships and so on, came right in front of our house. The pilot ship was always there, anchored there, and they used to stay there for quarantine. We saw all the ships that came into the harbor, it was very thrilling.
>
> (Wetherell 1980: 436–7)

CHRISTINA STEAD

Stead's place among Australian literary historians is controversial but fits quite nicely within the context of this book. Her reputation over the years has flitted between the poles of fame and obscurity. Although the author's first two novels, *The Salzburg Tales* (1934) and *Seven Poor Men of Sydney* (1935), were published in London, England, during the inter-war years, it was only in the 1960s that Stead, like her Australian peers, began to attract an audience in her own country and receive attention as one of the most international of Australian writers (Sheridan 1988). The first to be so elevated was Patrick White, followed to a lesser extent by Stead. Thanks to their work and that of compatriots such as Eleanor Dark, Sumner Locke Elliott, Miles Franklin, Louise Mack, Ruth Park, Kylie Tennant and Ethel Turner, most successful Australian authors have experienced little difficulty in achieving respect in the international literary world since the 1970s. The pioneers, like White, Stead and others, faced a more indifferent world.

According to Holt (1984), and based on her readings of Colin Roderick, publication of Australian literary works during the period after 1930 was driven by the maturation of the Australian novel within a particular antipodal setting. Holt claims that this stemmed from the process of industrialization and urbanization which had made itself felt within the western world. Most people were now "set in cities, a much more familiar setting for Australians, American and Europeans than the 'bizarre' and curious landforms of the inner continent" (Holt 1984: 9) which had previously formed the backbone of mainstream normative fictional accounts of Australia.

Sydney occupied an important place in Stead's works of fiction. Although she spoke from England, she wrote about Sydney at a particular time and within a particular historical and social context (Bennett 1991). For Stead, Sydney represented "human living space, identity and socio-cultural position" (Bennett 1991: 11). Through Stead's eyes, Sydney in *Seven Poor Men of Sydney* is more than physical and geographical space: it is animated by an underlying economic dynamic that illuminates the economic, social and cultural processes of alienation and subjugation, a perspective grounded in astute observation during her years in Sydney and the surrounding area. Stead provides insight into the social and economic conditions that defined everyday existence for the working-class population of Sydney in the 1920s (Mitchell 1981). The novel's observations are bound up with the artifacts and attitudes of the local, and feelings are based on place. In this light, we can allude to a "Steadian" gaze that allows the reader to peer into the buildings, to wander the streets and harbor district and engage in the ongoing struggle of working men and women during this particular era – a time now 70 years removed. Through her gaze, we can home in on landscapes and cityscapes that provide a navigational map on the interplay between people and place on an emotional, mental, economic and social plane.

Although Stead received her first literary award in 1974 (named in honor of Patrick White, the more popular and internationally known Australian novelist), her image as an expatriate writer (due to forty years abroad in Europe, England and the USA) has been more well-known and criticized both in her homeland and abroad. In 1967 Stead was refused the Encyclopaedia Britannica Award on the grounds that "she is not an Australian in terms of the award" (Green 1968: 150) because she had spent too much time abroad and too little time on Australian soil (Smith 1980: 212). In all likelihood, this expatriate image, often affixed to Stead, was a deciding factor when there was talk of a nomination for the Nobel Prize before her death in 1983: she did not receive the nomination (Sheridan 1988).

What are the details of Stead's life? In 1928 after spending her first twenty-six years in Sydney and the surrounding area, Stead embarked for London, England. While working as a clerk for Strauss and Company, a grain merchant firm in London, and fearing she might die from illness and exhaustion, she wrote *Seven Poor Men of Sydney* (1935) as only "something to leave behind" (Dutton 1985: 170). While in the employment of this firm, she met William James Blake, an American writer and Marxist economist. It was Blake who became Stead's husband soon after and who, throughout her career, was to have an important influence on the political and ideological tone of her literary works.

During her marriage with Blake, she traveled to and lived in Europe and America (Antwerp, Brussels, London, Paris, New York, Los Angeles, Switzerland and various locations in Spain during the civil war). After Blake's death in 1968, Stead returned to Australia in 1969 for four months. In 1974 she returned to Sydney to take up permanent residence. From approximately 1974 until her death in 1983, Stead lived in Sydney, a place that had been a formidable force in her formative years.

During her literary career, she wrote thirteen novels and numerous short stories. *Seven Poor Men of Sydney* (1935) is set wholly in the city of her youth. A later work entitled *For Love Alone* (1944) uses Sydney as the setting for its first half, and in another work, entitled *The Man Who Loved Children* (1940), her childhood environment is "translated into its American equivalent" (Green 1968: 151). Fisherman's Bay in *For Love Alone*, and in *Seven Poor Men of Sydney* in particular, is the Watson's Bay of Sydney to which her family moved in 1917. Many of her other works, although a "reflection of the mental, if not physical landscape of the author's youth" (Green 1968: 151), were based on American and European settings.

It is important to stress that, although Stead tended to focus upon political issues, she wrote within the social-realist genre. She had a gift for lyric prose and displays a sensitive perspective to the "range and minuteness of observation" (Green 1968: 151) on the "thwarted lives of working class Australians" (Lidoff 1982: 6). Stead's fictional works are based on a "careful observation of people and life situations ... from the selection and recording of the actual" (Wilding 1983: 150).

In *Seven Poor Men of Sydney* her political observations examine the destructive effects of economic exploitation and alienation in the post-World War I period and the spatial segregation of the working class in Sydney. Her attitudes on the political are well summed up by Wilding (1983) in Stead's obituary which appeared in *Australian Literary Studies*. On this somber occasion, Wilding notes that "her opposition to exploitation pervades all her work, whether the exploitation is economic, sexual or familial ... the capitalist, imperialist and military exploitations were the primary exploitations that had yet to be effectively confronted, resisted and removed" (Wilding 1983: 150–1).

SEVEN POOR MEN PLUS ONE IN CAMDEN TOWN

Stead is first and foremost a character writer who is not interested in plot but in what the cast of the novel "do with their lives and [what] their lives do with them" (Wetherell 1980: 444). Through the novel's eight characters (seven

men and one woman), Stead demonstrates a "deep awareness of contemporary socio-economic and political issues" (Segerberg 1987: 122); she provides
the reader with an image of a city gripped by forces of exploitation asserted
against society. As the fictional text unfolds, the characters' life-stories
become intermeshed in a tableau of economic hardship, personal struggle, the
creation of a class-divided society and alienation, including withdrawal and
death. It is through the "poverty-stricken and intellectually strenuous"
(Green 1968: 152) young men and women in the novel that the reader is given
a pungent, powerful whiff of the city's character. Her insight is grounded in
a basic sense of charity towards human relationships. Characters come to
their own ends through "real plots that occur in life ... [and how people] are
manipulated by society or surrounding circumstances or other people"
(Wetherell 1980: 444).

It has been conjectured that *Seven Poor Men of Sydney* was the first
Australian novel to reflect Sydney as a "world city." There is little doubt that
the city provided Stead with fertile ground for her account of the impoverished conditions within which the working person must function and
struggle. The sense of marginalization that colors her characters stems from
the capitalist system, which contributes to the creation of poor and wealthy
areas and, for the poor, a climate of alienation where the ultimate escape is
suicide.

This relationship between the working man's plight and the prevailing
capitalist economic system is vividly unearthed by Baruch Mendelssohn (a
printer), when he summarizes the connubial relation between Gregory
Chamberlain, owner of the press, and his employees:

> We're married to Chamberlain, or we're his concubines. He pets us,
> snarls, he sees to the general supplies and we get no pin-money at all
> ... he looks at it as housekeeping ... I object to living in a harem, first,
> from natural jealousy, second, because I carefully surveyed myself in
> the glass this morning and I can certify I'm not an odalisque.
>
> (Stead 1935: 105)

The capitalist system is "that detestable and degrading economic prison"
(p. 237) which forces the working classes to live like a bunch of "niggers, glad
of [their] bonds, and licking the hand that whips [them] and singing to the
Lord to make [them] meek" (p. 238).

The sense of alienation created by cities is not new to geographers who
have examined the impact of *laissez-faire* economics on Victorian society in
general and the city in particular. The expansion of the market system, the
unprecedented growth of cities and the subsequent increase in poverty
(Polanyi 1957) precipitated spatial demarcation in the city based on economic
and social class, and hence contributed to the growing character of alienation
in the modern city. Spatial and social dislocation along with discrimination
in the employment and housing markets presented obstacles for the laboring

poor in both residential location and job opportunities (Ward 1989).

The emergence of low-income inner-city residential areas in Sydney is the most visible consequence of *laissez-faire* economics. The shortage of alternative residential housing and the need to be in close proximity to the place of employment enhanced the spatial enclosing of low-income populations. Here the linkage between employment and residence was pronounced due to the lack of adequate transportation systems and because of the seasonal nature of employment. Ward (1989: 5) states that these urban slums and enclaves symbolize the "social costs of laissez-faire policies during the expansion of industrial capitalism."

As with many other writers who have chosen to write about particular cities, Stead offers the reader insight into the heterogeneity of Sydney and the close proximity of wealthy and not-so-wealthy neighborhoods. The inner city (cityscape) and the harbor district (landscape), so important to her novels and other Australian writers, is delineated by following the lives of the various characters in the book. She looks at the city as a place of employment and interaction. As a social realist, she is able to shed light on the multi-faceted character of Sydney and the places where the common working person scrounges out a meager existence. The reader is allowed to wander through the slums and travel the most alienating and dehumanizing sections of the city where people are ill-housed. We are given vivid descriptions of the twilight areas where the unemployed (tramps, striking seaman) and employed (printers) reside.

Harvey's Manchester (1973) and Burgess's Chicago (Park et al. 1925) would in all likelihood be no different from Stead's Sydney; the crucial point here is that Stead's themes were characteristic of literature emanating in the early twentieth century which openly condemned

the dislocating effects of industrial and technical developments within Australian urban life – effects such as the growing disparity of interest between classes and the frequent incidents of corruption and venality among wealthy sections of society, the growing crime of the cities, the loneliness and solitude of city life as distinct from the manifestations of these qualities in the bush, the hopelessness and abject poverty of the city's poorer classes.

(Woodward 1975: 120)

Catherine Baguenault (a rebel), perhaps one of the key figures in the novel, exists between the poles of home comfort and security, and the down-trodden lodgings of outcast Sydney. During periodic bouts of confusion, she seeks emotional resurrection from her "afoot and lighthearted" journeys on the streets of the city; the answers she seeks can only be found in the lowest of places. In one scene, she walks

by street, lane, lamp-post, policeman, picket fence, suburban church, dark shut factory and green graveyard, through several sleeping suburbs until she had lost herself.... She got her bearings at last, and after walking another half-hour came to a shelter in a very poor section.... There was a dirty lavatory on a landing, a wash-basin, and a bench with a gas-ring on which the vagrant women there could cook anything they liked, if they had anything to cook. In the morning they could work for the charitable institution and thus earn their breakfast; there was no reason for them to starve.... With the lowest and lost, with the degraded, unambitious and debauched, Catherine reviewed her life.

(Stead 1935: 247)

And while Michael Baguenault (a ne'er-do-well) walks with Tom Withers (a printer) from Elizabeth Street to the Raymond Terraces, through the Domain onto the Loo (Woolloomooloo) and up into Paddington, we are afforded a glimpse of the hopelessness and displacement of the city's lower classes in the face of unrelenting urbanization. As they walk, they come across old women "without underclothes" sitting on the gutters and children playing who are all "skinny and covered with sores" (p. 276). In this scene, the reader is afforded a view of a slum terrace housing built about 1850 by a builder

whose affection for the old country had induced him to reproduce on a new free soil the worst slums of Camden Town.... Each front had one dark door and one curtained window. Mingily, dingily, the little houses looked on the squalling street.

(p. 276)

In this section of the city, women hitch up their skirts only to reveal their "drawers, grey with age" (p. 277). Their teeth are absent and the roots "blackened."

Continuing this explorative journey, we are coaxed into less-frequented regions as Joseph Baguenault (a printer) turns out of Paddy's Market, where most shops are closed and young men play leap-frog in the streets:

a pool of blood on the pavement, with several clots, made them look around: opposite were two streets in which were houses of ill-fame – a fight between bucks, a girl's having a baby, a bleeding nose?

(p. 139)

Developing the theme of spatial segregation in the city, we are afforded a view of the area surrounding Cambridge Street, historically an area of vice, decadence and immorality for Sydney in the 1900s (Kelly 1986: 48). In this area is Castlereagh Street. While Michael and Withers traverse this area, they come across a young sailor

with two street-walkers on his arms, one old and one young, both haggard and both drunk.... Farther along a toothless young woman

with painted cheeks hung with heartrending tenderness on to the arm
and coat of a vain pimp.

(p. 262)

Adjacent to zones of blight is the area surrounding Macquarie and Pitt
Streets, a district known for its elite clientele. With hard-earned money in
hand, Joseph notices "how splendidly the shop-windows were lighted, how
richly dressed in crumpled silk and polished ashwood" (p. 135). On
Elizabeth Street, near Macquarie Street and Hyde Park, Catherine takes up
residence in a room in an old and dilapidated building. And juxtaposed to
these locales, we travel through another area with Joseph in which

> there were old men by themselves carefully comparing the prices of
> shoes in half-closed shoe-shops; there was a huge "boot-emporium"
> for the populace, decorated with hideous cartoons in green and red, of
> tramps with red noses and their toes out of their shoes, and women in
> laced boots with bursting breasts and big behinds.
>
> (p. 137)

Although tenement dwellings have an inherently dark and gloomy side, these
neighborhoods exhibit an underlying character uniquely their own. It is from
Baruch's window "fourth floor back, in a side street in Woolloomooloo Flat"
that the reader is provided with the following vista:

> His window commanded the Inner Domain, the Art Gallery, the spires
> of St. Mary's Cathedral and the Elizabeth Street skyline. On the right
> hand, as he looked from his window, were the wharves of the German,
> Dutch, Norwegian and Cape lines. In the backyard was a wood-and-
> coal shed covered with creepers, pumpkins, old tires, kites'-tails,
> buckets and old scrubbing brushes. There was a clothes-line across the
> yard ... and upon the line the tenants' garments, washed by the woman
> on the ground-floor.... The house backing to theirs was only three-
> storeys high. A cheap chop-suey restaurant occupied the ground-floor
> ... while the rooms above were for sleeping and for letting out. A
> couple of ladders reached from the windows of these upper floors to
> the roofs of lean-to sheds and outhouses.... On Saturdays and Sundays
> the whole neighborhood swarmed with children, and everybody was
> out of doors with sleeves rolled up.
>
> (pp. 158–9)

This neighborhood is fascinating in its heterogeneous character. What is
registered is an intense feeling of what Sydney would feel, look and smell
like:

> Sometimes colored boxers, cheerfully dressed, paraded past with belles
> of the neighborhood, and there was a stream of girls, Australian and
> Italian, large, bright-colored, buxom, with high heels and transparent

blouses. Opposite Baruch's back window, poor Chinese, sailors and loafers could be seen paying attention to females in the chop-suey gambling and lodging-house.

(p. 161)

In a sordid netherworld between Sydney's residentially segregated neighborhoods, there exists a segment of the population which lives on the streets and sleeps in the parks: this homeless population reflects the ultimate effects of rapid urbanization in Victorian society. In the northern/central portion of Sydney, near the Inner Domain overlooking Garden Island, the tramps of the city customarily take up residence for the evening in the caves along the coast. "Dead leaves ... bad fruit fallen out of trees ... shadowy emanations from the ground, abortions. They [are] not alive" (p. 308).

On the avenues leading into the university, drunks and beggars sleep on the lawns and

round the paddocks at one side, in the bushes, hanging on the picket fence, children of the slums of Golden Grove and Darlington found bloody rags and torn clothes displayed by larrikins. There the regulars of the "University Arms" slept off their spirituous heaviness. In this place Baruch always found his passions rising, because it was dark, brutal, strange.

(p. 165)

When Withers does not get paid he sleeps in the park with the

other deadbeats: only they're smarter than I am; they don't sweat their guts out for a chap who buys his daughter a fur coat and himself a new car [Chamberlain].... The tales they told me, enough to make a monkey bite its mother. They're a lot of philosophers, and remittance-men, sons of belted earls and what not.

(p. 91)

Spirits lost to the sea

Although the harbor occupies a symbolic and esthetic stronghold in the context of the book in the sense that all social and economic life clings to its shores, the actual proximity of the city to the harbor induces a sense of the "metaphysical realms of nature" (Ducker 1986: 160). Water has a penetrating power in the novel. The ever-present waterways provide a wandering zone for the city's population where people refresh themselves in times of pleasure and of pain. It is important to recognize the fact that until 1880 approximately 80 per cent of Sydney residents lived "within daily walking distance of the harbor" (Kelly 1986: 48). People lived, worked and traveled on the harbor contours. Interestingly enough, Stead, in an interview on February

24, 1980, said "the sea is a continent with no passports; it's a country in itself" (Wetherell 1980: 432).

Perhaps more significantly, the harbor affords a sense of escape. Loss of direction in a strange and distant antipodal environment during a chaotic era encourages flights of fantasy, especially for Michael, one of the novel's central characters. He would prefer to be at sea with the people who are the "lowest of the low," traveling the oceans to other worlds. He says

> can you imagine them eating together, sleeping together? The berths below teeming with lice, the food stinking in this weather, rations of rum served out to keep 'em happy till they clear the Heads ... no responsibilities and absolutely not wanted here: exiles ... it must be a relief to be with a lot of perverse dummies, whose backs aren't always bristling with righteousness.
>
> (Stead 1935: 229)

Michael seeks refuge in lands geographically distant:

> Samoa, Hong-Kong, Shanghai, Singapore, Cochin-China. It's lovely there in spring, the cosmopolitan crowds, the gutter like canals, the street-sellers, the little girls in tea-houses, the blue sky, the dogs, leprosy, tigers, temples.
>
> (p. 285)

Escape for him means retirement to a

> monastery or a cave in the desert, which is what [he] always wanted to do, if it weren't so bloody out of date. If [he] could ... join a desperate expedition, one to the North Pole ... or one to Central Africa.
>
> (p. 232)

Joseph also seeks escape. While he is employed in Chamberlain's printing establishment, he mentally flees from work to think about his future. During this feat of escapism,

> his mind floated out over the harbor and wove invisible skeins in the invisible fine air. He was busy fitting together his future like a jig-saw puzzle. He thought of sailing outside the heads and going to the old countries, where the morning sun gilded domes, palaces, royal parks and hives of cities, bigger ports, and where men had a history that looked through millenniums.
>
> (pp. 95–6)

Of course, the harbor locales depicted in *Seven Poor Men of Sydney* are more than just metaphorical and symbolic. They are "drawn from the specific, the actual, the material" (Wilding 1983: 150). For example, Watson's Bay and The Gap are concrete geographical places in this city. The plummeting cliffs of The Gap, above Watson's Bay, is where the terminally alienated person falls to his or her death. Just as The Gap helps give life and character to Sydney, The Gap takes life away. Suicide at The Gap is a "commonplace affair." And, in the context of the book,

> everyone knew why a person committed suicide: if it was a man, because he couldn't pay his bills or had no job; if a woman, because she was going to have a baby.
>
> (Stead 1935: 81)

In one scene, Baruch is having a conversation with Joseph and mentally wanders to another night at Fisherman's Bay where he saw "a rowing-boat tied to one of the piles of the wharf. In the rowing-boat a tarpaulin [covered] the body of the Gap's latest victim, a bankrupt shopkeeper" (p. 221). In the ultimate tragedy of the book, Michael, one of the poor men, commits suicide by plunging from the cliffs of The Gap. It is only at the seashore that he can find what he lost – the "notion that life was worth living" (p. 241). Michael is never truly able to create a place for himself in the world (Lidoff 1982). He perceives the world as rotten and his friends "are too dull to know what's going on under their noses" (Stead 1935: 243). Michael has an intuitive sense of his own death due to this "total awareness." The price of "total awareness is utter inaction, a kind of death, as he instinctively knows" (Green 1968: 155). The ranges of human experience go beyond belief for Michael and, in recalling everyday life, he summons the following scene:

> the cloud of dust was full of people, rushing past with songs and kickings, old mutterers singularly and angularly breaking into yells, bad children, fairies, old professors, confessors, aiders and abbesses, two-legged palsied palimpsests, clerks, sharks, narks, shades, suspicions, university janitors, spiral-horned rams, stock exchange rampers, rabbits, whorlie-whorlies, willy-willies, whories, houris, ghosts, knouts, ghouls, walking-gourds, grimalkins, widdershins and withering wights but in such a horried, enclaced, perplexed, twisted and lolloping rhythm as I shuddered to look upon.
>
> (Stead 1935: 312)

As Lidoff (1982: 129) explains, the central element in Michael's suicide represents a blending of "inner and outer worlds" and the "total loss of identity in death." There can be little doubt that in the following passage, Stead, in a somewhat grotesque manner, attempts to blend Michael's personality with the sea:

> He wrapped his coat around him as he wished to wrap the deep sea round him and its sleep fathoms down. He wished to sleep, to have the water sing as now for ever in his ears, and the inextinguishable anguish in his mind to be hushed. So he stood fixed, with fixed and troubled look cleaving the sea.... The wind sways him like the rooted plants and

grasses ... he is already no longer a man but part of the night. The pine trees crowd him to the ledge, the light wheels, down underneath is the howling parliament of waters deciding on his fate. The gusts on rock and ledge as spirits hold his heart in their shadowy hands and squeeze the blood out of it; darkness only runs through his veins now. He takes a step nearer the edge.... He falls into the sea, the wave a moment later cracks his skull against the submerged pediment of the cliff, and his brains flow out among the hungry sea-anemones and mussels. It is done; all through the early morning the strings of the giant mast cry out a melody, in triumph over the spirit lost.

(Stead 1935: 289–90)

Michael has a troubled mind and it is in the sea, both literally and figuratively, that he is able to find more "repose than in any human heart.... [The sea] understood his miseries through its own rages and revolts, his inconstancies through its tides, his longings through the bottoms" (p. 289). Michael's suicide can be attributed to geographical and historical circumstances; chaos is the only adequate way to describe how he feels and experiences the world.

Kol Blount (a paralyzed young man) elaborates on this when he says "these deaths and youthful suicides show the fearful tension in which we live" (p. 359). Tom Winter (a librarian and the seventh poor man in the novel) is a victim of the social fatalism that can engulf the human psyche: his sacrificial single-mindedness accomplishes nothing. Like Blount, Michael is paralyzed and armless and his own death represents a nihilistic form of success, "since he wills it and consigns himself deliberately to the sea" (Green 1968: 155).

Femme s'echappe de la Forêt

The last theme warranting examination is Stead's portrayal of women and their "proper" place in Victorian society (Mitchell 1981). She addresses the issues of female economic dependence and oppression in a male-dominated culture. Within the novel, women's place is prescribed by society. Baruch explains to his friends that women are members of the dependent and exploited classes. He says

their peculiarities are imposed on them to keep them in order. They are told from the cradle to the grave, you are female and not altogether there, socially or politically: your brain is good but not too good, none of your race was ever a star, except in the theatre.

(Stead 1935: 236)

When it comes to earning a livelihood, Stead's women do not care about such matters since they only want to marry money, not make it. Michael says to Catherine, "You girls are all the same: you don't care whether the chap loves you, you only care to get a house and furniture" (p. 55).

In another scene, Michael accuses Mae Graham, the woman with whom he is infatuated, of marital fraud and accuses her of marrying an "insurance policy." To him, the proper place for a woman is in the home, waiting on her husband and children and arranging knick-knacks. Housewives are especially singled out for his scorn:

> You housewives are absolutely ignorant of the world; you don't know how stories are fabricated in newspapers or in scriptures, how the house is put together, how cloth is made or dyed. All you know is religion, home, fashion, some painted mechanical creatures that come all made into the world. The one bit of creation you can and must do, does itself unconsciously.
>
> (p. 36)

According to Michael, women are enough to make a man turn into a "homosexual." Withers claims that women will drive men mad. He says to Michael

> This collection of skirts will drive you dippy before they've finished: one with her religion, the other with the reds. What a woman's man are you. Why don't you get a real woman and taste their real goods, not all this eyewash?
>
> (p. 45)

Perhaps the only "real woman" in Stead's eyes is Catherine. This character is not like the other women in the novel. Walker (1983) says she is identified and explained in the image of the "other": the feminine, alternative value system, the "rebel who can find no place for herself in society" (Lidoff 1982: 128). She is seen as mad in a sane world, a wandering gypsy who is subjected to the "jim-jams." Although highly gifted, she is a tormented and torn individual of circumstance. As a woman who has reached maturity in Australia, she has yet to find a purpose in this patriarchal society. Her brother tells her, "You are not like the rest of the world, you're mad. With you one can be free, one can say anything, nothing is absurd or horrid" (p. 232). On a symbolic level, Stead infuses the character of Catherine with the feminine counterpart of the male psyche: that of the "creative anima" (Walker 1983: x). Catherine herself admits her uniqueness. She says that she has fought all her life for

> male objectives in men's terms. I am neither man nor woman, rich nor poor, elegant nor worker, philistine nor artist. That's why I fight so hard and suffer so much and get nowhere. And how vain ambition seems when you look at it, unambitious.
>
> (Stead 1935: 248)

In this chaotic context, she is a chaotic nun in reverse. It is not religion and order but disorder and irreligion that drives her mad.... She'll have your soul in any case, the female devil. She's like one of those female spiders that eat their husbands.

(p. 44)

The true nature of woman's situation in Sydney's restrictive Victorian society, that of economic and social bondage, is summed up by Baruch in his dialog on a drawing entitled "La Femme s'echappe de la Forêt". In this image we are shown a

> naked woman with agonized contortion of body and face bursting through a thicket, tearing her thigh on a splintered tree, while a boa constrictor and a tropical vine loaded with large lilies hung before her and impeded her.

(p. 178)

According to Baruch, this symbolizes woman's escape from the forest; it means "the middle-class woman trying to free herself, and still impeded by romantic notions and ferocious, because ambushed, sensuality" (p. 178). But what is more interesting is the fact that Stead based the character of Catherine on a friend of hers by the same name: "She was eccentric, but she came from a very middle-class, sedate family, she was the freak" (Stead 1974: 241). The novel's Catherine refuses to conform and, so as not to risk her own self-destruction, she aligns herself with the mad and voluntarily admits herself to the Forestville Insane Asylum in the hope that a solution will present itself. She says, "It's the best thing," since the insane are "kind people ... easy to live amongst, easy to humor, and if cranky, no worse than plain men" (Stead 1935: 303). In an interview in 1973, Stead commented that the individual from which the character of Catherine was based on was the kind of character you meet in old Bohemia.

CONCLUSION

Seven Poor Men of Sydney reflects the dynamic nature of Sydney during the 1920s. The urban setting and life of the characters that Stead presents is an observed Sydney, a literary extension of Dickens's London, and as real as "Dostoyevsky's St. Petersburg" (Green 1968: 152). Her Sydney is a city which could have been any urban center in the western world during the onslaught of industrialization and urbanization (Woodward 1975). In a more colorful vein, Stead likened Sydney to Manhattan:

> It has the same narrowing right down to the waterfront. It's very confined there where the big city, where the real city is, just like Sydney, and it has many waterways and all the back-country, Brooklyn and Queens and all those places like our south-eastern suburbs.

Manhattan has got the beautiful North Shore which we have.

(Wetherell 1980: 434)

Stead in her own unique way provides the reader with an image of city life, especially as seen from the perspective of the lower classes. The landscape and cityscape presented by Stead is constructed from ideology, place, space and time and delineated in the form of language. Like other Australian writers, her sense of Australianness cannot be defined in a narrow national sense but must be looked at from a wider historical and geographical vantage-point. Like other fiction writers from "down under," it is more a question of an "acceptance of a common environment, both human and natural, and an Australian use of language" (Dutton 1985: 8). Stead is not the only – or brightest – beacon in this distinct and fascinating literary tradition: other writers illuminate different vistas and dialogs on place and space from different temporal vantage-points. They too are worthy of critical study from a geographical and literary perspective.

REFERENCES

Bennett, B. (1991), *An Australian Compass: Essays on Place and Direction in Australian Literature*, South Fremantle: Fremantle Arts Center Press.

Briggs, A. (1963), *Victorian Cities*, London: Odhams Ltd.

Buckley, V. (1986), "Unequal twins: A discontinuous analysis," in J. Davidson (ed.), *The Sydney–Melbourne Book*, Sydney: Allen & Unwin.

Ducker, J. (1986), "Sydney versus Melbourne revisited," in J. Davidson (ed.), *The Sydney–Melbourne Book*, Sydney: Allen & Unwin.

Dutton, G. (1985), *The Australian Collection: Australia's Greatest Books*, London: Angus & Robertson.

Eaden, P., and Mares, F. (eds.) (1986), *Mapped but not Known: The Australian Landscape of the Imagination*, Adelaide: Wakefield Press.

Green, D. (1968), "Chaos, or a dancing star? Christina Stead's 'Seven Poor Men of Sydney'," *Meanjin Quarterly*, 27/2: 150–61.

Harvey, D. (1973), *Social Justice and the City*, London: Edward Arnold.

Holt, P. (ed.) (1984), *A City in the Mind: Sydney Imagined by Its Writers*, Boston: George Allen & Unwin.

Hughes, R. (1986), *The Fatal Shore: The Epic of Australia's Founding*, New York: Random House.

Kelly, M. (1986), "Nineteenth-century Sydney: 'Beautiful certainly; not bountiful'," in J. Davidson (ed.), *The Sydney–Melbourne Book*, Sydney: Allen & Unwin.

Kiernan, B. (1971), *Images of Society and Nature: Seven Essays on Australian Novels*, New York: Oxford University Press.

Kramer, L. (ed.) (1981), *The Oxford History of Australian History*, New York: Oxford University Press.

Lidoff, J. (1982), *Christina Stead*, New York: Frederick Ungar Publishing.

Mitchell, A. (1981), "Fiction," in L. Kramer (ed.), *The Oxford History of Australian History*, New York: Oxford University Press.

Moore, I. I. (1971), *Social Patterns in Literature*, Los Angeles: University of California Press.

Park, R. E., Burgess, E. N., and McKenzie, R. D. (1925), *The City*, Chicago: University of Chicago Press.

Polanyi, K. (1957), *The Great Transformation: The Political and Economic Origins of Our Time*, Boston: Beacon Press.

Segerberg, A. (1987), "Getting started: The emergence of Christina Stead's early fiction," *Australian Literary Studies* 13/2: 121–45.

Sheridan, S. (1988), *Christina Stead*, Indianapolis: Indiana University Press.

Smith, G. K. (1980), *Australia's Writers*, Melbourne: Nelson.

Spearritt, P. (1986), "Statistical tables – Melbourne and Sydney at the Census," in J. Davidson (ed.), *The Sydney–Melbourne Book*, Sydney: Allen & Unwin.

Stead, C. (1934), *The Salzburg Tales*, London: D. Appleton-Century Company.

—— (1935), *Seven Poor Men of Sydney*, London: D. Appleton-Century Company.

—— (1940), *The Man Who Loved Children*, New York: Simon & Schuster.

—— (1944), *For Love Alone*, London: Harcourt Brace.

—— (1974), "Christina Stead: An interview," *Australian Literary Studies*, 6/3: 230–48.

Walker, S. (ed.) (1983), *Who Is She?*, New York: St. Martin's Press.

Ward, D. (1989), *Poverty, Ethnicity and the American City, 1840–1925: Changing Conceptions of the Slum and the Ghetto*, New York: Cambridge University Press.

Wells, A. (1986), "Cities of capital," in J. Davidson (ed.), *The Sydney–Melbourne Book*, Sydney: Allen & Unwin.

Wetherell, R. (1980), "Interview with Christina Stead," *Australian Literary Studies*, 9/4: 431–48.

Wilding, M. (1983), "Christina Stead," *Australian Literary Studies*, 11/2: 150–1.

Woodward, J. (1975), "Urban influence on Australian literature in the late nineteenth century," *Australian Literary Studies*, 7/2: 115–29.

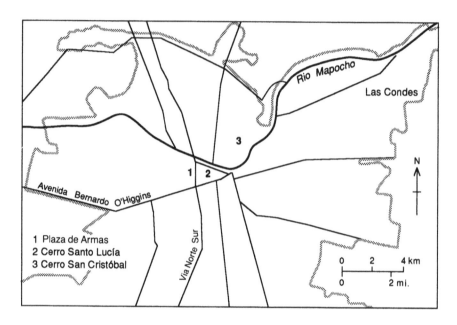

1 Plaza de Armas
2 Cerro Santo Lucía
3 Cerro San Cristóbal

Map 16 Santiago

16

TRANSGRESSING BOUNDARIES

Isabel Allende's Santiago de Chile

Emily Gilbert

The history of Chile is the history of its capital, Santiago de Chile. Santiago was founded by the Spaniard Pedro de Valdivia in 1541 but it was not until the end of the eighteenth century that it began to emerge along a European-style city plan (Loveman 1988: 98).[1] De Valdivia arrived at what is now Santiago from the Mapocho River, which he used as one of the reference-points for the city. Today the river forms one side of the roughly triangular-shaped downtown, bounded on the other sides by the Via Norte Sur in the west and the Avenida del Liberator Bernardo O'Higgins to the south. At the center of the triangle is the Plaza de Armas, the downtown core. This central area is congested by the modern traffic that struggles through the inadequate narrow streets and by the pedestrian activity that gravitates around the abundant shops, restaurants and expensive hotels on the streets emanating from the Plaza de Armas. There are few high-rises in this city prone to earthquakes; thus for the most part the city's expansion has been lateral. The urban sprawl extends in an indefinite extension of the colonial grid-system (Violich 1987: 257). Insufficient housing has been the outcome of Santiago's rapid urban development. The downtown area has heterogeneous residential areas with high-density low-, middle- and upper-class housing (Violich 1987: 286). Squatter settlements, known in Chile as *poblaciones callampas*,[2] have developed both in and around the city. They generally can be found in marginal land-use areas such as the banks along the Mapocho River which are prone to flooding (Blouet and Blouet 1982: 242–3).

Santiago, and the country as a whole, was unique in South America for its relatively stable and democratic governments; the image of Chile held by Chileans was that of a peaceful nation.[3] Perhaps Santiago's unique location inland, in a country so dominated by its oceanic borders, had facilitated its insular protectionism. However, the image of a democratic nation dissolved when, on September 11, 1973, General Augusto Pinochet, leader of the military, headed a coup against President Salvador Allende of the Popular Unity Party. The elite, perceiving Allende's communist ideas as a threat to their status, had incited the military to depose him. Support for the junta was also anticipated from the traditional bastions of the Catholic Church, western foreign governments and Chile's own democratic political parties which formed the opposition (Arriagada 1988: 9). Undeniably, the military action pleased many among these groups, but others were shocked by the excessive violence and bloodshed, and support was not readily forthcoming. However, it was only in 1989 that democratic elections were again held in Chile, and while Pinochet is no longer officially President, he is still commander-in-chief of the military. The structure of the government may have changed, yet many of the practical realities have not.

In 1973 over a third of Chile's population was concentrated in Santiago; thus the effects of the coup were perhaps the most obvious in the capital. The post-1973 changes to the city's morphology typified a government whose priority was the comfort of the elite. Economic problems were not solved but hidden, and an illusion of wealth and prosperity was maintained for the benefit of the upper classes. The physical changes to the capital, while significant, merely reflected the more insidious transformations of the daily routines and attitudes of the city's inhabitants. People no longer mingled nor dawdled, conversing, in the streets, curfew was imposed, and, with the exception of church services which experienced a notable increase in attendance, gatherings were prohibited. The thousands murdered or disappeared prompted those haunted by their political backgrounds to flee into exile.[4] Isabel Allende, niece to President Salvador Allende, was one of those who escaped from Pinochet's militarized Chile. Fifteen months after the coup she, her husband and their two children fled to Venezuela, fearing imprisonment. Allende, born in Peru, was raised and educated in Chile; she called it home. Her flight from the coup marked the end of her innocence and her childhood (Manguel 1992: 622), yet it also marked the beginning of her career as a novelist. In Caracas she wrote her first two novels, *The House of Spirits* (in 1982) and *Of Love and Shadows* (in 1984), which narrate the transformation of Santiago at the time of the coup.[5]

THE NOVELS

Allende's first two novels can be treated as contiguous. The continuity is not maintained by the characters who, at least in name, are completely different, but by the point of view, theme and context, which remain consistent. The pivotal character in *The House of Spirits* is Esteban Trueba, patriarch. His reminiscences frequently interrupt the narrative of his granddaughter, Alba, who has pieced together the family's stories from her own memories and from her grandmother Clara's notebooks. Irene Beltrán is the protagonist of *Of Love and Shadows* and she could easily be an older version of the young, naïve Alba of the first novel. Both women are from upper-middle-class backgrounds and grew up in similar imposing, protective houses in an upper-class neighborhood not far from the city's downtown. Despite Allende's claim that neither novel is autobiographical (Pinto 1991: 35), it is curious to note that the common points that connect the two novels – the background, the city, the class, the writing – correspond to the few autobiographical details Allende has willingly revealed to her numerous interviewers.[6]

The House of Spirits began as a long letter to Allende's dying grandfather (Zinsser 1989: 42). When it reached 500 pages – her grandfather had by then died – she knew it was more than a personal letter; it was, as her husband noted, "a novel" (Pinto 1991: 28). Initially rejected by Latin American publishers, the manuscript traveled to Spain for publication and its subsequent popular success has been overwhelming, with translations in most of the languages of western Europe. Few Latin American authors, and no other Latin American women, have reached this level of international success. The novel – to summarize briefly 433 pages – is the story of Esteban Trueba and his family: his wife, Clara, their first child, Blanca, the twins, Jaime and Nicolás, and Blanca's daughter, Alba. The family saga of love, death and magic predominates throughout the first three-quarters of the novel. As the year of the coup approaches, and as political and economic tensions are exacerbated, these social concerns begin to subsume the family plot as well as pull the family apart. By the time of the novel's climax – the coup of 1973 – it is apparent that the story is no longer the story of just one family, but of the country as a whole and the effects of the coup on the generations of its families. Throughout the novel there is a growing awareness among all the family members, including the patriarch Esteban (though not until the end), of the one-sidedness of their understanding of the city, and the country, in which they live.

Just as it split Allende's life in half (Pinto 1991: 26), the coup is also the event that divides the two novels. *Of Love and Shadows* takes up where *The House of Spirits* left off: it is the story of Santiago five years later. While also commercially successful, the critical acclaim for this second novel was not as forthcoming, with criticisms levied at both the lack of suspense and the ineffectual love-story.[7] Paradoxically, *Of Love and Shadows* is based on an actual event: the 1978 discovery of a mine near Lonquen which contained the bodies of twenty-six civilians executed by the government (Pinto 1991: 32). From the information revealed in the newspapers in Caracas, Allende created a fictionalized narrative about the personal implications of this discovery. In the novel, Irene Beltrán, a writer for a glossy women's magazine, and her sidekick, photographer Francisco Leal, are the ones who discover the mine while on an apparently routine excursion to a "peasant" community, where the duo had gone to report on the story of Evangelina Ranquileo, a young girl who is daily overcome by seizures with healing powers. Drawn to the family when Evangelina mysteriously disappears, Irene and Francisco are directed to the mine, located in fictive Los Ricos about an hour away from the capital, by Evangelina's brother Pradelio, himself in hiding from the military. Endangering their own lives, they secretly leak this information to the church. Meanwhile they have also fallen in love, despite Irene's long-standing engagement to an army officer, Gustavo Morante. Irene's newly acquired political activism intensifies, and she is gunned down on the street in front of her office building late one afternoon. With help from their friend and political ally Mario, she and Francisco escape from the surveillance at the hospital, and finally go into exile in Argentina. The novel's descriptions of Santiago leave no doubt that even five years after the coup the regime's brutalities continue, though no one, not even its supporters, had expected the dictatorship to last so long.

WRITING THE CITY

Allende's city exists in time, beginning at the turn of the century and culminating in the late 1970s, suffering through political conflicts but also transcending them. In spatial terms, the city that emerges from the two novels is defined by the personal perspectives of the narrators. As the young women become more cognizant of the world around them, their conception of the city changes: the images become fractured and politicized; they become aware of the *cities* in which they live. In *The House of Spirits* and *Of Love and Shadows* the descriptions of Santiago fall into two categories. The first is a macro-level allegorical exegesis of Santiago's historical development. At this level the tension between rural and urban society is palpable.[8] The second level of representation is from within the city itself. This perspective includes both the city's public spaces – the streets, the buildings, the roads, the neighborhoods and the people who occupy them – and the private spaces, particularly the contrasts between the homes of the city's elite and those of the lower classes.

But to begin, is it even possible to talk about the city of Santiago in Allende's fiction? Santiago itself is not named in either novel; it is referred to as either "the capital" or "the city." However, descriptions of the physical realities such as the two hills that watch over the downtown, and the historical references such as the coup, firmly identify the city as Santiago.[9] Why this anonymity and what is its effect? From a North American perspective one logical reason might be the figurative and literal distance between Allende in exile and Santiago. Or perhaps time and space are intentionally ambiguous to protect both Allende and her chances of returning to the city.

Gabriel García Márquez's *Clandestine in Chile* (1987; orig. 1986) suggests an alternative reading. *Clandestine in Chile* is the story of Miguel Littín, an exiled film-maker, and his covert return to Chile under a false identity, as told to García Márquez.[10] Chapter Eight, "Two of the dead who never die: Allende and Neruda," offers particular insight into Allende's un-naming. In García Márquez's text, the Chileans do not refer to Salvador Allende and

Pablo Neruda by name but by their titles, "the President" and "the Poet" respectively. These are terms of respect and admiration which immortalize the two heroes. Similarly, Isabel Allende uses these allusions to refer to Salvador Allende and Pablo Neruda, but she also uses the same technique to refer to Santiago. Thus referring to Santiago as either "the city" or "the capital" is an indication of its status as primate city, the privileging of its place in Chile. Therefore the un-naming of the city, which on the one hand universalizes the particular, also suggests its primacy and familiarity.

THE DIALECTIC OF CITY AND COUNTRY

In *The House of Spirits* Isabel Allende infuses Santiago's identity with the origins of its wealth. Chile's affluence is located in its capital; it is there that the farm-owners reside, exploiting the land from a distance (Violich 1987: 258). The tripartite bases that underlay the colonial economy of "land, Indian or Negro labor, and the authority to exploit them" (Loveman 1988: 75) have informed contemporary economic practice. Out of the mining in the north, the agriculture in the south and the subjugation of the natives came what Brian Loveman calls Chile's history of "Hispanic Capitalism," though it was not until the nineteenth century that wage-labor was implemented in the country (Loveman 1988: 73–4). The land around Santiago was parceled out to Spanish owners in 1604 and by the eighteenth century the system of *inquilino* labor – which would last up until the 1960s – was already in place.[11] The financial resources of the feudal-like economy, which depended upon the rural natural resources of mines and land, were concentrated among the elites of Santiago (Loveman 1988: 97–8; Violich 1987: 258), thus creating a typically capitalist class-based society. The rich aggregated in the city but had access to their country villas, while the poor were relegated either to the country shanties or to the pockets of slums scattered throughout the city.

Chile's external economy mirrored its internal one; historically, it too was an economy of displacement with a dependence upon foreign economic ties with both Spain and Peru.[12] Today, Chile still relies upon international investments, largely those of the USA.[13] This economy of foreign involvement played a large part, as instigator and perpetrator, in the military coup in 1973, and later prompted the denationalization of industry by Pinochet's regime.

In *The House of Spirits* Chile's economic history is allegorically represented in the development of the Trueba family fortune: money is acquired in the mines in the north of Chile and is then invested in the agricultural land of the south. The novel opens at the turn of the century with the del Valle family at mass in the parish of San Sebastián, Santiago. The family's church attendance coincides with father Severo del Valle's political ambitions: he is out to buy the vote. Into this family Esteban, though of a more modest background, eventually marries. Using his mother's blue-blood maiden name, Esteban secures a concession in the mines in the hopes of earning enough money to marry Severo's daughter, Rosa the Beautiful. When she suddenly dies, he flees from the city which he now "hates" (Allende 1988a: 44), away from both his painful memories of her and his difficult mother and sister.

Esteban escapes to Tres Marías, the derelict family estate, located near San Lucas, south of the capital, and with the money he has acquired from his mining, embarks on a program of revitalization. Summoned to his mother's deathbed nine years later, Esteban returns to the city for the first time, and, as she is dying, promises her he will marry. Knowing no one but the Truebas he returns to visit them, and within the year he is married to Rosa the Beautiful's younger sister, Clara. The majority of the following twenty years passes peacefully at Tres Marías, and the family travels, at their convenience, between city and country.

The freedom to travel between city and country is a prerogative of the elite. Although the dynamic of city and country is more elaborated in *The House of Spirits* than in *Of Love and Shadows*, the tensions are still present. On the one hand, Irene and Francisco are able to motorcycle in and out of the city with relative freedom. On the other, Digna Ranquileo, Evangelina's mother, can only travel to the city to see Isabel and Francisco when she fears for the life of her son Pradelio. Digna's trip to the city is extraordinary in two respects: first, for practical reasons the long journey is near impossible since there was no one to look after the children; and, second, Digna personally distrusts cities and modernity, a feeling reinforced when her daughter is switched at birth in the hospital.

In *The House of Spirits* the country workers display similar antagonisms. The prosperity Esteban brought to their barren lands neither justifies nor compensates his merciless rule, nor his rape, mostly before his marriage with Clara, of the young girls of the estate. Despite the success of his farming techniques, Esteban's city methods are unable to deal with the swarm of ants which invades Tres Marías. Pedro García removes the ants within a day – like the Pied Piper he talks them off the estate – while the gringo from the city spends weeks analyzing their species to determine which poison to use. The workers at Tres Marías are, like the Ranquileos, cut-off from the city, and Clara's funeral in Santiago is for many their first chance to travel to the capital. Thus the social distance created by the historic and economic dialectic of city and country supersedes the spatial distance between Santiago and the rural communities.

Although a negative environment for the rural workers, for those outcasts who do not conform to the static values of the country the city can be liberatory. In *The House of Spirits* Tránsito Soto, a prostitute Esteban first encounters in the Red Lantern (a brothel at San Lucas near Tres Marías), always dreams of escaping to the city to get "rich and famous" (p. 69), and, with the extra money Esteban gives her in a rare moment of generosity, she makes it. Esteban encounters her later in the Christopher Columbus, a brothel in Santiago, where she is reputed to be the best in the house. Now rich and famous, she aspires to do away with her *patrón*, and establish a cooperative of "whores and fags" (p. 116). When Esteban later returns to the Christopher Columbus after Clara's death, it is a flourishing tourist attraction, well known abroad and transformed into a "social event and a historic monument" (p. 315). The city gives Tránsito Soto the opportunity to succeed, as it allows her to assume control of her body and its sale. Even after the coup, when the brothel is losing business, Tránsito Soto commands considerable power; she is able to repay Esteban's earlier monetary favor by having his granddaughter released from the detention center.

Mario, in *Of Love and Shadows*, is similarly an outsider in his family's mining-town: he is homosexual. His father and brothers ridicule his effeminate mannerisms and even threaten his life. His only alternative is the city's anonymity and receptiveness to difference, where he can live openly as a homosexual. In Santiago Mario finds work in a beauty salon and becomes one of the most respected stylists in the country. Despite his mining background the elite accepts him, and he joins their social circuit, yet he never forgets those who are less fortunate and works actively in the resistance movement. The stereotypical figures of Mario and Tránsito Soto – stereotyped both by their own communities, but also by the text – find freedom and acceptance in the heterogeneity of the city. However, their freedom is bought with their economic success, and their place in the capital, like that of the Truebas and the Beltráns, is secured by their wealth. Still, within the parameters of the text, it is a wealth they could not have achieved within the social and spatial boundaries of their own communities; for the marginalized, it is a wealth that is limited to the city.

For the Truebas, wealth transcends the boundary of city and country, and their ease of travel between the two provides a constant escape from wherever they do not wish to be. Perceptions of the city vary among classes, but even individuals experience conflicting responses to the city as their relationship to it changes. Clara anticipated her first trip to the country, but, as Férula (Esteban's sister) wryly notes, it was because to her it was "a romantic idea" (p. 103). Clara's naïvety vanishes when she realizes how much work needs to be done at Tres Marías, but she also discovers her "mission in life" (p. 105). From this point on, Clara is very much involved in helping the poor, whether it be in the country or in the capital when she returns to the conveniences of civilization to give birth to her twins. The Trueba women spend most of their time on the estate but when Esteban, in a fit of rage at Blanca's relationship with Pedro Tercero García, first beats his daughter and then lashes out at Clara, the city is still there to welcome these two refugees from the country who arrive "looking like disaster victims" (p. 207). The violence shatters the idyllicism of country life for mother and daughter and neither ever returns to Tres Marías. Soon after, Esteban joins them and settles permanently in the capital, but his wife refuses to ever speak to him again.

Irene Beltrán, like Clara, experiences an awakening in the country; there violence is more apparent, more personal. Irene, also oblivious to the contrasts in her city, merely passes through them on Francisco's motorcycle, past "the exclusive neighborhoods with their lush trees and lordly mansions, the gray, noisy middle-class zone, and the wide cordons of misery" (Allende 1988b: 47). Only when she becomes involved with the Ranquileo family does Irene acquire a social conscience. The police, who open fire on the Ranquileos and their guests in the hopes of curing Evangelina (though the firing was only into the air and may have been well intended) frighten Irene and she has her first taste of the danger of the military. When Evangelina disappears Irene finally awakens to the brutalities of the government. Withdrawn in its peaceful neighborhood, Irene has been hitherto protected from the outside world by the physical bearing of her house. In *Of Love and Shadows* it is not that the atrocities of the Pinochet regime are not present in the city, only that in the city there are more places to hide. In the small Los Ricos community, where everyone knows one another, there are no such safe houses in which to seek refuge.

Esteban's relationship to the city is much more problematic; his feelings are in constant flux as he moves back and forth between city and country, never liking where he is, and seldom liking where he has been. Moving to Tres Marías was an escape from his negative personal associations with the city, and though he laments the country's decivilizing effect he calls it "home" and attempts to bring to it a kind of quasi-civilization. However, when frustrated with his farm-workers, the city becomes his escape and he yearns to "go live like a prince in the capital" (Allende 1988a: 65). When Esteban returns to the capital for his mother's death the nine years away have rendered the city unfamiliar. It was:

> a jumble of modernity; a myriad of women showing their bare calves, and men in vests and pleated pants; an uproar of workers drilling holes, knocking down telephone poles to make room for buildings, knocking down buildings to plant trees; a blockade of itinerant vendors hawking the wonders of the grindstone, that toasted peanut, this little doll that dances by itself without a single wire or thread, look for yourself, run your hand over it; a whirlwind of garbage dumps, food stands, factories, cars hurtling into carriages and sweat-drawn trolleys, as they called the old horses that hauled the municipal transport; a heavy-breathing of crowds, a sound of running, of scurrying this way and that, of impatience and schedules.

(p. 83)

At this point Esteban concludes that the city is "a shithole" and begins to romanticize his memories of "quiet" Tres Marías. Approaching his mother's house, he notices the progressive decline of the neighborhood "ever since the rich had decided to move their houses farther up the hill from everyone else and the city had expanded into the foothills of the *cordillera*" (p. 84). The drastic superficial changes of the city – which Esteban describes as deterioration – are contrasted immediately to the inside of his mother's house where nothing has changed since he left almost a decade ago.

Despite Esteban's disgust with the city he spends a few years in Santiago, occupied with the preparations for his marriage to Clara, the building of his grandiose house on the corner, the wedding itself and the birth of their first child, Blanca. However, the family, now also including his sister, Férula Trueba, decide to spend the summer at Tres Marías. This begins a pattern of alternating between Tres Marías during the hot summers and the city during the school year, a phenomenon typical of the wealthy class. Yet Férula hates the country, and in all the years she lives at Tres Marías she never grows used to not having the amenities of city-living.

The children, who board at schools in the capital, also respond to Tres Marías differently. The twins, Jaime and Nicolás, having spent most of their time at a British-style boarding-school packed with planned activities and organized sports, are bored in the country. Nicolás in particular is bored: he passes his time, much like his father before him, seducing the young women of the area. Their sister Blanca, on the other hand, is only happy in the country, since Pedro Tercero García is there. Unhappy at her boarding-school run by nuns, she feigns an endless cycle of illnesses. Eventually her mother rescues her and takes her back to Tres Marías, but already Blanca has initiated a pattern of hypochondria that she never overcomes. Despite Blanca's yen for the country she is the only one, other than her father, who is impressed with the Count Jean de Satigny who comes to stay with them. For her, his presence is a chance for dressing up and resuming her city manners. Only later, when Blanca, pregnant with Pedro Tercero's child, is forced to marry the Count does she discover, alone in the north of the country, that his appearance is but a thin veneer of civilization that disguises his secret sexual appetites. These so disgust and terrify Blanca that she flees from their house in the north to her mother in the city, never to see her husband again.

Tres Marías remained astoundingly impervious to the social unrest developing in the rest of the country. During the closing of the nitrate fields, Esteban's farm prospered, while thousands became unemployed. These workers migrated towards the city and,

> as they approached the capital, were slowly forming a belt of misery around it. They settled in any way they could, under planks of wood and pieces of cardboard, in the midst of garbage and despair. They wandered the streets begging for a chance to work, but there were no jobs and slowly but surely the rugged workers, thin with hunger, shrunken with cold, ragged and desolate, stopped asking for work and

asked for alms instead. The city filled with beggars, and then with thieves.

(p. 134)

The urban conditions Allende describes are characteristic of the squatter settlements in many Latin American cities and emphasize the essential dynamic of rural versus urban prosperity. In Chile, as the demand for nitrates ran out, the workers were left with nothing – no money, no land and nowhere to go but to the city. Still, in the capital, there was little hope for acquiring anything but hand-outs. The social concerns prompted by the closing of the northern mines led to the growth of the labor movement and the filtration of its socialist ideals to the south – to the workers on estates such as Tres Marías. Owners such as Esteban emphatically opposed the dissemination of any kind of "communist propaganda" on their property. In the novel, Pedro Tercero García, the farm-manager's son and Blanca's clandestine lover, is barred from the estate for organizing the workers. Ironically, Esteban would later be hard-pressed to avoid Pedro Tercero's songs which ignited the country's resistance movement.

The threat of socialism and Salvador Allende's support for Agrarian Reform pushes Esteban into politics; he is elected Senator of the Republic for the Conservative Party. However, his title is useless in preventing Allende's government from implementing what Esteban calls "the Marxist cancer" (p. 306). Agrarian Reform was not unique to Allende's government. Land-reform laws had already been passed in 1962 under President Jorge Alessandri and were espoused by the Eduardo Frei government (Loveman 1988: 375). However, the threat to the elite was Salvador Allende's commitment to the urban and rural poor which depended upon this policy of the "expropriation and redistribution of agricultural land" (Loveman 1988: 287). In *The House of Spirits* Esteban's estate, by some chance, is one of the last properties to be seized for public benefit, but the workers eventually do gain control. In an embarrassing and fruitless confrontation with his laborers, Esteban tries to regain what he has lost, but his powers have already been stripped by the government and he no longer has the strength – political or physical – to face his ex-workers.

Chile's agrarian Reform not only applied to the internal redistribution of land, but was also a policy of repatriation of foreign-controlled business. In some senses Esteban's comment that "the Agrarian reform ruined things for everyone" (Allende 1988a: 52) is valid, though the land workers, many of whom received immediate benefits, might disagree. The Ranquileo family in *Of Love and Shadows* is unique in that they own their own land. Their lives are isolated, but they do not feel poor because they are property-owners. Digna Ranquileo, like Esteban, never believes that the land reforms can work, and she is proven right when the dramatic economic changes lead to an uncontrollable working class, an appalled upper class and the intervention of the USA. For the reasons iterated in the two novels, ultimately the implementation of the reforms provoked the 1973 coup. One of Pinochet's first decisions was to reverse the process of Allende's Agrarian Reforms and "return" many of the foreign companies to the foreigners.

The land returned to the elite who were satisfied with the emphasis on privatization, but again the workers suffered. In *Of Love and Shadows* Digna Ranquileo, despite her hardships – the difficulties of tending the land without her husband's help (he is a circus performer and is absent most of the year) and the increasing difficulty of defending the land from the big enterprises – adamantly insists that the family stay where they are rather than move to one of the new agricultural villages "where every morning the *patrónes* chose the number of laborers they needed, saving themselves the problem of tenant farmers. That was poverty within poverty" (Allende 1988b: 20).

The importance of the ownership of land, to both upper and lower classes, was an integral aspect of survival in Chile both before and after the coup. Land assured some sense of economic security. Digna Ranquileo echoes the thoughts of Esteban Trueba who affirms once at the age of 25 and again in his later years that "land is all you have after everything falls apart" (Allende 1988a: 309). In the city, this rural concern with the ownership of land is translated into the privatization, and thus the control, of public space.

INSIDE THE CITY

The concepts of private and public space endure subtle modifications with the changes in government. To the protagonists, private spaces primarily comprise their own residences. The elite can escape to the country but they also have an escape in the city – the insularity of their houses ensconced in quasi-private neighborhoods. The houses of the Beltráns and the Truebas are their protection: "a cloistered world," segregated from the other, poorer, areas of the city (Allende 1988a: 266). Esteban had his house – "the big house on the corner" as it was commonly referred to – built in the High District, one of the prestigious neighborhoods, in time for his wedding in the early 1900s. Its model is the house of Esteban's father-in-law, Severo del Valle, built with old money. But Esteban wants more than just its aristocratic bearing, various wings, courtyards and service areas; he intends to surpass the house of his youth, where his mother now "spends her mornings immobile in her chair, looking out the window at the bustle in the street, and observing the gradual decline of the neighborhood that in her youth had been so elegant" (p. 40). His house must have:

> two or three heroic floors, rows of white columns, and a majestic
> staircase that would make a half-turn on itself and wind up in a hall of
> white marble, enormous, well-lit windows, and the overall appearance
> of order and peace, beauty and civilization, that was typical of foreign

peoples and would be in tune with his new life.

(p. 93)

The building is completed around the turn of the century (the time of Chile's rapid urbanization), spurred by developments within the cities, the advancement of the communication networks of rail lines, telegraph and steamship and the organization of a national labor market (Loveman 1988: 208). The houses of the del Valles and the Truebas identify them as both the participants and the benefactors of this urban renewal, and the houses, in their design and location, are expressions of their privilege.

The Beltráns' house in *Of Love and Shadows* is quite similar to that of the Truebas in both its physical appearance and its symbolic function as sanctuary. Like the Trueba "cloister," this house also protects Irene, who "had lived surrounded by the gales of hatred, but remained untouched by them behind the high wall that had protected her since childhood" (Allende 1988b: 118). There are two significant differences, however, in the Beltrán household. The first is the absence of the father, or any male figure. Irene's father, Eusebio, whom she loved unconditionally – though as she was to mature she realized he could be "despotic and cruel" (p. 144) – had walked out without explanation four years earlier. The second difference arises from their subsequent financial difficulties. The women install the Will of God Manor, a retirement home, on their first floor. Although Irene seems willing to move, "Beatriz would rather work herself to death and perform all manner of juggling acts than reveal her reduced circumstances. To leave the house would be to publicly acknowledge poverty" (p. 43). When making the outer changes to the house, Beatriz insures that from the street they are invisible and that the house maintains an appearance of affluence.

Irene's house fascinates Francisco, although he immediately recognizes her mother's "class prejudices and ideology" (p. 39). He describes his own home, designed by his father, as "small, old, modest, greatly in need of repair" (p. 24), yet far from the supposed squalor that Beatriz conjures up about middle-class neighborhoods. Since the coup and the tough economic times for those who, like Francisco's father, Professor Leal, are on the lists of undesirables because of their political ideas, an addition was built to accommodate Francisco's brother, Javier, and his family. The house, cramped and in disrepair, is still a home, and as Professor Leal notes: "everything else depended on largeness of spirit and liveliness of intellect" (p. 24).

This home is contrasted to Mario's penthouse apartment, located in a fashionable downtown district, "decorated in black and white, in a tasteful, modern, and original style. Geometric lines of steel and crystal were softened by three or four very old baroque pieces and by Chinese silk tapestries" (p. 89). Despite its comforts, the apartment offers none of the warmth of the Leal house nor early childhood memories such as those Francisco reconstructs of his kitchen whenever he needs to feel secure.

At the other end of the spectrum of residential space are the slums in the city, which endure both before and after the coup, often immediately adjacent to the wealthy areas. When Férula is kicked out of her brother's house she moves to a run-down tenement, the ground covered in garbage, not far from Father Antonio's old church:

> The tenement was a long passageway of ruined houses, all exactly the same: small, impoverished dwellings built of cement, each of a single door and two windows. They were painted in drab colors and their peeling walls, half eaten by the damp, were linked across the narrow passageway by wires hung from side to side, which were used for laundry but this late at night swung empty in the dark. In the center of the little alley there was a single fountain, which was the only source of water for all the families who lived there, and only two lanterns lit the way between the houses.
>
> (Allende 1988a: 149–50)

Then there is the rooming-house of Pedro Tercero García who, having come to the capital to continue his political struggle, lives on the last street of the working-class district where the comrades gather for their Socialist meetings. Conditioned to the comforts of the upper class, Pedro's lover, Blanca, can never accept his living arrangements; even when he moves to a middle-class apartment building closer to downtown, she thinks of it as "sordid, dark and narrow and the building overcrowded" (p. 312).

Amanda, Nicolás's lover, and her younger brother, Miguel, who later becomes involved with Alba, occupy an apartment which had survived a succession of occupants. It is described as a:

> moldering old house that fifty years earlier had probably boasted some ostentatious splendor but had lost it as the city gradually expanded to the foothills of the cordillera. At first it had been occupied by Arab merchants who encrusted it with pink friezes; later, when the Arabs moved their business to the Turkish quarter, the owner had turned it into a boarding house, subdividing it into poorly lit, sad, uncomfortable and awkwardly constructed rooms for tenants of little means. It had an impossible labyrinth of dark, narrow halls, in which the stink of cauliflower soup reigned eternally.
>
> (p. 231)

The Bohemian quarter, where Amanda later moves, offers another example of living conditions in the city. Here, "in the center of town, only yards away from the modern buildings made of steel and glass, streets of painters, ceramists, and sculptors had sprung up on the side of a steep hill" (p. 336); the men have time for leisure, the women are beautiful and the children play happily in the street. Yet beneath the superficial gaiety lie hidden social problems: hidden away in her studio apartment, Amanda's arms and thighs are covered with needle-marks.

Not only are the houses in Santiago private, but neighborhoods, separated by social and economic class, are privatized. The social barriers create a social distance where often no spatial difference exists, as often neighborhoods of extreme economic differences are adjacent. Frances Violich comments on this alienation:

> youngsters growing up in low income sections such as Barrancas may actually reach adulthood before learning of even the existence of upper-class districts like Las Condes or Vitacura, let alone know anything of the ways there. Likewise, and perhaps more critically, upper-class youth grow up totally ignorant of or insensitive to life in the poor, working-class *comunas*. Both live apart, in separate worlds. Community facilities are rarely shared in a way that might stimulate mutual awareness and understanding.
>
> (Violich 1987: 280)

Both Allende's novels describe this fragmentation of urban space. However, the situation is much more extreme after the coup; there is not only a fragmentation of space, but a fragmentation of those who occupy that space. Francisco Leal daily crosses the boundaries within the city:

> The city was divided by an invisible frontier that [Francisco] was regularly obliged to cross. The same day that he photographed exquisite dresses of muslin and lace, in his brother José's barrio he treated the little girl who had been raped by her father, then carried the latest list of victims to the airport where, after reciting the password, he delivered it to a messenger he had never seen before.
>
> (Allende 1988b: 83)

Francisco instinctively protects Irene from finding out about these cities which she knows nothing about, from facing "the unresolvable misery, the injustice and repression that he experienced daily" (p. 77). However, Irene's experience with the Ranquileos alerts her to this boundary and she becomes "filled with a new determination born of her desire to know the truth; she felt that she must cross that threshold" (p. 115). To cross the threshold means to delve into both the emotional and the political dimensions of the city, as well as to physically transgress into the areas of poverty.

The Trueba women, however, enjoy a long history of delving into the poor neighborhoods. Nívea, Clara's mother, the first in a long line of women activists, hangs suffragette posters on the city walls at night, and during the day campaigns for women's rights to education as well as for children's rights. Accompanied by her daughter, Clara, she delivers food and clothes to the slums. Later Clara does the same with her daughter, Blanca, with the acknowledgment that what the poor really need is justice and that her charity works only "assuage our conscience" (p. 136). Similarly, Férula partly has her own salvation in mind as she says the rosary in the tenements of the Misericordia District. Yet by no means is it only the women who engage in public service: Jaime Trueba, Pedro Tercero García, Francisco Leal and Mario are examples of men who actively participate in social work and political resistance. However, the women in the novels also challenge conventional notions of woman's place in the domestic/private world, and thus blur the traditional distinctions between public and private spaces. Moreover, the private spaces of their homes are made public. For example, Irene and Beatriz open a retirement home on their first floor, Clara encourages an endless procession of clairvoyants, Blanca gives ceramic classes for mongoloid children on the patio and Alba hides the victims of persecution in the empty rooms of the big house on the corner. Thus the women both occupy public space and make their private spaces public.[14]

Conversely, public space under Pinochet's regime is privatized through the imposition of rules and regulations such as the curfew.[15] The street scenes of day and night are vividly contrasted. While the Plaza de Armas bustles during the day, at night the streets are eerily quiet. Irene and Francisco, stuck in their office building late one evening during curfew, gaze out at the city below them: "They saw no sign of life; below them spread a deserted city; it was like science fiction, as if some cataclysm had erased every trace of humanity" (Allende 1988b: 106). Morality orders were imposed dictating short hair for the men and skirts for the women. In the streets the army would trim men's long beards or slice women's pants into skirts. Public congregations were also outlawed. *Of Love and Shadows* details one of the few mass-demonstrations when people gathered in Los Ricos, many journeying from the barrios outlying the capital, to identify the dead bodies as they are extricated from the mine. Still others remain in the city and protest with a sit-in outside the Vicariate: "A procession was organized to pray for the victims, and before the authorities realized what was happening, an unmanageable crowd was marching through the streets carrying banners and placards demanding liberty, bread, and justice" (p. 231). The protest mirrors an earlier and smaller demonstration, detailed in *The House of Spirits*, at the death of the Poet. At that time, the protesters swamped the morgue and followed his coffin to the cemetery, chanting solidarity slogans, protected by the international journalists who filmed the procession.

The novels also record the changes in the diurnal habits of the people under the Pinochet government. The Plaza de Armas, where Esteban first caught sight of Rosa the Beautiful, is where Professor Leal marches with a placard that reads: "AT THIS VERY MOMENT, MY SON IS BEING TORTURED" (Allende 1988b: 208). The city of Esteban's youth at the turn of the century includes places such as the Hotel Francés where he indulges his first earnings in a magnificent cup of hot chocolate and where Clara and Férula later seal their friendship over tea and pastries. Exuberance characterizes the places in the city, from the street of Jewish jewelers where Esteban runs to buy his wife a diamond brooch, to the honeymoon suite of the best hotel of the city where Blanca stays her wedding-night, to the market where Alba is taken for fried fish with her real father, Pedro Tercero García, to the park in the center of the city where Alba and Miguel go for romantic walks. However, after the coup, the city becomes sinister. Irene, when looking for Evangelina, discovers the under-side of the city: the detention centers, the police lock-ups, the morgue and even the off-limits section of the psychiatric hospital. Parks, which used to be city's oases, are now tainted with the memory of Javier's body, which was found hanging from a tree. The military converted even the sports stadium into a detention camp.

The perception of the Church, before and after the coup, is one of the most radical changes in the city. Before the coup the upper classes, like the Truebas and the del Valles, periodically attend religious services downtown at the Church of San Sebastián, though for the most part in *The House of Spirits* Clara's spiritualism and magic replace formal religion. It is not therefore surprising that Esteban considers religion a "feminine affair" (Allende 1988a: 136), nor is his implication that religion is thus apolitical – that the Church's alms for the poor can be seen therefore as a social project, not a political statement. However, in *Of Love and Shadows*, the Church plays a significant political role against the dictatorship. Part of the Church's importance arose from the facts that it was the one place left where people could legally congregate and that the "priests and nuns were forced to postpone their spiritual tasks in order to minister to the earthly needs of their lost flocks" (p. 385). José Leal, Francisco's brother, both a priest and an activist, lives and works in the poor district. It is to this brother that Francisco tells of the bodies in the mine. José informs the Cardinal, who then organizes a delegation to fully open the mine, in opposition to the government. It is significant that to protest the executions, the relatives of the victims and their supporters gather at the Cathedral, perhaps their only safe haven, and initiate a hunger strike, while others pray outside the Vicariate. Even the limited spiritual magic in this novel – Evangelina's mysterious and healing seizures are somewhat saintly – become political when the military interferes.

Despite the horrors which took place after the coup there is no question that Chile was in trouble before 1973. During the months after Allende's election, as described in *The House of Spirits*, women take to the streets with pots and pans complaining about the scarcity of food. The streets seem dark and deserted at night, unlike the thronging masses of earlier days. The garbage is not collected, the street lights are smashed and posters of the opposing political parties cover all the available wall space. The city's dependence on cars becomes apparent when Allende imposes rations on gasoline and:

the lines of automobiles could last two days and a night, constricting the city like a gigantic motionless boa tanning itself in the sun. There was not enough time to stand in so many lines, and since office workers had to get around the city on foot or by bicycle, the streets filled with panting cyclists that looked like a frenzy of Dutchmen.

(Allende 1988a: 348)

The economic hardships during President Allende's government were largely generated by the US government's intervention, coupled with the sabotage of members of the upper classes engaged in covert activities like Esteban. This opposition, coupled with the President's lack of political support in the city,[16] made it difficult for him to enact his urban-planning ideas. Allende disapproved of the urban phenomenon of sprawl and thus supported an increase in urban density. He also condemned segregation and encouraged a greater mix of the social classes (Robin 1972: 10).

These controversial ideas and the promise of economic restructuring mean that Allende's election is not celebrated in the High District. In *The House of Spirits* the champagne bottles are left unopened and the window shutters are closed to the people gathering in the center of the city. Alba runs from her grandfather's big house on the corner to join the celebration downtown. There she mixes with the residents of the shanty towns and the working-class districts, other students and workers, all "shouting in a single voice that the people united would never be defeated" (Allende 1988a: 340). That night the poor march through the private neighborhoods for the first time, seeing a part of the city previously inaccessible to them. The problem of class divisions, depicted by the march on the elite neighborhoods, was not resolved by Allende's three years in government; paradoxically, in some ways these problems were exacerbated.

However, Pinochet's government further entrenched these class divisions. Five years after the coup, the omniscient narrator of *Of Love and Shadows* comments that "Through an unwritten but universally known law of segregation, two countries were functioning within the same national boundaries: one for a golden and powerful elite, the other for the excluded and silent masses" (Allende 1988b: 177). When freed from the detention center where she was raped and tortured, Alba is dumped somewhere near the Misericordia District, in "an empty lot full of garbage, with rats scampering among the refuse … [near] a wretched slum, with houses made of cardboard, planks and corrugated metal" (Allende 1988a: 428). On her way home the following day after curfew, Alba

could see the city in all its terrible contrasts: the huts surrounded by makeshift walls to create the illusion that they do not exist, the cramped, gray center, and the High District, with its English gardens, its parks, its glass skyscrapers, and its fair-haired children riding bicycles. Even the dogs looked happy to me. Everything in order,

323

everything clean, everything calm, and that solid piece of a conscience without memory. This neighborhood is like another country.

(p. 429)

The division of the city is economically determined and the social distancing creates a spatial alienation.

The morphological changes to the city which Allende describes reinforce class divisions and offer another example of the privatization of public space. In both novels Allende describes the walls built by the Pinochet regime to keep the poor in their place, and, as described in *The House of Spirits*, to protect those who do not want to see: "Cement walls were erected to hide the most unsightly shantytowns from the eyes of tourists and others who preferred not to see them" (Allende 1988a: 381). José Leal, in *Of Love and Shadows*, lives in a shack situated "in a large, densely populated neighborhood that was invisible from the road, hidden behind walls and a row of poplar trees with naked branches stretching towards the sky – a place where not even vegetation thrived" (Allende 1988b: 217). On the other side of the wall the main city streets have never looked so attractive:

In a single night, as if by magic, beautifully pruned gardens and flower beds appeared on avenues; they had been planted by the unemployed, to create the illusion of a beautiful spring. White paint was used to erase the murals of doves and to remove all political posters from sight.

(Allende 1988a: 381)

Poverty had not gone away: it had only disappeared. In an inversion of the ancient city walls, whose function was to defend citizens from their outside enemy, these inner-city walls protect the elite from the enemy within.

Littín notes that, rather than solve the country's problems when he overthrew the government, Pinochet sought to change the appearance of the city:

To give an immediate and impressive appearance of prosperity, the military junta denationalized everything that [President] Allende had nationalized, selling off almost anything of value to private capital and multinational corporations. The result was an explosion of flashy luxury goods and decorative public works that created an illusion of spectacular wealth and economic stability.

(García Márquez 1987: 42)

As Allende describes, those with money were delighted. The champagne bottles of the upper classes, left unopened at the 1970 election, were uncorked at news of the coup while "all that night helicopters flew over the working-class neighborhoods, humming like flies from another world" (Allende 1988a: 372). The narrator notes that "the city had never looked so beautiful. The upper class had never been so happy: they could buy as much whisky as they wanted, and automobiles on credit" (p. 381). However, although the stores filled with staple goods, some of which had not been seen in years, they cost three times more than usual, no one could afford to buy and "three days later the smell of rotting meat infected every shop in the city" (p. 373). However, for those with money the pretense of prosperity could be, and was, maintained.

Appearances are deceptive. After the coup, the streets appear full of people during the day, yet it was a "stage-set normality and operetta peace" (p. 389); this bustle protects those who, at night, have to be hidden. Then the streets empty, but for the trucks filled with bodies and detainees and for the roving police cars. No one witnesses the soldiers abducting Alba from her grandfather's house, despite the uproar in broad daylight and flames leaping from the burning books. Beatriz Beltrán ignores this aspect of the city. She believes that the women marching in the Plaza de Armas every Thursday protesting against the *desaparecidos* are being paid by Moscow. Beatriz also ignores the news about the bodies found at the mine, although everyone else was talking about it, even the bourgeoisie. She can not even believe that the shots aimed at her daughter were deliberate, and maintains, until the very end, that it was only an accident.

Allende frequently employs this theme of stripping away the appearances, or the difference of night and day, to uncover what lies hidden beneath, whether with regards to a character's appearance,[17] the events at Los Ricos mines or the city itself. However, as Miguel Littín notes, the superficiality of the city cannot disguise the changes in the people:

> As we approached the center of the city, I stopped admiring the material splendor with which the dictatorship sought to cover the blood of tens of thousands killed or disappeared, and ten times that number driven into exile, and instead concentrated on the people in view. They were walking unusually fast, perhaps because curfew was so close. No one spoke, no one looked in any specific direction, no one gesticulated or smiled, no one made the slightest gesture that gave a clue to his state of mind. Wrapped in dark overcoats, each of them seemed to be alone in a strange city.
>
> (García Márquez 1987: 16–17)

Dressing up the city so that it appears to be prosperous is Pinochet's idea of urban planning. But the city is never just what it appears to be.

CONCLUSION

For a North American reader the themes that have been touched upon – particularly the transgression of boundaries whether between country and city, public and private space, appearance and reality, or fact and fiction – perhaps evoke an allusion to post-modernism. Similarly, Allende's novels also problematize the writing of history and, tangentially, writing the city. The naming of places in the novels is for the most part fictionalized (for example, the Church of San Sebastián and Los Ricos). The neighborhoods also, though described in detail, are given general labels such as "the High District" or "the Misericordia District," which have no actual application. However, other places like the Plaza de Armas exist both in the novels and in Santiago. Fact becomes fiction, and fiction becomes fact.

This confused sense of place elicited by these ambiguous zones effectively conveys the narrator's own sense of unfamiliarity with the city. It is also evocative of a government which, as Allende describes, sought to change history, politics and concepts of space:

> They adjusted the maps because there was no reason why the North should be placed on top, so far away from their beloved fatherland, when it could be placed on the bottom, where it would appear in a more favorable light; and while they were at it they painted vast areas of Prussian-blue territorial waters that stretched all the way to Africa and to Asia, and appropriated distant countries in the geography books, leaping borders with impunity until the neighboring countries lost their patience, sought help from the United Nations, and threatened to send in tanks and planes.
>
> (Allende 1988a: 383)

Allende's descriptions of Santiago parallel the above quotation in their transgression of borders. While specific details of places are provided, their relative location in the city is not. Thus the various neighborhoods become labyrinthine, without fixity, without orientation. For the reader who wishes to create a map of the city, the text proves a frustrating guidebook; but the reader more interested in perceptions of the city will feel at home.

In the novels, time and space are ambiguous. *The House of Spirits* challenges the presumed linearity of narrative – and thus history – by piecing together the story from various perspectives. Alba begins to tell the story inwardly to herself when a vision of her grandmother comes to her in the dog-house at the detention center. This storytelling keeps her alive. When released, Alba writes the story with the help of her grandmother Clara's notebooks. However, they are problematic sources since they are organized by subject, not chronologically.[18] Esteban also writes pages of his own memories, which are interspersed throughout the narrative. A further complication is that Allende has been criticized for the appropriation of the voices of torture victims – in both novels – since she herself was never tortured (Martin 1989: 354).

Time and space have some basis in fact, but are used allegorically. Reviewers have found the second novel to be more problematic because of Allende's long absence in exile and her distance from Chile.[19] The fictionalization of the discovery of the mine at Lonquen in *Of Love and Shadows* aroused contempt. Allende gathered information from the newspapers in Caracas and filled in the details that the government had suppressed, but also introduced a whole string of invented characters. Allende's own mother claimed that the details around the discovery of the mine were "unrealistic." Ironically, when Allende returned to Chile in 1990 a man went to visit her and told her that he had been the one to discover the mine and that all the details she described were true. Allende tells her interviewer Alberto Manguel that "I suppose that if you tell a story exactly as you feel it, truth will take over. I thought that writing creates reality. It's the other way round: reality dictates your writing" (Manguel 1992: 625).

Allende's novels are overtly political, with a particular consideration of economic inequalities. Thus the initial correspondence to post-modernism is questionable. Rather, Allende's novels are more appropriately identifiable as magic realism, a literary style typical of Latin American writers and one that has been around much longer than post-modernism.[20] Yet Allende also transgresses the literary history of the representation of the Latin American city. Her description, neither entirely negative or positive, emerges out of a context of pessimistic descriptions of the city which have focussed on the city's repressive elements rather than recognizing its liberatory potential (Martin 1989: 119).

What clearly emerges from Allende's novels is a sense of hope. Gerald Martin claims that the difference is gender-related and that Allende's emancipatory ideals are entwined with her feminist goals (Martin 1989: 351). In her chapter "Writing as an act of hope," Allende herself makes this connection. She also claims that the true political literature of our time is writing that gives "both women and men a chance to become better people and to share the heavy burden of this planet" (Allende 1989: 54). Thus in her novels the patriarchy of the system is not blamed on the father of the family, whether Esteban Trueba or Eusebio Beltrán. In itself, this rewriting history is an act of hope, a way of infusing memories with significance. Thus Allende's Santiago, fractured by its inequalities of wealth and cowering under a military dictatorship, is, in the end, a city of hope.

NOTES

1 Many thanks to Adriana Benzaquén, Paul Simpson-Housley and Jamie Scott for their comments and critiques, and most of all for their support.
2 *Poblaciones callampas* means "mushroom towns," which refers to the way these Chilean squatter settlements can literally appear over-night.
3 It is not that Chile was ever without conflict: for example, just six months after the city was founded the Mapuche Indians attacked. Recent history includes the 1891 civil war, which many have compared to the 1973 coup, as well as a minor coup in 1924. The one real period of democracy was between 1932 and 1973. However, despite these instances the perception of Chile was that it was above all this fighting, and Allende has noted that "We had had many years, many

generations, of democracy in Chile, and it always seemed to us that these things happened in other countries, not in ours" (Pinto 1991: 26).

4 Examples of the Chilean diaspora are plentiful: some examples include the film-maker Miguel Littín, and the literary critic and playwright Ariel Dorfman.

5 Allende has since published two other works of fiction: one novel, *Eva Luna*, and one collection of short stories, *The Stories of Eva Luna*. Other writings by Isabel Allende include four unpublished plays (Manguel 1992: 622), a children's book *La gorda de porcelana*, and a collection of satirical short stories on the war between the sexes: *Civilice a su troglodita* (Hart 1989: 11). This chapter will only deal with the first two novels, since they provide the most detailed account of the historical transformation of the city.

6 Though Allende has granted many interviews, there appears to be little autobiographical material available about her, at least in translation. For further information see the interviews in Manguel 1992, Pinto 1991 and Zinsser 1989.

7 Patricia Hart's *Narrative Magic in the Fiction of Isabel Allende* (1989) makes numerous negative comments about Allende's novels; particularly *Of Love and Shadows*, though the overall tone of her text is positive. For details with regard to other critical reviews see Manguel 1992: 624 and Zinsser 1989: 50.

8 The classic text on the country/city dynamic is Williams, *The Country and the City* (1973). For a Latin American perspective on this subject see Martin's *Journeys Through the Labyrinth: Latin American Fiction in the Twentieth Century* (1989).

9 Alberto Manguel, in an interview with Isabel Allende, notes that "however universal they [Allende's stories] may seem, [they] are firmly rooted in the terrible history of her country over the past two decades" (Manguel 1992: 621). Doris Meyer makes a similar comment *vis-à-vis Of Love and Shadows*: "Like *La casa de los espíritus*, *De amor y de sombra* is a novel set in an unnamed country, which is unmistakably the author's homeland, under the dictatorship of an unnamed general, obviously Augusto Pinochet, in approximately 1978" (Meyer 1988: 151).

10 *Clandestine in Chile* chronicles the twelve years in Chile since the coup. The information was gathered through interviews and personal observations. The events in *Clandestine in Chile* are more sinister than in either of Allende's novels. One example is Littín's description of the morning of the coup:

> Twelve years ago, at seven o'clock one morning, an army sergeant had let go a burst of machine-gun fire over my head and ordered me to fall in with a group of prisoners being herded toward the Chile Films building where I worked. The whole city was shuddering with the reverberations of dynamite blasts, gunfire, and low swooping planes.
>
> (García Márquez 1987: 21)

11 The *inquilinos* were tenant agricultural laborers. For details regarding the controversy of the *inquilinos* among Chilean historians see Loveman 1988: 89–91.

12 Gerald Martin takes a literary approach to this issue of displacement. Martin writes that the place of Latin America in western myth means that it may be destined to always be "country" to Europe's "city" (Martin 1989: 105).

13 "By 1970 over one hundred United States corporations had investments in Chile, among them twenty-four of the top United States-based multinationals" (Loveman 1988: 282).

14 Another, more subtle theme of women's public resistance which permeates the novels is the transformation of traditional women's crafts (i.e., Rosa the

Beautiful's embroidery, Blanca's sculptures and Alba's paintings) into political activism. Throughout her collection of articles entitled *Women of Smoke*, Marjorie Agosin likens the politicization, and thus publicity, of traditional crafts to the women leaving their private spheres and emerging into the streets: "visibility and publicity were powerful weapons of protest" (Agosin 1989: 40). Furthermore, Agosin, in that she refers to "sewing as a form of writing a text" (Agosin 1989: 93), alludes to a historical continuum of the narratives of Latin American women.

15 Curfew was implemented directly after the coup. During the Pinochet regime curfew was often reintroduced, for example in *Of Love and Shadows* in 1978 and in 1985, when Miguel Littín returns to Chile.

16 Salvador Allende won the overall vote in the 1970 election but had placed third in the city. Allesandri, the most conservative of the candidates, received the most votes: 460,146 votes to Allende's 416,854 (Robin 1972: 28).

17 While dressing is often considered a private affair, the audience is the public. The post-coup ethos demands that appearances be maintained. Beatriz's character manipulates her personal appearance to communicate her status. Pinochet's ordinances on public morality also insist upon righteous superficial appearances. Thus, in some sense both the junta and the elite need to believe that the exterior appearance reveals, or can modify, internal qualities. For a brief discussion on appearance, reality and clothing see Hart 1989: 139.

18 For a feminist analysis of the narrative strategy see Meyer 1990. Meyer claims that the narrative represents a community of women who can transgress time and space and thus be empowered.

19 For other discussions of the effects of exile see Doris Meyer's 1988 article, which details the psychological, mythological, historical and feminist reasons for a feeling of exile among Latin American writers (Meyer 1988: 153); see also "Political code and literary code: The testimonial genre in Chile today" in Dorfman 1991, which discusses the role of testimonial for Chileans in exile and for those who survived the "concentration camps."

20 This chapter touches upon only a few instances in the two novels of magic and spirituality, two of the many categories often attributed to magic realism. Yet my concern has not been to insert the author within a literary style, whether postmodernism, magic realism or any other "ism." I introduce the topic in the conclusion only to make it clear that there are alternative ways to reading the text than solely from the standpoint of North American or European literary theory. For further references with regards to magic realism see Hart 1989, Gordon 1987 and Earle 1987.

REFERENCES

Agosin, Marjorie (1989), *Women of Smoke*; trans. from the Spanish by Janice Molloy, Stratford, Ontario: Williams-Wallace.

Allende, Isabel (1988a; orig. 1982), *The House of Spirits*, trans. from the Spanish *La casa de los espiritus*, by Magda Bogin, Toronto: Bantam Books.

—— (1988b; orig. 1984), *Of Love and Shadows*, trans. from the Spanish *De amor y de sombra*, by Margaret Sayers Peden, Toronto: Bantam Books.

—— (1989), "Writing as an act of hope," in William Zinsser (ed.), *Paths of Resistance: The Art and Craft of the Political Novel*, Boston: Houghton Mifflin Company.

Arriagada, Genaro (1988), *Pinochet: The Politics of Power*, trans. by Nancy Morris with Vincent Ercolano and Kristen A. Whitney, Boston: Unwin Hyman.

Blouet, Brian W., and Blouet, Olwyn M. (1982), *Latin America: An Introductory Study*, Toronto: John Wiley & Sons.

Dorfman, Ariel (1991), *Some Write to the Future: Essays on Contemporary Latin American Fiction*, Durham and London: Duke University Press.

Earle, Peter G. (1987), "Literature as survival: Allende's *The House of Spirits*," *Contemporary Literature*, 28/4: 543–54.

García Márquez, Gabriel (1987; orig. 1986), *Clandestine in Chile: The Adventures of Miguel Littín*, trans. by Asa Zatz, New York: Henry Holt & Company.

Gordon, Ambrose (1987), "Isabel Allende on love and shadow," *Contemporary Literature*, 28/4: 530–42.

Hart, Patricia (1989), *Narrative Magic in the Fiction of Isabel Allende*, London and Toronto: Associated University Presses.

Loveman, Brian (1988; orig. 1979), *Chile: The Legacy of Hispanic Capitalism*, second edn., New York: Oxford University Press.

Manguel, Alberto (1992), "Conversation with Isabel Allende," *Queen's Quarterly*, 99/3: 621–6.

Martin, Gerald (1989), *Journeys through the Labyrinth: Latin American Fiction in the Twentieth Century*, London and New York: Verso.

Meyer, Doris (1988), "Exile and the female condition in Isabel Allende's *De amor y de sombra*," *International Fiction Review*, 15/2: 151–7.

—— (1990), "'Parenting the text': Female creativity and dialogic relationships in Isabel Allende's *La casa de los espíritus*," *Hispania: A Journal Devoted to the Interests of the Teaching of Spanish and Portuguese*, 73/2: 360–5.

Pinto, Magdalen García (1991), *Women Writers of Latin America: Intimate Histories*, trans. by Trucy Balch and Magdalen García Pinto, Austin: University of Texas Press.

Robin, John P., and Terzo, Frederick C. (1972), *Urbanization in Chile*, New York: International Urbanization Survey, The Ford Foundation.

Rojas, Sonia Riquelme, and Rehbein, Edna Aguirre (eds.) (1991), *Critical Approaches to Isabel Allende's Novels*, New York: Peter Lang.

Violich, Francis, in collaboration with Robert Daughters (1987), *Urban Planning for Latin America: The Challenge of Metropolitan Growth*, Boston: Oelgeschlager, Gunn & Hain.

Williams, Raymond (1973), *The Country and the City*, New York: Oxford University Press.

Zinsser, William (1989), *Paths of Resistance: The Art and Craft of the Political Novel*, Boston: Houghton Mifflin Company.

17

EDEN, BABYLON, NEW JERUSALEM

A taxonomy for writing the city

Jamie S. Scott and Paul Simpson-Housley

INTRODUCTION

As Kevin Lynch has pointed out, to study the city is not always a matter of simple observation. "We must consider not just the city as a thing in itself," Lynch writes, "but the city as being perceived by its inhabitants."[1] Writers are among the most perspicacious of urbanites, and so their observations of the city will be of especial interest. In western letters, literary representations of city life are haunted by biblical figures. In the Hebrew scriptures, for example, Deutero-Isaiah envisions the time when the Israelites will return from exile in Babylon to a restored Zion, "an exalted city, one which resembled the fabled garden of Eden ... a holy city which would fulfill its original destiny as the dwelling place of Yahweh and the center of religious truth."[2] In an analogous way, the first urban Christians look forward to the time that the *urbs Romana* will be transformed into "the New Jerusalem, that comes down from heaven to earth."[3] In western literature, Saint Augustine's *City of God* most fully develops this tension between the threat of Babylon and the promise of a New Jerusalem into "a marvellous paradox" of civilized alienation.[4] For Augustine, we are alien citizens because

> the earthly city, if dominated by Satanic self-love, is also the place wherein the city of God is at work recruiting its citizens-elect, transforming the children of Cain into fellow travellers with Abel: men and women who pray for the peace of Babylon, but who live there only in body, being in heart and mind pilgrims en route to the heavenly Jerusalem.[5]

In western letters, then, urban realities are located over against the assumed innocence of an Eden now lost, but somehow promised again in a New Jerusalem, which itself may only be understood in the contrast of a sinful Babylon.[6] In this light, the city is an ambiguous place, "a sign of human estrangement from God ... the hope of an ultimate if distant reconciliation."[7]

Because those who dwell in the city dwell in alienation from God, they dwell in alienation from nature, from others, and from themselves. William Wordsworth's "Home at Grasmere," for example, a poem at once Augustan and Romantic in inspiration, configures themes of lost innocence and future hope by contrasting rural topography with images of urban estrangement:

> He by the vast Metropolis immured,
> Where pity shrinks from unremitting calls,
> Where numbers overwhelm humanity,
> And neighbourhood serves rather to divide
> Than to unite. What sighs more deep than his,
> Whose nobler will hath long been sacrificed;
> Who must inhabit, under a black sky,
> A City where, if indifference to disgust
> Yield not, to scorn, or sorrow, living Men
> Are ofttimes to their fellow-men no more
> Than to the Forest Hermit are the leaves
> That hang aloft in myriads, nay, far less,
> For they protect his walk from sun and shower,
> Swell his devotion with their voice in storms
> And whisper while the stars twinkle among them
> His lullaby. From crowded streets remote
> Far from the living and dead wilderness
> Of the thronged world, Society is here
> A true Community – a genuine frame
> Of may into one incorporate.[8]

The mood here is elegiac, lamenting the dehumanizing effects of William Blake's satanic mills, themselves a late eighteenth-century figure of Babylon. But Wordsworth also celebrates the realities of "a true community," itself a figure of Eden rediscovered in Grasmere, suggesting the possibility, at least, of a New Jerusalem.

For other writers, the possibility of a New Jerusalem becomes a paradisal probability. Take, for example, *The Cashier*, a novel by the French-Canadian writer Gabrielle Roy. The protagonist, Alexandre Chenevert, a bank clerk, has ventured into the country, to Lac Vert, to rest nerves frayed by the turmoil of life in the city:

> One evening he happened to be on the shore of a sheltered cove. This spot had become a place of escape for him, whence he could plunge with avid heart into the past. And there, on that evening, Alexandre found the city again. In place of dusky banks, he perceived the swarm of lights by which cities reveal themselves in the fullness of the night. Homesickness for the crowded lives there, for the intermeshed lives,

startled him, more compelling than any longing he had ever felt in all his days, like a longing for eternity.

He thought of the shop windows bursting with provisions ... He dreamed also of the newspapers and magazines piled high on the sidewalks, bearing tidings of all the world. There was life, that endless, stirring, brotherly interchange.[9]

Like Wordsworth, Roy invokes the figure of a lost Eden rediscovered in a rural environment. The spiritual rejuvenation afforded by this contrast between rural topography and Babylon then enables Chenevert to transform images of transience and alienation into images of eternity and universal community. For Chenevert, who suffers from inoperable cancer, a hospital placard advising citizens "Take care of yourself in time/ At the outset, cancer is curable" reveals how "the streets of the city" bear witness to "man's solicitude."[10] Drawing upon the evolutionary philosophy of heterodox Roman Catholic theologian, Teilhard de Chardin, Roy sees in medicine one of the more optimistic outcomes of the effects of science and technology upon urban life. Rural simplicities help Chenevert focus anew on expressions of the way in which Babylon becomes the New Jerusalem, "the eventual reconstruction of an actual redeemed city situated amidst a renewed paradisiacal earth."[11]

EDEN, BABYLON, JERUSALEM: A TAXONOMY FOR WRITING THE CITY

Taking Wordsworth and Roy together, we have a picture of the city as at once the site of alienation and the site of human efforts at community. Writing the city is thus an exercise in critical mediation. Drawing upon the contrasts afforded by rural topography, writing the city is an attempt to articulate our "persistent awareness of the otherworldly unreal cities of Revelation, New Jerusalem, and Babylon ... that frame this world."[12] *Sub specie aeternitatis*, writing the city has to be concerned with the spaces of a middle march, to adapt George Eliot's famous title. It is the work of portraying a community in all its social, economic and political dimensions, in an effort to avoid the realization of Babylon and to progress further along the road from a lost Eden to the realization of a New Jerusalem.

For all their geographical and historical differences, the representations of city life discussed in the chapters in this volume exemplify such tensions between nostalgia for a lost Eden and the unrealities of Babylon and a New Jerusalem. The yearning for a lost Eden informs this fiction in two ways. Sometimes the ambiguities of city life are contrasted with the simpler certainties of life in the country; on other occasions, aspects of the rural landscape are introduced into the urban environment in an effort to recover some more comforting sense of human identity in an otherwise alienating world. For example, in "Transgressing boundaries: Isabel Allende's Santiago de Chile," Emily Gilbert emphasizes an ambiguous dialectic between city and country. Isabel Allende's *The House of Spirits* revolves around the Trueba family, whose wealth enables them to travel freely between city and country, enjoying the sophistication and excitement of Santiago when it pleases them and retreating to the calmer atmosphere of Tres Marias when city life begins to overwhelm. A less ambiguous relationship between things rural and things urban appears in "Contrasting the nature of the written city: Helsinki in regionalistic thought and as a dwelling-place." Pauli Karjalainen and Anssi Paasi point out how Pirkko Lindberg's *Saalis* refers to a summer home in Spring Point, which acts as a sanctuary for Inna's family. This rural retreat reminds Inna herself of "what the world is like outside the city." Whatever their differences, both of these instances contrast urban life with life in the country. When she returns to Helsinki, however, Inna keeps a stuffed lynx in her flat to remind her of "the untamed forests of the Lake of Truth ... a piece of wild country in the midst of all this urban, artificial stuff." Here we have an instance of a character attempting to moderate the dehumanizing effects of the urban environment by introducing aspects of rural life into the city. In this vein, though on a more public level, Lafontaine Park functions as recreational space in Michel Tremblay's "Cronique du Plateau Mont-Royal," as Pierre Deslauriers indicates in his chapter, "Very different Montreals: pathways through the city and ethnicity in novels by authors of different origins." Or again, Jonathan Crush's chapter, "Gazing on apartheid: post-colonial travel narratives of the golden city," refers to the bucolic atmosphere characterizing white life in the Johannesburg of Adam Hochschild's *The Mirror at Midnight*, where we find "quiet winding streets lined with plane trees or the brilliant blue flowers of jacarandas."

Such nostalgia for a lost Eden is not common in the fiction under consideration in these essays, however, and to a large extent, this absence should not surprise us. Obviously, the city exists by explicit difference from the country, and nostalgia for a lost Eden may even be seen as a dangerous distraction from the real business at hand, which is to assume responsibility for urban realities in the here and now. In what economic, social and political realities, then, does the fiction considered in these essays envision the threat of Babylon in today's cities taking shape? Let us address the representation of the twilight zone, notably urban decay and the deterioration of living conditions. If we return to the origins of urban decay in Victorian England, we discover in Peter Preston's chapter, "Manchester and Milton-Northern: Elizabeth Gaskell and the industrial town," the insanitary conditions of working-class housing. Preston quotes Alexis de Tocqueville's view that this housing, usually located in marshy land, could be designated as the last refuge between poverty and death. Nearer our own time, Gary Brienzo's chapter, "Belfast: Bernard Mac Laverty's heart of darkness," considers personal and public suffering in the capital of Northern Ireland. In "St. Paul

Could Hit the Nail on the Head," the outward realities of decaying and demolished buildings in Belfast resonate with the psychological depression of Mary, who feels torn between her Protestant husband's intolerance of the Roman Catholic church and Father Malachy's marginalized loneliness. Deslauriers also describes soot and filth as striking features of East Montreal. These living conditions are vividly portrayed in Leonard Cohen's novel, *The Favorite Game*. In yet another instance, Yossi Katz, in his chapter "Jerusalem in S. Y. Agnon's *Yesterday before Yesterday*," refers to the stench of rotting garbage and other detritus in Jerusalem under the hot glare of the middle-eastern sun.

In another vein, the threat of Babylon occurs in various images of alienation, expressed in a spectrum of suffering ranging from homelessness through gender differences to urban violence, even impending economic, social and political chaos. Deborah Carter Park's chapter, "Tinsel town: Sydney as seen through the eyes of Christina Stead," refers to homelessness and vagrancy in the streets and back alleys of this paradise of perpetual sunshine. Drunks and beggars sleep on the lawns of an avenue leading to the university, itself a symbol of enlightenment and human dignity. Gender alienation occurs in Susan Rosowski's chapter, "Willa Cather as a city novelist." In Cather's novel, *The Song of the Lark*, Thea Kronborg moves to Chicago, a city in which ideas and business are made by men, causing Thea to think she had moved to a foreign place. In another North American urban context, the fading industrial lights of Detroit, once the automobile capital of the world, provide a setting of almost unbelievable violence. Lorne Foster, in his chapter "City primeval: high noon in Elmore Leonard's Detroit," chronicles countless instances of criminal inhumanity in a city where the normal constraints of civil behavior are all but ignored. In particular, the sheer number and horror of homicides in Leonard's Detroit seem to recapitulate the mythical dimensions of the first primeval act of civilization, Cain's murder of his brother Abel. Similarly, Crush refers to the brutality of Soweto, where the per capita rate of murder is four times that of New York. In an analogous vein, Cather's *Alexander's Bridge*, another novel Rosowski discusses, transports Alexander through the great cities of Boston, New York, Paris, Liverpool and London, linking him to ideas of urban progress in an ironic invocation of the false aspirations of modernity. Perhaps it is but a short step from these sorts of alienation to the breakdown of urban order altogether. Alec Paul, in his chapter "The St. Petersburg of *Oblomov* and *Crime and Punishment*," discusses how Fyodor Dostoyevsky depicts St. Petersburg as a city teetering on the verge of economic, social and political chaos, a threat of Babylon brought about by wave after wave of rural immigrants trying to find sustenance in the alien urban environment.

But writing the city is not always a matter of the sad but inevitable threat of Babylon; images of the New Jerusalem occasionally intrude. In particular, the essays in this collection and the fiction they discuss refer to two kinds of hope contained in the promise of a New Jerusalem: the hope of progress, and beyond this aspiration, the hope of perfection. In obvious disagreement with Cather's ironic condemnation of modernist notions of urban progress, we find a simple statement of the belief in progress in the chapter by Karjalainen and Paasi, who discuss Alpo Ruuth's novel, *The Last Autumn*. In this portayal of Finnish urban life, stately new blocks of bourgeois housing are perceived as marks of progress, especially when compared to the wooden shacks surviving alongside them. But perhaps this sense of progress is somewhat too optimistic, an anachronistic hearkening-back to simpler times. Certainly, Rosowski's chapter demonstrates how Cather's novel *Shadows on the Rock* somewhat nostalgically portrays seventeenth-century Quebec as a setting of almost idyllic perfection, a city protected from intruders who, in the name of modernity, might have destroyed the natural irregularities of its urban form, which follows the flowing contours of its headland. In a similar way, John Radford's chapter, "A place of tombs: the Charleston of William Gilmore Simms," refers to the antebellum perfection of Charleston. Radford comments that in 1806, the year of Simm's birth, Charleston enjoyed the greatest per capita wealth in the USA and ranked sixth in that country's urban hierarchy. But it hardly needs pointing out that this state of urban utopia depends upon the slave labor of the rural plantations. Similarly, we should not need reminding that when Gilbert notes Irene Beltrán, in Allende's *The House of Spirits*, passing through the lush trees and lordly mansions of exclusive neighborhoods, this state of earthly perfection is the domain of the fortunate few and in stark contrast to the sprawling *poblaciones callampas*. Similarly, Anna Makolkin's chapter, "City-icon in a poetic geography: Pushkin's Odessa," emphasizes how flowing wine, plump oysters, Rossini and Hellenic theater created a state of bliss in the Odessa of Alexander Pushkin's *Eugene Onegin*, though, again, this urban idyll belongs only to an aristocratic elite. If we are looking for images of the hope of perfection, perhaps the closest we will come is in Katz's exploration of Agnon's Jerusalem. Ironically, for all its rotting garbage and other detritus, the old Jerusalem expresses for Agnon something of the promise of a New Jerusalem, a sense of spiritual invigoration, holiness, knowledge and under-standing.

These chapters on urban writing and the fiction they discuss thus explore the ways in which various authors portray efforts to escape the implications of Babylon and to realize the promises of a New Jerusalem. In these efforts lie implied critiques of contemporary urban realities. Adapting Michel Foucault's notion of "heterotopia," we might say that to write the city is to hold up a critical mirror to present urban realities.[13] But the mirror is a kind of utopia, a place, in which

> I see myself there where I am not, in an unreal, virtual space that opens up behind the surface; I am over there, where I am not, a sort of shadow

that gives my own visibility to myself there where I am absent.[14]

At the same time, however, this mirror possesses its own reality:

> [I]t makes this place that I occupy at the moment when I look at myself in the glass at once absolutely real, connected with all the space that surrounds it, and absoutely unreal, since in order to be perceived it has to pass through this virtual point which is over there.[15]

Foucault coins the term "heterotopia" to refer to this sort of unreal reality. Reiterating the structure of Foucault's argument, we might say that fictional representations of city life are heterotopias, real other places which are yet unreal, mirrors held up to the realities of contemporary urban existence, offering implied critiques of the economic, social and political conditions we encounter in the Babylon of today's cities, yet unwilling or unable to dictate specific remedies, clear visions of a New Jerusalem.

CONCLUSION

We have not yet referred to two essays in this collection: Jacqueline Gibbons's "Chikamatsu's Osaka" and Rana Singh's "Modern Varanasi: place and society in Shivprasad Singh's *Street Turns Yonder*." These chapters stand out for at least one clear reason: each essay treats of fiction portraying a non-western city. So far, we have couched our reading of the chapters in this collection and of the fiction they discuss, in terms of the Judeo-Christian tropes of the lost Eden, the threat of Babylon and the promise of the New Jerusalem. The question thus arises: how, if at all, may these tropes be applied to a reading of the two remaining chapters and the representations of urban environments they consider?

In response, we may say that it is difficult to find any sense of a lost Eden in either essay. Though there might appear to be instances of a rural landscape within Osaka and Varanasi, the worlds of nature and humankind seem to flow more seamlessly into one another in the writing of Chikamatsu and Shivprasad Singh. Gibbons speaks of Chikamatsu's Osaka being situated in a fruitful plain on the banks of a navigable river, with bridges built of cedarwood, but there is little sense of the city being an imposition upon the natural scene. Similarly, Rana Singh writes of the central role the River Ganga (Ganges) plays in the personal and public life of Shivprasad Singh's Varanasi, but, again, we have little sense that the worlds of nature and humankind are at odds. Rather, in both chapters, the urban environment provides a stage for the playing-out of a wide variety of human activities, understood not in terms of a lost Eden, but in terms of the richness and diversity of life in the here and now. This richness and diversity include the kinds of economic, social, and political discrimination we see in the chapters on western fiction. Gibbons notes, for instance, that a love-affair in the pleasure quarter of

Tokugawa Osaka could affect the whole extended family. Yet she stops short of describing this affair in terms of gender alienation rooted in an apocalyptic plunge into a kind of Babylon. Rather, this affair is portrayed as another strand in the multicolored weave of urban life in Osaka, and moral considerations are couched in practical economic and social terms: to allow the heart to rule the head in such a matter is to risk the loss of the family's livelihood and, hence, its social position. Or again, though the banning of women from playing men's parts and the elimination of homoerotic elements in *kabuki* plays is certainly a function of the ethical values of the Tokagawa Shogunate, this intermingling of gender roles is frowned upon because it threatens the economic, social and political hegemony of men over women, not because the Osakans located their city within some transcendent vision of eternal life and death. Similarly, even when Rana Singh discusses the twilight zone of rowdies, black marketeers and even corruption among the ritual priests at the *ghats* in Shivprasad Singh's Varanasi, it is with a sense of acceptance or, at least, an understanding that the scale of urban existence makes the presence of such people inevitable. Even if Varanasi is the home of sinners, this judgment does not depend upon some dialectical vision of a future city transformed into either a Babylon or a New Jerusalem. As Rana Singh writes, "Sacramental glory and manipulative culture go side by side."

The writings of Osaka and Varanasi thus serve to relativize and render ambiguous the images of the lost Eden, the threat of Babylon and the promise of a New Jerusalem. It is not that Gibbons and Rana Singh do not identify analogous portrayals of urban life in the work of Chikamatsu and Shivprasad Singh. Rather, the urban landscapes of Osaka and Varanasi are contextualized more within the vertical than the horizontal plane. In a sense, even writings of non-western cities perpetuate the dialectic between rural and urban topographies on the one hand, and, on the other hand, the dialectic between Babylon and the New Jerusalem, but once we look beyond the Judeo-Christian tradition and western history, these dialetics are relativized and far more ambiguous with respect to today's cities than the tropes themselves allow. But we would like to suggest that this development is not simply a matter of the tropes being alien to non-western cultures; after all, in their broadest reach, the images of the lost Eden, the threat of Babylon and the promise of the New Jerusalem were intended as transcendent, universally applicable readings of urban realities. Rather, we suggest that this relativizing and ambiguity is a two-way street. On the one hand, it is true that that the images of Eden, Babylon and a New Jerusalem arise out of the context of the Judeo-Christian tradition and more broadly western history, and as such fail to take into account the influence of urban images and realities from non-western cultures. On the other hand, however, what we find in Osaka and Varanasi we also find in the cities of Europe, the Americas and elsewhere. We live in a multicultural situation, not only in the sense of a global village, in which all cultures are aware of one another's presence and impinge upon one another economically, socially and politically, but also in the sense that every great city incorporates within its domain the cultures of the world. The diverse realities of this decentered urban landscape transcend and explode images of the city as a place in quest of a lost Eden in the form of a New Jerusalem and in flight from the threat of an encroaching Babylon. Today's cities are polychrome, not monochrome, not leaning monolithically towards the threat of Babylon or towards the promise of a New Jerusalem, but diverse and decentered, at once centripetal, focussed on downtown, and centrifugal, localized in the vibrant variety of ethnic neighborhoods. Also, the increasingly mutlicultural character of urban life today further fragments and relativizes personal and public identity in the city. As Burton Pike has argued, the contemporary urban environment is characterized less by *angst*, by a personal and public sense of the tension arising out of the threat of Babylon and the promise of a New Jerusalem, and more by anomie: "[W]hat happens happens, and nothing seems to matter."[16] Indifference, not anxiety, seems to characterize much urban life today. We are faced less with the need to resist Babylon while we pursue the evanescant promise of a New Jerusalem, and more with the challenge of discovering coherence and significance at all amidst the relativities and ambiguities of the urban environment.

Perhaps the images of the lost Eden, the threat of Babylon and the promise of the New Jerusalem still persist. As Peter Hawkins points out, in novels like Italo Calvino's *Imaginary Cities*, Mark Helprin's *Winter's Tale* and Donald Barthelme's *Overnight to Many Distant Cities*, we do find hopeful visions of Babylon transformed into a New Jerusalem, even if in these writings of the city we discover a secularized and much more variegated and diverse vision of the urban environment.[17] But for the most part, where images of the New Jerusalem intrude, they are invariably ambiguous, and, if not ambiguous, then expressive of a minority situation, established and maintained at the economic, social and political expense of a multifarious majority. Where our analytical goals have to do with identifying kinds of writing that are likely to invigorate critical discussion and, if possible, bring about real changes in the economic, social and political structures characterizing life in the city, we are left, ironically, with works of literature that problematize rather than fulfill the promise they offer. If writing the city expresses a post-lapsarian attempt critically to mediate the unrealities of Babylon and a New Jerusalem, what is written is itself another unreality, a fiction posing as a realistic portrayal. These efforts at critical mediation know no reliable standpoint. As Georg Lukacs has noted of many nineteenth-century novels, what appear to be objectively realized, critically independent visions of cultural experience are in effect culturally conditioned, fictional idealizations of contemporaneous realities.[18] The creators of realistic fictions are themselves rooted in an historical matrix of economic, social and political conditions. Insofar as the fictionalized city is only recognizable in terms of actual urban realities, the former runs the risk of recapitulating many of the shortcomings characterizing the historical conditions of which the latter purports to be a critique. Every effort to portray the city realistically thus expresses implicitly the ideological agenda sustaining the economic, social and political structures the city writer would transform in an effort critically to mediate between the unrealities of Babylon and a New Jerusalem. To put it another way, we might say that realistic portrayals of city life always contain an implicit ideological element that forestalls the possibility of ideological critique. Drawing upon these ideas, we may say that writing the city represents an effort to awaken the possibility of ideological critique in imaginative fiction by ironizing, and hence problematizing, the dialectic between fictional realism and implicit ideology. For the purposes of this collecion of essays, this ironizing and problematizing takes place in and through contrasts between rural and urban topography, between a lost Eden and the alternatives of Babylon and a New Jerusalem. In the last analysis, however, we must recognize the ironic, and hence problematic, ideological interpenetration between things Judeo-Christian and western and things otherwise.

NOTES

1 Kevin Lynch, *The Image of the City* (Cambridge: Harvard University Press, 1960), p. 3.
2 Robert R. Wilson, "The city in the Old Testament," in Peter S. Hawkins (ed.), *Civitas: Religious Interpretations of the City* (Atlanta: Scholars Press, 1986), pp. 10–11.
3 Wayne A. Meeks, "Saint Paul of the cities," in Hawkins (ed.), *Civitas*, p. 16.
4 Rowan Greer, "Alien citizens: A marvellous paradox," in Hawkins, (ed.), *Civitas*, pp. 39–56.
5 Peter Hawkins, "Nightmare and dream: The earthly city in Dante's *Commedia*," in Hawkins (ed.), *Civitas*, p. 74.
6 In this regard, see also James Dougherty, *The Fivesquare City: The City in the Religious Imagination* (Notre Dame: University of Notre Press, 1980); Jacques Ellul, *The Meaning of the City*, trans. Dennis Pardee (Grand Rapids: William B. Eerdmans, 1970); Harvey Cox, *The Secular City* (New York: Macmillan, 1965); Gibson Winter, *The New Creation as Metropolis* (New York: Macmillan, 1963); John S. Dunne, *The City of the Gods* (New York: Macmillan, 1965); and Charles Williams, "The redeemed city," in *The Image of the City and Other Essays* (London: Oxford University Press, 1958), pp. 102–10.
7 William Chapman Sharpe, *Unreal Cities: Urban Figuration in Wordsworth, Baudelaire, Whitman, Eliot, and Williams* (Baltimore: Johns Hopkins University Press, 1990), p. 1.
8 William Wordsworth, "Home at Grasmere," in *The Poems*, 2 vols., ed. John O. Hayden (New Haven: Yale University Press, 1977), vol. I, p. 713.
9 Gabrielle Roy, *The Cashier* (Toronto: McClelland & Stewart, 1955), pp. 148–9; originally published as *Alexandre Chenevert* (Montreal: Beauchemin, 1954).
10 Roy, *The Cashier*, p. 172.
11 James L. Machor, *Pastoral Cities: Urban Ideals and the Symbolic Landscape of*

America (Milwaukee: University of Wisconsin Press, 1987), p. 22.

12 Sharpe, *Unreal Cities*, p. 15.
13 Michel Foucault, "Of other spaces," *Diacritics*, 16/1 (Spring, 1986), p. 24.
14 ibid., p. 24.
15 ibid.
16 Burton Pike, "The city as cultural hieroglyph," in Hawkins (ed.), *Civitas*, p. 131. See also Burton Pike, *The Image of the City in Modern Literature* (Princeton: Princeton University, 1981).
17 Peter S. Hawkins, "Introduction," in Hawkins (ed.), *Civitas*.
18 Georg Lukacs, *Studies in European Realism* (New York: Grosset & Dunlap, 1964).

INDEX